Engineering Practice in a Global Context

Understanding the Technical and the Social

Editors

Bill Williams
Polytechnic Institute of Setubal, Portugal

José Figueiredo
IST, University of Lisbon, Portugal

James Trevelyan
The University of Western Australia, Crawley, Australia

CRC Press
Taylor & Francis Group
Boca Raton   London   New York   Leiden

CRC Press is an imprint of the
Taylor & Francis Group, an **informa** business

A BALKEMA BOOK

CRC Press/Balkema is an imprint of the Taylor & Francis Group, an informa business

© 2014 Taylor & Francis Group, London, UK

Typeset by MPS Limited, Chennai, India
Printed and Bound by CPI Group (UK) Ltd, Croydon, CR0 4YY.

Library of Congress Cataloging-in-Publication Data

Engineering practice in a global context : understanding the technical and the social / editors, Bill Williams, Polytechnic Institute of Setubal, Portugal, José Figueiredo, IST, Technical University of Lisbon, Portugal, James Trevelyan, The University of Western Australia, Crawley, Australia.
 pages cm
 Includes bibliographical references and index.
 ISBN 978-0-415-63696-4 (hardback)
  1. Engineering—Social aspects. 2. Engineering—Study and teaching (Higher)
I. Williams, Bill, 1947– editor of compilation.
 TA157.E6225 2014
 338.4'762—dc23

                                                                      2013020251

Published by: CRC Press/Balkema
                P.O. Box 11320, 2301 EH, Leiden, The Netherlands
                e-mail: Pub.NL@taylorandfrancis.com
                www.crcpress.com – www.taylorandfrancis.com

ISBN: 978-0-415-63696-4 (Hbk)
ISBN: 978-1-315-87936-9 (eBook)

Table of contents

Foreword

Are you a working engineer? An engineering educator? Are you satisfied with the dominant image of engineering work as the application of engineering sciences to design new technologies for the benefit of humanity as a whole? Does this image paint a sufficiently complete picture for you of what engineers do? If so, you may have opened the wrong book. I daresay you're also pretty unusual among engineers these days.

This volume is, to my knowledge, the first scholarly collection designed for working engineers and engineering educators that dedicates itself to making visible the full range of actions that engineers actually perform on the job. If you find the dominant image of engineering knowledge and work to leave out some things, perhaps some really important things, you will find the contributions to this volume both provocative and appealing. And okay, if you happen to be a researcher, engineer, or co-worker with a scholarly or personal interest in what this book characterizes as engineering practice, you should read it too.

By writing for working engineers and engineering educators, these authors offer experiments in what I think of as critical participation. They work to make visible what is hidden by the dominant, linear image of engineers drawing on the engineering sciences to design new technologies. But these chapters do not stop there. The authors frame their studies in ways to persuade you not only that engineering is much more, and much more interesting, than the straightforward application of sciences to technological design. They also want to help you use their findings and insights to rethink and potentially transform relatively standardized ways of teaching and learning engineering. Critical participation includes both knowledge production and engagement.

By offering themselves as critical participants in the making of engineers and doing of engineering, these authors are also competing with one another. This competition is unusual, however. It is not about being right in an absolute sense, about discovering the single truth that will diffuse across engineering communities and gain acceptance by all through the sheer force of its intrinsic value. The competition is rather to make a difference for you and among you. It is to identify pathways for challenging the dominant image of engineering work in ways that have a chance of scaling up and gaining acceptance for new images and practices, in some places, among some engineers, sometimes. For any one of these initiatives to be successful, even to the point of scaling up across engineering communities more broadly, the others do not have to be wrong or incorrect. The scholarship of critical participation in engineering education and work is not a zero-sum game.

In a compelling introduction to the volume, James Trevelyan names the central purpose and contribution of this book with the provocative claim that engineering is not simply what engineers do. Rather, he and others in this volume argue, engineering is better seen as activities, or practices, through which engineers mobilize their knowledge to influence outcomes that are in fact produced by larger groups of people. Engineers influence outcomes. A key feature of this characterization is that engineers rarely, if ever, work alone. Furthermore, engineers typically influence outcomes within enterprises of one sort or another. These enterprises may be formal organizations, frequently commercial firms, or they may also be projects that reach across organizations and communities of people.

The implications of viewing engineers as influencing outcomes produced in concert with others are profound. Key contents of engineering practice lie in the communications and other interactions that take place with others. The so-called boundary between the technical contents of engineering work and its nontechnical contents dissolves away as the work of influencing outcomes necessarily includes both. In fancier language, key contents of engineering practice lie in the normative, or value, commitments of engineers and the pathways, or directionalities, they select to implement those commitments and exert influence. But the fancy language is not necessary to realize that routinized problem solving in the engineering sciences with single answers combined with brief introductions to uncertainty in technological design does not even begin to capture what is involved when engineers work to influence outcomes.

The studies that populate this volume rely on interviews, surveys, and the direct observation of engineers to make visible networks of relationships that include clients, other professionals, colleagues and teammates, legislation and standards, finance, contracts, technologies, production facilities, communities, and environments (Chilvers/Bell). Since these studies are not predictive in the sense of "sum of the forces equals zero," you may be inclined to resist their findings as knowledge. I urge you to think of these studies as providing maps of engineering practice.

How do you judge the value of a map? Since you cannot dial an 800 number directly to Reality to check, you have to compare a new map against other maps and mapping practices that you already accept, such as counting steps while walking or pulling out the GPS. So your challenge in assessing the findings in this volume is to judge them against your own knowledge of what the dominant image of engineering knowledge and work hides. Judge these studies according to a standard of plausibility. Given what you know, is this account plausible? What does its map include? Does it leave out something you consider important? And remember that you too constantly engage in mapping practices that you consider authoritative, from explaining to new colleagues how your organization came to be what it is today to explaining to a supervisor just who knows what about what and can be depended upon to do what by when. These studies have value.

Here's a short list of some things you can learn by reading about engineers influencing the outcomes of work undertaken with others. Many engineers develop a strong emotional attachment to mathematics. An overly strong attachment can inhibit one from working well with others, in part since the mathematical thinking typically used on the job to estimate and confirm multiple solutions differs from the objectivist, one-answer-per-question mathematics taught in school (Goold & Devitt). Effectively working with others, especially non-engineers, can depend upon skillful acts of redefinition until participating stakeholders converge on a definition or definitions of the problem that all can accept (Itabashi-Campbell/Gluesing). The amount of time engineers spend calculating on their own pales in comparison to the time spent coordinating the work of others (Williams/Figueiredo).

When engineers move from the routine work of design and problem solving to embrace an entirely new design endeavor, they like to try reusing previous solutions, aligning themselves with other established academic worlds, and searching for general, simplified models to apply. These efforts frequently fail, however, and it becomes difficult for engineers to sustain the self-image of a rational actor or genius craftsman in favor of the strategist (Kaplan/Vinck). When engineers at all levels of expertise, from novice to experienced, participate in design teams, they always pay attention to both the social interactions that take place on the team and broader contextual implications of the design process (Borgford-Parnelf/Deibef/Atman). Engineers who participate with other specialists in cross-disciplinary projects react differently to the experience. For some, it's just a matter of working with new people, learning some new things, or building some new alliances. For others, it a profound opportunity to reflect critically on their own commitments and practices, as well as to understand wholly new perspectives (Adams/Forin).

The engineering work of shepherding a new artifact from virtual worlds into the physical world involves complex scaffolding of people and things. From designing the artifact to developing new infrastructures to mobilize it or perhaps regulate its use, engineers must figure out how the artifact may be attached to, or constrained by, a multitude of other human and non-human entities (Hubert/Vinck). It may not be easy for engineering advocates of sustainability to integrate that commitment into routine practices of building design and construction. Motivated engineers might find themselves at odds with actors who are otherwise engaged and committed, such as architects. Such encounters make visible power relations on the job, but power here has technical contents. It is more than just politics (Chilvers/Bell).

Engineers trained and working in different countries may differ dramatically in their abilities to influence outcomes produced with other people. Engineers may be positioned differently in social hierarchies and vary in their identities, forms of knowledge, and commitments (Trevelyan). Histories of engineers and engineering are unlikely to enable contemporary engineers to avoid the mistakes of the past. For one thing, these tend to be hidden. For another, conditions have changed. The present is not the past. Learning histories of engineers and engineering may be more helpful in enabling engineers to understand how they gained the identities, forms of knowledge, and commitments that define them today (Figueiredo). And, finally, a site walk through an engineering project, a collaborative explication of what is taking place in a collaborative enterprise, demonstrates that informal learning on the job can be as important for engineers, if not more so, than formal engineering learning. Engineering education does not happen only in the classroom (Rooney/Willey/Gardner/Boud/Reich/Fitzgerald).

A common response to proposed changes in dominant engineering pedagogies is the challenge: What do you want to drop? Might the issue be better framed as: How can we teach engineering differently to enable engineers to see, understand, and anticipate the sorts of practices these authors make visible? The answer necessarily turns on looking for and analyzing the technical contents of seemingly nontechnical work done by engineers, and then adjusting pedagogies to accommodate the expanded visions.

Collaborative engineering judgment is a practice of communication with those whom engineers seek to influence, as well as with those affected by the outcomes of engineering influence. Learning about and listening to others is essential to participating effectively in a design exercise, site walk, problem-solving experience, mobilization of software, negotiation with architects and contractors, and so on.

Each author or set of authors in this volume offers a distinctive pathway for not only making visible arrays of practices involved in engineering work but also for formulating and scaling up alternative images and practices. In addition, each recognizes that no alternative images and practices scale up without the active participation of working engineers and engineering educators. That's why they wrote this book primarily for you.

Gary Lee Downey
Department of Science and Technology in Society,
Virginia Tech, Blacksburg, Virginia, USA

About the editors

Bill Williams originally trained as a chemist at the National University of Ireland and went on to work in education in Ireland, UK, Eritrea, Kenya, Mozambique and Portugal and to run international distance courses for the International Labour Organization in various African countries.

He is a lecturer at the Barreiro School of Technology of Setubal Polytechnic Institute in Portugal, where he teaches technical communication to civil engineering and construction management undergraduates. He has been an invited lecturer at IST, University of Lisbon and the Technical University of Madrid.

He is an associate member of the Engineering Management and Management Science Research Centre at IST, Lisbon (CEG-IST), has been an active member of the European Working Group on Engineering Education Research of SEFI since its inception in 2009 and is a founder member of the Portuguese Society for Engineering Education (SPEE).

José Figueiredo is a Professor in the Engineering and Management Department of IST – University of Lisbon and a member of the Engineering Management and Management Science Research Centre at IST, Lisbon (CEG-IST). He is an Electronics Engineer with an MBA in Information Management and a PhD in Industrial Engineering. He currently teaches project management and communication skills. His papers in conference proceedings, international journals and international edition book chapters have focused principally on project management and on Actor Network Theory.

In addition to his long-standing involvement with university teaching he also set up two small companies in the information technologies sector. He has been involved in consultancy work with a number of Portuguese and international companies.

Professor James Trevelyan is a Winthrop Professor in the Mechanical and Chemical Engineering School at The University of Western Australia. His main area of research is engineering practice, and he teaches design, sustainability, engineering practice and project management.

He and his students produced the first industrial robot that can be remotely operated via the internet in 1994. The robot has been controlled by a conservatively estimated 500,000 people in dozens of countries.

He was presented with the 1993 Engelberger Science and Technology Award in Tokyo in recognition of his work, and has twice been presented with the Japan Industrial Robot Association award for best papers at ISIR conferences. His teaching has also been well recognised: he was presented with the ASME Award for Mechanical and Mechatronics Teaching at the 2003 AAEE Conference in Melbourne. He has earned four distinguished teaching awards at UWA, and received a further best paper award at the 2004 International Conference on Engineering Education Research conference in the Czech Republic.

From 1996 till 2002 he researched landmine clearance methods and his web site is an internationally respected reference point for information on landmines. He was awarded with honorary membership of the Society of Counter Ordnance Technology in 2002 for his efforts, and was also elected a Fellow of the Institution of Engineers Australia.

His recent work on engineering practice aims to understand how engineering work is actually performed, an aspect of engineering that was not previously researched in great depth. This research has helped explain why engineering services in the developed world seem to cost much more than they should, often much more than in industrialised countries, a significant factor inhibiting poverty reduction. Professor Trevelyan is working on education initiatives to enable engineers everywhere to benefit from the insights emerging from this research.

Professor Trevelyan's web page is: http://www.mech.uwa.edu.au/jpt/ and this has a large amount of supplementary information on his research and teaching.

Acknowledgement

The preparation of this book has been supported by a grant from the Fundação para a Ciência e a Tecnologia (FCT) of the Portuguese Ministry for Education and Science as part of the project *What Engineers Do* PTDC/CPE-PEC/112042/2009. In addition Bill Williams has been supported by a PROTEC grant ministered by the Instituto Politécnico de Setúbal.

We are grateful to António Câmara, CEO of YDreams, for his early encouragement of our work. We would like to express our thanks to our proof-reader Ursula Rutherford and research assistants Nuno Delgado and Carlos Tiago for their invaluable assistance. Finally, on behalf of all the authors we would like to thank the many engineers in various parts of the globe who took time to willingly participate in our studies of their practice and without whom this book would not have been possible.

Introduction

This volume aims to provide the reader with a broad, global cross-section of the empirical research being carried out into engineers at work. The chapters provide pointers to other relevant studies over recent decades – an important aspect, we believe, because this area has only recently begun to coalesce as a field of study and up to now relevant empirical research has tended to be published in a wide variety of academic disciplines.

The book addresses the following groups of readers:

- researchers and students with an interest in engineering practice,
- professional engineers, particularly those looking for knowledge on engineering practice,
- engineering educators,
- people who employ, recruit or work with engineers.

The book is not designed to be read linearly, although it is useful to start with Chapters 1 and 2. A glossary of less common terms appears on page 281.

CONTRIBUTIONS IN THIS BOOK

The first chapter by Antonio Dias de Figueiredo opens a historical perspective on the ways that engineers and engineering have been portrayed through the eyes of philosophers, historians, writers and researchers over more than 2000 years. This chapter sets the stage for the rest of this book, helping us see the research presented in each chapter in the light of historical accounts of engineering. It helps to highlight the present difficulties experienced by engineers when they try to explain to others what engineering is, how it is different from science, and what it is for. The paper explores four threads of these portrayals of engineering in detail: craft, science, society and design, and how these different aspects impinge on contemporary debates on engineering education. There are many references that enable the reader to explore these histories in more detail. The sparseness of our knowledge, at least in English-language media, is an open invitation for historians of technology to dig deeper and enlarge our knowledge.

James Trevelyan has built on several research studies by himself and his students to argue the need for a theoretical framework for engineering practice. He demonstrates the need by pointing out the loss of practice knowledge in both academic and professional arenas. He proposes that the scientific basis for engineering, currently largely restricted to physics, chemistry, mathematics and related sciences, is in need of revision. He suggests the inclusion of human sciences and other disciplines that now offer informative explanatory insights into the many factors that shape the landscapes of practice; for example, factors that determine the relative ease of implementing different engineering solutions. The physical sciences cannot, by themselves, offer such informative views of the landscapes of practice. He points out some of the challenges that will be encountered along the way, not the least of which may be resistance from traditional academic interests.

Taking a widely adopted perspective within science, technology and society studies, Matthieu Hubert, Frédéric Kaplan and Dominique Vinck have presented us with two chapters that focus on the evolution of particular technologies. Hubert and Vinck explain how much is heterogeneous in engineering work at various levels: whether it be when the international technology roadmap for semiconductors (ITRS) is defined (which exerts an overarching influence on the micro-world of nanotechnology at the heart of information technology) or inside R&D projects, for instance when engineers are developing a new sensing technology in an attempt to measure the 'naturalness' of materials. Kaplan, also working with Vinck, explores what is going on when engineers are confronted by new challenges; they explore another aspect of information technology, the development of 'digital humanities', in the shape of an electronic book that evolves in quite unexpected ways into a new research tool for traditional humanities scholars. They demonstrate the application of actor network theory (ANT) to construct the technical objects of their studies, the roadmap, the sensor and the digital humanities research tool, as networks of intersecting influences that span nations, disciplines, space and time. We see engineers playing their parts without necessarily understanding the wider context in which their limited personal influence plays a role.

The evolution of these technological objects reflects the networks that take us beyond the traditional engineer's view of technical rationality. Engineers often express frustration at decisions made in organisations that, they think, are derailed by 'office politics', a term they use to describe any influence that they think is inconsistent with technical rationality. These accounts can help engineers understand and appreciate the complex social dynamics that shape the technical constraints impinging on their technical work – timescale, budget, resources, technology choices and instrumentation.

In their chapter on design teams, Jim Borgford-Parnell, Katherine Deibel, and Cynthia Atman bring us an analysis of an engineering design meeting, part of a much larger international collaborative study on a common data set. Their diagrams and quantitative approaches will no doubt come as a welcome relief for many engineering readers from the extended prose in other chapters like this one. They show how different engineers participate in different ways in a design meeting, how the flow of conversation gradually moves from higher level, broader issues of requirements to close-in details of implementation. The value in this chapter lies in creating awareness among engineers of process in meetings and social interactions. Readers can understand ways in which they too can observe events at a higher level and see them in a new light. While the focus of their chapter is on design, there are close parallels with other chapters that explore the working together theme.

Robin Adams and Tiago Forin bring an education and learning science perspective to a similar situation, investigating the differences and similarities in the way that people experience cross-disciplinary practice in engineering contexts – what they've come to understand about cross-disciplinary work, how and why they approach this kind of work, and how they see themselves as cross-disciplinary professionals. They studied 22 participants including engineers, biologists, computer scientists, software engineers, electronics engineers, and veterinary scientists. They used phenomenography, an approach in which they sorted the different narratives of their participants into categories that reflect different ways of experiencing cross-disciplinary practice. Some experienced it simply as 'working together'. Others experienced intentional learning to accommodate the perspectives of the other disciplines. Some exercised strategic leadership as they saw opportunities for combining discipline perspectives. The last group found themselves challenging and transforming their own disciplinary practice, for example an engineer who abandoned the familiar conventional built-test-fail process in favour of a new form of analysis-led discovery process. This group reported encounters with deeply embedded social values in what they previously thought was value-free objective science, and transformed their practices into more participatory and socially constructed perspectives. Adams and Forin conclude with suggestions for those who engage in cross-disciplinary work as well as those who seek to enable successful cross-disciplinary projects.

Working within an organisation science framework, Rachel Itabashi-Campbell and Julia Gluesing have explained how a detailed empirical study of engineering problem-solving in the automotive industry has led them, like others before them[1], to frame problem solving in a social context. They discovered that successful problem solving can be distinguished from unsuccessful attempts by differences in social interaction patterns. Success came from aligning the stakeholders' conceptions of the problem. Working from the narratives of their participants, engineers with extensive experience, Itabashi-Campbell and Gluesing propose the notion of 'ba'[2], a shared experiential space that allows the stakeholders to learn from each other. In this space, participants are able to move outside the 'bounded rationality' that characterises their instinctive reaction to the problem situation, and move toward a shared understanding or 'cognitive synchronicity'.

For engineers, a problem that can be solved without bringing other stakeholders on board is a rare event. Many engineers have lamented how they see perfectly viable, elegant technical solutions cast aside, often ascribing this to the limited vision of stakeholders such as accountants. This study provides suggestions that engineers could apply, helping them navigate the complex turbulence of competing stakeholder agendas in real organisations.

Bill Williams and José Figueiredo have presented results from an extensive mixed methods study of engineers in Portugal, further strengthening evidence for the significance of technical coordination as a prominent aspect of engineering practice. They build on earlier work by James Trevelyan and his colleagues, and have extended

[1]Korte et al., "Early Engineering Work Experiences," 2008. Jonassen et al., *Everyday Problem Solving in Engineering*," 2006.
[2]Nonaka, I. and Takeuchi, H. (1995), The Knowledge-Creating Company, Oxford University Press, New York, NY.

Trevelyan's framework of engineering practice. Their chapter opens, appropriately, with John Law's notion of heterogeneous engineering that emerged from his study of Portuguese maritime technologies in the 15th century. They build on the idea of an actor network to develop a model of practice by individual engineers finding workable solutions in a variety of different settings. Through quantitative and qualitative analysis, they have drawn attention to the importance of the firm as a setting for practice, with a focus on issues such as reputation, competitive bidding for work, regulations, codes and standards, and client satisfaction. The network developed at the conclusion of the chapter helps to show how an appreciation for engineering can be built with a much richer understanding of the social landscape in which engineers practice.

Andrew Chilvers and Sarah Bell have taken an empirical ethnographic approach in their chapter that tells the story of an engineering firm responding to the needs of their architect client, working on behalf of a local council. At the same time, they are attempting to enact the sustainability vision of their firm, a vision that conflicts with the exigencies of meeting the client's requirements within a constrained budget and time span for design. Their study reveals the detailed interplay between these different stakeholders, some separated by commercial interest, others by time and space. We see the complexities of building relationships with clients who have their unique character and personal agendas. In contrast to earlier chapters that, to the reader, are distant from the actual events providing the data, in this chapter we are taken right up close to the actual conversations, both those taking place centre stage, and those on the periphery of the social interactions. This chapter confronts us with the social and technical details that characterise the daily life of engineers.

James Trevelyan's second chapter is an attempt to compare engineering practice in the first world with the challenges faced by engineers in the third world, on the back-streets of Asian mega-cities. We see engineers faced with the necessity of personally persuading householders to pay their water bills, sometimes with the necessity of parallel technical intervention, deliberately blocking their sewerage connections to increase persuasive pressure. We see confusion on the manufacturing shop floor, where engineers try to train uneducated day labourers, only to find them replaced with more untrained labourers after just a few months. The pressure of immense disparities in social power and wealth dictates the actions of engineers, no matter how firmly they espouse technical and economic rationalities. By contrasting three different settings, a water supply utility, a metal manufacturing plant, and telecommunications engineering, we can begin to recover an appreciation of the social and economic significance of different engineering practices. South Asia presents us with an environment in which contemporary water engineering practice imposes a daily struggle for the minimal safe drinking water needed for survival, accounting for more than 10% of a society's economy. However, the unexpected success of telecommunications engineering stands as a beacon of hope, raising questions on the differences in engineering practices between different settings, and their influence on economic prosperity, health and welfare. The author concludes that engineering practice issues provide at least a part of the explanation for poverty that endures, despite improvements in economics, governance and education.

Mathematics forms the foundation for university studies in engineering and most engineering science is formed on a scaffold constructed in the languages of mathematics. Eileen Goold and Frank Devitt have described a study of the ways that practicing engineers apply the mathematics that framed their formal education. They explain how tacit understandings of mathematical ideas underpin thinking by engineers in the workplace. They have also managed to describe the diversity of ways in which mathematical thinking finds its way into engineering practice, ranging from simple applications of secondary school mathematics to complex computer modelling developed on a foundation of advanced mathematical analysis techniques. Presenting both quantitative and qualitative findings, they have provided us with a much clearer understanding about the ways that mathematical thinking supports engineering practice, such as enabling engineers to justify conclusions using mathematics, taking into account missing and uncertain information. While the study was performed in Ireland, the results seem to be widely generalizable, and invite further studies in different countries and engineering disciplines to strengthen and elaborate the findings.

Donna Rooney and her colleagues have provided insight into workplace learning in civil and construction engineering through a 'practice' lens. They review theoretical contributions that help us conceive workplace learning as something richer than the conventional notion of attending formal courses. Rather than learning as being something that an individual achieves through, for example, studying books, Rooney explains how we can conceive of learning that results when engineers work together on a project with multidisciplinary stakeholders. In other words, they present a view of practice that is learning: learning occurs through the participation in engineering practice in a social context. Knowledge, often new knowledge, is constructed between participants and exists as a distributed network of different understandings. The chapter shows how recent changes in how we perceive practice and learning have influenced the development of these new perspectives on practice. A site walk by engineers provides a real life setting through which the chapter illustrates practice learning and knowledge construction. The authors suggest that these perspectives could open up new and richer ways to recognise and foster workplace learning by engineers, and help young engineers better manage their transition from formal education measured by individual grades to a 'learning practice'.

Each of these chapters provides different insights into engineering practice, some with a broad canvas that presents a high level perspective, some with close-up details that reveal practices of individual engineers. Each article emphasises different aspects of the social, technical, and behavioural aspects of practice, helping to expose some of the knowledge of practice that has remained hidden for too long. Each provides informative reading for future students studying engineering practice, and useful material for future authors who will build on this work to develop texts to provide accessible learning materials for young engineers.

Chapter I

On the historical nature of engineering practice

Antonio Dias de Figueiredo
CISUC, Department of Informatics Engineering, University of Coimbra,
Coimbra, Portugal

I INTRODUCTION

In her influential book about the engineering profession, *Retooling: A Historian Confronts Technological Change*, Rosalind Williams, an eminent historian and dean at the Massachusetts Institute of Technology, claims that "the establishment of an autonomous engineering profession oriented toward ideals of broad social responsibility ... has not happened and is not going to happen" (Williams, 2002, p. 80). Her book was written from the point of view of someone who participated intensively in the organisational and intellectual transformation that MIT was going through at the time, and the shaping of her arguments reflects the controversies, communication barriers and dead ends that emerge when engineering academics discuss the future of their discipline. The disagreement about the nature of engineering practice is, however, quite old, and not confined to the walled gardens of the academy. The identity crisis of the engineering profession, which Williams refers to when she points out that present-day engineers follow the flow of innovation rather than commitment to an organisation (Williams, 2002, p. 63), is one of a number of identity crises that punctuated the evolution of engineering in the last two centuries. These crises have never been solved, and my conjecture is that much could have been gained if those concerned with their solution had expanded their insights by looking into the valuable legacy of engineering practice.

According to Williams (2002, p. 30), engineering is undergoing a process of "expansive disintegration" that has been sparked by its own success. The number of people calling themselves engineers is increasing and engineering-like activities are expanding, but the emergence of technoscience is blurring the borders between disciplines, so that both engineers and non-engineers are now engaged in multidisciplinary projects that replace the disciplines as the organising principles of techno-scientific activity. In Williams' opinion, this is causing engineering to disappear as a coherent and independent profession. Also, in her view the master of the new (disintegrated) engineering should be not the state, the army, the corporations, or the market, but democracy (Williams, 2002, p. 87).

My personal view, as an engineer, is that what Becker & Carper (1956) describe as 'occupational identities' is essential for the qualified performance of any techno-scientific (and social) activity. I also believe that, whatever the outcome of Williams' prediction, the abstract acceptance of democracy as a guide for the construction and evolution of an occupational identity is too vague to be of any use. As put by Nelson Foote, who first called our attention to the importance of identity for motivation:

> "Doubt of identity, or confusion, where it does not cause complete disorientation, certainly drains action of its meaning, and thus limits mobilization of the organic correlates of emotion, drive and energy which constitute the introspectively-sensed 'push' of motivated action" (Foote, 1951, p. 19).

My view is that those wishing to call themselves engineers, those responsible for the education of the engineers of the future, and those engaged in producing new knowledge in the fields related to engineering will have much to gain if they look into the precious heritage of millennia of engineering practice.

The aim of this chapter is to offer an introductory description of this heritage and to carry out an exploratory interpretation of its relevance to the clarification of the nature of engineering practice. The chapter begins with a description of the evolution of engineering from its early beginnings to the end of the Renaissance. As the history of engineering accelerated enormously since then, through the industrial and information revolutions, it would be impossible to synthesise such an intensive expansion into a single chapter. So, the period between the Renaissance and the present is analysed through a brief digression into the history of engineering education. From this historical examination, the chapter proposes an interpretive framework where engineering is seen as the transdisciplinary combination of four key disciplinary dimensions. The chapter concludes with the proposal of the concept of whole engineer to describe the organic aggregation of the four dimensions.

2 THE EARLY BEGINNINGS OF ENGINEERING

The first manifestation of engineering practice must have occurred two and a half million years ago (Roberts, 2002, p. 9), when our prehistoric ancestors learned that they could sharpen the edges of the stones they used as tools and realised that the exploration of the improved tools could in turn lead to the discovery of new practices. This meant that practice in the use of artefacts led to knowledge about how to produce and improve them, in the circular relationship between practice and knowledge described by John Dewey as 'intelligent action', which lies at the root of engineering as a discipline and a profession. In Dewey's expression, 'intelligent action', unlike conventional trial and error, uses thinking to carry out the "rehearsal (in imagination) of various competing lines of action" (Dewey, 1930, p. 179). Imagination is, indeed, an essential ingredient of engineering practice. Much of the evolution of engineering resulted from individual and collective dreams about things that never existed. And the practice of engineering can be seen to a large extent as making these dreams come true.

Very little is known about the evolution of engineering practice before the invention of the written word, but early Mesopotamian clay tablets in cuneiform writing show that Babylonian engineers were concerned, as early as 3000 BC, with sophisticated practical problems in mathematics, algebraic equations, right triangles, areas of land, volumes of masonry, the cubic content of canals to be excavated (Kirby *et al.*, 1990, p. 10), and approximations to algorithmic rules (Chiu, 2011, p. 175). They used a base 60 number system, which inspired our current system for measuring angles and telling the time (Kirby *et al.*, 1990, p. 10). The Egyptian civilisation also left evidence of the use of mathematics in engineering, and they actually established a non-positional base 10 number system. As most of their writing was done on fragile dried papyrus, it did not survive to our times. However, the few writings found to date, such as the Rhind and Moscow papyri, do offer clues about the role played by mathematics in Egyptian life (Imhausen, 2006, p. 19) and suggest the extent to which it contributed to the impressive temples and monuments, huge obelisks and pillars, advanced hydraulic engineering, and sophisticated shipbuilding that characterise the achievements of Egyptian engineering since 3000 BC (Kirby *et al.*, 1990). The evidence from ancient Mesopotamia and Egypt, not to mention identical evidence from ancient India and China, show that from a very early age our engineering ancestors supported their intelligent action with mathematical and abstract thinking.

Besides intelligent action and mathematical thinking, they also resorted to drawings and graphical means of representation. As suggested by Madsen & Madsen (2011, p. 10), the prehistoric cave drawings and carvings that gave shape to the artistic appeals of our ancestors were also early forms of the graphic communication that became essential to the progress of engineering (Ferguson, 1992). Remains found in ancient Mesopotamia suggest that drawings of the planned structures, such as the ziggurats, were already in use by 2000 BC (Kirby *et al.*, 1990, p. 10) and evidence from the history of early Egypt points to the use of drawings, both in plan and elevation, and to the construction of models to scale (Kirby *et al.*, 1990, p. 24). This can be understood if we bear in mind that the stones for the Egyptian buildings, many of them of huge dimensions, had to be shaped at the quarries before transportation to the construction sites, which required the engineers to develop detailed plans of their structures and predict rigorously the places the stones would occupy in the finished constructions (Kirby *et al.*, 1990, p. 24).

This concern with project management points to yet another characteristic of early engineering, besides the exercise of intelligent action, the application of mathematical tools, and the use of drawings and graphical representations: that of managing their own projects, including managing the people involved. In Mesopotamia, the Code of Hammurabi addressed explicitly the rules, responsibilities, and acceptable standards of workmanship of the master builders (Chiu, 2011, p. 35), who had to manage within these standards their whole activity and that of the people who worked for them. In ancient Egypt the master builder, who was also an architect, engineer, and builder, usually held a very high status and was responsible for managing the whole process of quarrying the stone material, transporting it to the construction site, overseeing the whole construction, and organising the immense labour force (Chiu, 2011, p. 59). Project management in ancient Egypt improved considerably over the years, sometimes resorting to large-scale prototyping as in the case of the Bent Pyramid at Dahshur, which was used as an experimental setup to try out different slope gradients for the gigantic Giza pyramid (Miroslav, 2001) and shows the extent to which, from a very early age, the management of engineering projects could be deeply woven into their whole fabric.

Primitive engineering began in the realm of the crafts, inspired by utilitarian and artistic reasons. In ancient Mesopotamia, engineers were master builders who had started as craftsmen and then become experienced builders in the service of the expanded urbanisation and civic construction projects that made Mesopotamia the cradle of civilisation. In Egypt, engineers were also master builders who had started as craftsmen, many of them coming from the priesthood but enlarging their ranks as the construction and supervision of temples, irrigation projects, aqueducts, and roads expanded and required a growing number of skilled people (Chiu, 2011, p. 59). This brief description of the early beginnings of engineering is sufficient to point out a few threads or dimensions of engineering practice. First of all, engineering practice originated in the activity and culture of the crafts. Secondly, it developed around the use of plans, drawings and models, corresponding to what is described today as engineering design. It also rested upon rigorous measurement, calculation and, to some extent, abstract thought–the early manifestations of science. Finally, it recognised the need to organise and administer projects and people; the dimension of the social sciences (including business and management).

3 GREEK SCIENCE AND ROMAN ENGINEERING

Although the Greek and Roman civilisations were strongly influenced by the Mesopotamian and Egyptian cultures, their attitudes toward knowledge and practice evolved in radically different directions, causing the first fundamental split between the cultures of science and engineering. Up until the sixth century BC, when the ancient Greeks started conjecturing about the existence of general 'laws of nature', all human knowledge had developed from experience. The radical change introduced by the ancient Greeks was to recognise the existence of a hidden order in nature and hypothesise the existence of general natural laws that governed this order and could be discovered by man (Kirby *et al.*, 1990, p. 42). This explains why the ancient Greeks are said to have created science – the ability to build knowledge by uncovering the theories of natural phenomena and the mysteries of mathematics. The Mesopotamians and Egyptians were able to solve practical problems of geometry, but the real breakthrough was the invention of abstract geometry by Thales of Miletus, whose general theory of the relations and properties of lines, angles, surfaces, plans, and solids was developed independently from any concrete objects to which they might refer (Kirby *et al.*, 1990, p. 42). Three hundred years later, in the Hellenistic period, Euclid wrote his extensive Elements, a 13-book collection of definitions, axioms, propositions, and proofs in geometry, algebra, and number theory that would project the influence of Greek science up to our times. In a similar way, Aristotle's physics, an inspiring treaty on the philosophy of nature, became an important foundation of science for many hundreds of years.

Curiously enough, the Greek engineers did not find much use for those theories, so the history of ancient Greek engineering was mostly based on the skilful adaptation of the practices of earlier civilisations. There was, however, a significant difference: the beauty and sense of proportion of their realisations, namely in architecture (Kirby *et al.*, 1990, p. 43). The planning of their buildings, based on broad specifications from the city-state, only outlined general dimensions – volume, area, wall thickness, window sizes – so the *architekton*, whose role was identical to that of today's engineer, was free to use his creativity and artistic mind as the construction progressed and thus ascribe to his buildings the astounding architectural effect we witness today (Kirby *et al.*, 1990, p. 43). The *architekton* was regarded as a craftsman and building contractor, engaged by the state or by wealthy clients, who designed the buildings, hired and managed personnel, and made sure the construction was completed within an agreed time scale and budget (Chiu, 2011, p. 79). The status of the *architekton* does not seem to have been much prized in ancient Greece, and even those who are recognised today as geniuses, such as Iktinos, the main *architekton* of the Parthenon, do not seem to have been famous in their time (Chiu, 2011, p. 79).

In the Hellenistic period, when Greek culture was propagated to the non-Greek world following Alexander the Great's conquests, the Greek contempt toward the practices of engineering was moderated somewhat to meet the challenges of the new civilisational development. Archimedes of Syracuse was an example of this change. He was simultaneously a prominent scientist and a brilliant practicing engineer, who contributed to the discovery of the principles of statics, hydrostatics, mechanical power, and mensuration, but also to the invention of the compound pulley, the screw, and a wide variety of machinery, levers, cranes, and catapults. In any case, the enduring mark of the ancient Greek culture was that its most brilliant minds created abstract science and disregarded the practicalities of engineering. Unlike the Roman civilisation that followed, the Greek civilisation emphasised knowledge and demoted practice (Kirby *et al.*, 1990, pp. 95–96).

Roman engineering was the leading example of engineering of classical antiquity, leaving its mark all over Western Europe and propagating its influence across the centuries, well into the Renaissance. However, no first rank Roman scientists emerged (Kirby *et al.*, 1990, p. 56). The Romans had a strong genius for management and administration. They were not keen on theorising but they were fast learners and remarkably competent at adapting and improving the ideas and practices of others (Kirby *et al.*, 1990, p. 57). They had a strong interest in engineering and they appreciated its value for both military and civil affairs. They also encouraged and actually subsidised the systematic training of young and promising apprentices (Kirby *et al.*, 1990, p. 59). The combination of these factors, including their eagerness for management and their sense of effectiveness and economy, explains to a large extent the success of Roman engineering. One of their major discoveries seems to have happened by chance, the sort of chance that favours those who insistently explore the realm of practice with a prepared mind, the discovery of concrete. The huge dome of the Roman Pantheon which, two thousand years after it was built, is still the largest unreinforced concrete dome in the world, is both a tribute to the technical competence of the Roman engineers and an historical example of how knowledge can be wiped out when a civilisation collapses. Indeed, the technology of concrete disappeared with the demise of Roman civilisation and only in the early 19th century, almost two millennia after the completion of the Pantheon, would it be entirely recovered (Pacey, 1992, p. 5).

A major shortcoming of Roman engineering was the lack of knowledge of mathematics, despite the efforts of Vitruvius, the Roman engineer and author of The Ten Books on Architecture, a detailed treatise of engineering methods, who tried to penetrate the scientific legacy of the Greeks and extract some empirical rules for his books. Vitruvius exhorted his fellow engineers to master the knowledge of mathematics, the sciences, and the arts, but his advice had no effect. Indeed, the Roman engineer, or *architectus*, had no formal knowledge of statics and did not understand Greek trigonometry. As he mastered no systematic theory for the computation of stresses, thrusts, and weight distribution, and did not understand theoretically the behaviour of materials under tension, compression, bending, or shearing (Kirby *et al.*, 1990,

p. 80), he had to resort to frequent empirical tests and adopt very generous factors of safety.

The Roman engineers, or master builders, acted simultaneously as designers and builders and managed their large-scale projects in hierarchies much inspired by the political hierarchies of Roman society. They divided the work into multiple sections, to make possible the simultaneous construction of many parts of the buildings at the same time, while holding overall control (Chiu, 2011, p. 191). According to Vitruvius, they should also be cultivated enough to understand the symbolic meaning of what they were constructing (Oleson, 2008, p. 255), which meant being familiar with mathematics, geometry, philosophy, music, law, astronomy, and medicine, and be "men of letters who could produce a lasting record of their treatises that would become part of the future record of the state" (Chiu, 2011, p. 113). In other words, they should be very well cultivated people and not just mere craftsmen.

The four threads, or dimensions, of engineering practice identified for the early Mesopotamian and Egyptian civilisations – the crafts, design, science, and the social sciences (including business and management) – can also be recognised in the engineering practice of classical Greece and Rome. The Greek *architekton* and the Roman *architectus* also generally started their practice as craftsmen – or as artists, in the case of the Greek *architekton* – and they progressed in the complexity and responsibility of their tasks as their skills improved through apprenticeship and practice. Their robust ability as designers, with design understood as the ability to conceive novel solutions, can be appreciated in the abundant Greek monuments we can visit in the regions that were once covered by the Greek Empire and in the wealth of monuments, bridges, aqueducts, canals, dams, and other ancient Roman constructions scattered all over the large geographical area that belonged to the Roman Empire. In contrast, the dimension of science was not particularly strong in the engineering practice of classical Greece and Rome. On one hand, the Greek contributions were so advanced that they failed to propagate to the engineering practice of the time and would only be fully appreciated two millennia later. On the other hand, in spite of the strong endorsement of mathematics by Vitruvius, the most prominent representative of Roman engineering thought, Roman engineers were not keen on exploring abstract thought in their projects. Unlike the component of science, the component of the social sciences (including business and management) was particularly developed, thanks to the Greek invention of Western democracy and the Roman management and organisational excellence, which permeated their engineering practice. Indeed, although Socrates defined management as "a skill separated from technical knowledge and experience" (Pindur *et al.*, 1995, p. 60), the Greek *architekton* combined the roles of craftsman, artist, builder, and manager of his own projects, which included control, inspection, approval, and authorisation of payments to labourers and contractors (Chiu, 2011, p. 86). These management skills also included the practice of delegation, the recognition of labour division and specialisation, an understanding of worker motivations, and the willingness to lead and manage in democratic environments (Chiu, 2011, p. 188). Roman management approaches, inspired by Greek democratic tradition, were, in turn, strengthened by Roman hierarchical models and practices of organisation, planning, and optimised use of materials and workers. In addition, Roman democratic traditions also required the *architectus* to act as a leader capable of mobilising his workforce and reaching agreements with them (Chiu, 2011, p. 192).

4 THE REFOUNDATION OF ENGINEERING IN THE MIDDLE AGES

When, in the later Middle Ages, the expansion of trade, industry, and church architecture triggered the re-foundation of engineering, particularly visible in the invention of non-human power systems and the construction of cathedrals, it emerged from the Roman legacy of engineering, rather than from the abstract science of the Greeks (Kirby *et al.*, 1990, p. 95). The abstract science of the Greeks had emigrated to the countries of Islam, where its most representative works had been translated into Arabic, and to India, where it arrived through the Middle East Syriac writers and the sea route that connected Alexandria with north-west India (O'Leary, 1949). Thus, before the 12th–14th centuries, in the absence of formal science, the evolution of engineering was through intelligent action. When problems occurred, and they kept occurring in ambitious constructions such as the medieval cathedrals, they were solved thanks to the skill and genius of engineers who attempted lines of action no one had ever tried before. The resulting solutions were then used to inspire the solutions to similar, but more complex, problems in the future. Using Kuhn's terminology, the solutions they devised for their problems were 'exemplars' that encapsulated the tacit knowledge acquired through practice (Kuhn, 1970, p. 44; Kuhn, 1977), tacit knowledge that was learned by doing rather than by acquiring rules for doing (Kuhn, 1970, p. 253). This was how, from the 8th century onwards, the European engineers progressively learned to build bigger and bigger churches; replaced the traditional wooden roofs with fireproof stone vaults; extended the use of vaults to entire naves and aisles, progressing from barrel vaults to groined vaults, to ribbed vaults; learned statics from experience; and invented and improved the elegant flying buttresses that support the huge lateral loads transmitted from ceiling to walls in Gothic churches (Kirby *et al.*, 1990, pp. 102–103; Pacey, 1992, pp. 17–21). To assist in the development of their constructions they also resorted extensively to modelling, making constant use of full-scale templates and detailed drawings (Kirby *et al.*, 1990, p. 105).

The specialisation of the engineer in a single occupation is a very recent phenomenon (Hill, 1996, p. 7). Throughout classical and medieval times, and well into the 18th century, the engineer accumulated technical functions with many other functions. Sometimes an engineer was merely an artisan who had reached the top of his profession to become a master mason or a master metalworker (Hill, 1996, p. 9) but an engineer could also be a clergyman, or a nobleman, or a monk who, at the request of a patron, took a major role in designing and producing buildings, machines, and artefacts. The engineers of many of the cathedrals of the Middle Ages were clergymen. The first sponsor of the cathedral movement was Abbot Suger, of the monastery of St. Denis, in Paris; the pioneering cathedral of Durham was erected, even earlier, by a French bishop; and Sens cathedral in Burgundy, one of the earliest Gothic churches, was built by its bishop, Henry de Sens. Most of these clergymen belonged to the Cistercian Order, which worked as an efficient social network for the exchange of technical ideas (Pacey, 1992). Their approach to engineering was a mix of idealistic and religious values, aesthetic innovation, artistic symbolism, and enthusiasm for experimenting. This explains why the mind of the medieval engineer was shaped by a culture that transcended pragmatic solutions to contemplate, through experimentation, the ideals of creativity, beauty, and an aspiration to perfection that could come as close as possible to the imagined divine ideal (Pacey, 1992, pp. 8–9).

The predominance of engineering over science had continued through the Middle Ages, but as the religious orders became widespread and their culture influenced the growth of an educated population of merchants, physicians, bankers, lawyers, and other secular professions for whom literacy and numeracy were becoming valuable, the search for more formal kinds of knowledge started to make sense. The emergence of the universities – Paris (around 1150), Oxford (1190), Bologna (1200) – was closely related to this environment. Robert Grosseteste, a scholastic who taught theology at Oxford in the 13th century and had a major influence on the thoughts of Roger Bacon, is described as the medieval thinker who founded modern experimental science through the revival of Aristotle's methodological tradition of validating scientific claims by combining analysis and synthesis, 'resolution and composition'. The big split between science and engineering, developed in 600 BC with the divergence of abstract Greek science from the empirical traditions of its ancestors and conquerors, was now about to close with the flow into Europe both of Islamic science and philosophy, strongly enriched with the Greek scientific and philosophical traditions and translated from Arabic, and of the contributions coming from Indian scholars (Kirby et al., 1990, p. 96). The main outposts of this flow were Sicily and Spain, namely Toledo where an intensive activity of translation from Arabic to Latin developed following the conquest of the city by the Christians. The Italian maritime republics also contributed to this flow, and the recognised superiority of Islamic science and technology motivated, between the 11th and 14th centuries, frequent travels of Christian scholars to Arab countries where they learned the sciences and attended the main Muslim centres of higher learning. The solid experimental engineering culture of the time, now combined with the abundance of Greek and Islamic literature translated into Latin and the drive to pursue philosophical and theoretical inquiry at the monasteries and universities, explains to a large extent the course that led, through the Renaissance, to the rise of modern science in the 17th century.

The four threads of engineering practice identified in the previous sections – the crafts, design, science, and the social sciences (including business and management) – can also be recognised in the Middle Ages. The importance of the crafts was certainly a characteristic of this historical period, closely associated with the spread of craftsmen's guilds and confraternities that strengthened technical competence, consolidated apprenticeship as a regular way of reproducing knowledge, and boosted respectability. Associations of craftsmen are known to have existed in Egypt in the Hellenistic period, and identical organisations became widespread all over the Roman Empire from the 3rd century BC. However, the medieval guilds played a pivotal role in creating widespread communities of craftsmen who exchanged techniques across Europe, preserved secrets, protected collective interests, promoted training, and actively contributed to the quality of the professions. The differentiation of hierarchical levels that rewarded the experience, merit, and achievements of their members became so respectable that it is still recognisable in the degrees and academic rituals of our universities. Religious ideals also played a role in strengthening the status of craftsmen in Europe, namely since the 12th century under the influence of the Cistercian monks, who attached superior moral value to manual labour. An identical religious influence was felt again in the 17th century, inspired by the Protestant work ethic that prevailed in the nations where the industrial revolution was brewing (Pacey, 1992). The religious and philosophical thinking of St. Augustine (354–430), which so much influenced the medieval worldview, to a large extent drove the Gothic cathedral movement of the 12th century and contributed to a unique combination between design and science. Augustine saw aesthetic perfection in the 'proportions' emerging from the 'perfect ratios' and claimed that artistic creation could not exist unless the 'laws of numbers' were observed (Simson, 1956, p. 23). He scorned the architect who acted as a mere practitioner and valued the architect who applied mathematical rules and thus acted as a 'scientist' of his art (Simson, 1956, p. 23). For Augustine, the creative process could not exist in the absence of proportions dictated by mathematical laws, these alone being capable of leading the mind "from the world of appearances to the contemplation of the divine order" (Simson, 1956, p. 24). Apart from the central concern with the consonance of parts, or proportion, Gothic architecture also attached exceptional value to luminosity, which symbolised the enlightenment of the human mind by the divine intellect (Simson, 1956, p. 52). This led to the daring replacement of solid walls by stained glass windows and contributed to the intricacy of the relationship between design and science in Gothic architecture and to the creation of "a new relationship between function and form, structure and ornament" (Simson, 1952, p. 6). Unlike the dimensions of science and design, which had such a unique development in the engineering practice of the cathedral movement, the dimension of the social sciences (including business and management) of medieval engineering practice does not seem to have much to add to the earlier practices in the construction of large buildings. The introduction of non-human power systems must have had a significant impact in other engineering sectors of the Middle Ages, and maybe even in the construction of cathedrals, but little could be found in this respect in the scarce historical literature on the topic.

5 RENAISSANCE ENGINEERING AND THE POWER OF DESIGN

At the peak of the 15th century, a time of massive expansion of trade, fast growth of the merchant class, intensive urbanisation, and great rivalry between the Italian city-states, artist-engineers were much valued and respected figures of Italian court life (Bjerklie, 1998). On one hand, they imagined the technical solutions essential to support the offensive and defensive military ambitions of their wealthy lords. On the other hand, they designed the magnificent buildings and public works sponsored by those lords and created the agricultural and industrial improvements needed for their success and reputation (Galluzzi, 1987; Bjerklie, 1998). Four prominent artist-engineers illustrate the genius and singularity of Renaissance engineering: Filippo Brunelleschi, Mariano di Iacopo, Francesco di Giorgio, and Leonardo da Vinci (Bjerklie, 1998). Their most revolutionary achievement, which changed radically the practice of engineering, was to use graphical representation in ways that enabled far more powerful modes of reasoning, much improved descriptions, and much better communication of ideas across the barriers of tacit knowledge – which are so inscrutable when it comes to imagining and building something that does not exist. This revolution included the use of the sketch as an intellectual tool, the discovery of the laws of perspective, and the exploration of the techniques of cutaway, exploded and rotating views (Bjerklie, 1998).

Filippo Brunelleschi (1377–1446), better known today as the engineer who conceived and built the dome of the Cathedral of Florence, an impressive engineering achievement even by today's standards, was prominent in the Renaissance for his contribution to the invention of the mathematical laws of linear perspective and for the construction of a large number of elaborate machines. His insights and empirical experiments on linear perspective, which were later codified and published by one of his followers, Leon Battista Alberti (1404–1472), became for Western culture "a style of thought, a cultural perception, a way of imagining the world" (Romanyshyn, 1989, p. 69). Indeed, much earlier than Descartes (1596–1650), Brunelleschi and his followers made us see the world from the outside, as spectators looking out of a window, and offered us the virtues and evils of the epistemological assumption that we can know the world better if we look at it from a distance (Cypher & Richardson, 2006). In fact, even the way we see the world today on our television sets and computer screens, a culturally assimilated convention, is a consequence of this invention (Cypher & Richardson, 2006).

Unlike Brunelleschi, who saw himself as a craftsman and not as an author, Mariano di Iacopo (1382–1458), also known as Taccola, and Francesco di Giorgio (1439–1501), both from Siena, became famous not just for their impressive accomplishments as engineers but also for their writings. The revolutionary feature of their treatises was the invention of cutaway and exploded views and the use of abundant illustrations to express graphically what would be difficult or impossible to express in words (Bjerklie, 1998). Taccola pioneered the systematic use of personal sketch books to register ideas and observations, and di Giorgio developed an extensive hierarchical classification of machines based on the identification of principles that led him to produce more advanced designs (Bjerklie, 1998). In other words, he demonstrated that we could improve knowledge and practice by abstracting principles from empirical evidence and that we could then depart from these principles to imagine and validate better solutions.

Leonardo da Vinci (1452–1519), often described in the literature as a solitary genius, was, on the contrary, both a consequence and a culmination of the creative engineering environment of the Italian Renaissance. He relied extensively on the contributions of his predecessors and pushed forward the patterns of practice and knowledge they had attained as artists-engineers. His notebooks which, like those of Taccola and di Giorgio, combined text and pictures, are eloquent illustrations of the extent to which the intelligent combination of drawings and text can support reasoning and become powerful forms of intellectual experimentation. In this sense, they represent a major breakthrough in the practice of engineering: the exploration of sketching, drawing, and modelling – that is, of design – as a powerful aid to thought (Buxton, 2007, p. 105). The use of the notebook as a systematic personal recording of the progress of an engineer in his projects was another innovation in the engineering practices of the Italian Renaissance, which spread to science practice as well.

Leonardo da Vinci's constructivist method of developing new ideas and solutions was also a major contribution to the practice of engineering, the advancement of science and the progress of the philosophy of knowledge (LeMoine, 1999). Formulated as a second thought in one of his notebooks, it stated:

"But first I shall do some experiments before I proceed farther, because my intention is to cite experience first and then with reasoning show experience is bound to operate in such a way. And this is the true rule by which those who speculate about the effects of nature must proceed." (Ms. E, folio 55r) (Capra, 2007, p. 161).

More than a century before Galileo, Leonardo da Vinci had thus stated the foundations of experimental science, a fact only discovered recently, as the recovery, compilation and study of Leonardo's scattered works progressed. It is interesting to notice that Leonardo's constructivist approach was strengthened by his motto, *ostinato rigore* (obstinate rigour) (Wilson, 2000, p. 97) which establishes persistent rigour for its own sake, a typical craftsman's value, as a key element of his method.

Of the four components of engineering practice discussed in the previous sections – the crafts, design, science, and the social sciences (including business and management) – the one that stands out in the Renaissance for its originality and contribution to the progress of engineering practice is design.

6 ENGINEERING EDUCATION

The history of engineering in the last three centuries runs approximately parallel to the history of engineering education. Three main traditions have inspired engineering education in the Western world. The oldest one developed in the centralised bureaucracy of the French monarchy in the 16th century when the French royal government integrated in its army a permanent body of engineers dedicated not only to the techniques of warfare (artillery, fortress building, and road and bridge construction for troop deployment) but also to the main constructions of the centralised nation-state (road and bridge networks, canals, river and harbour improvements, and water supply systems) (Reynolds, 1991, pp. 7–8). Building on this tradition, and in order to increase its workforce of qualified engineers, the French monarchy established in 1747 the *École des Ponts et Chaussés* (School of Roads and Bridges), a turning point from the traditional apprentice-based strategies of engineering education towards more systematic schooling practices inspired by the principles of science and mathematics. This new approach was so successful that it led in the following years to the creation of other formal technical schools and culminated, in 1794, already under the auspices of the French Revolution, with the foundation of the *École Polytecnique*, still recognised today as an icon of formal theoretical and mathematical engineering education in France (Reynolds, 1991, p. 8). The industrial revolution, which had taken place in the United Kingdom, and quickly spread to Central Europe, widened massively the need for engineers and led to the emergence of a new category of engineering that largely transcended military needs. For this reason, it was called civil (meaning non-military) engineering. Civil engineering was the mother of the future branches of engineering – mechanical engineering, chemical engineering, electrical engineering – which gradually emerged as the evolution of technology increased specialisation. By 1800, engineering was a high-status and well-established professional occupation in France, attracting a significant proportion of its members from the lower nobility and upper-middle class (Reynolds, 1991, p. 8).

Another European tradition that influenced engineering education emerged in the United Kingdom as a result of the industrial revolution. The subsequent rapid commercial and industrial expansion soon led to the creation of corporations and partnerships which launched several ambitious projects requiring engineering manpower (Reynolds, 1991, p. 8). Unlike its French counterpart, British engineering education was still strongly based on apprenticeship and the profession attracted people from all classes. It favoured a practical and empirical approach to problem solving and tended to view mathematical and theoretical approaches with suspicion (Reynolds, 1991, 9). It was in this context that John Smeaton (1724–1792), often described as the father of British civil engineering, made a decisive contribution to strengthen the foundations of British engineering. On the one hand, he encouraged in 1771 the formation of a "Society of Civil Engineers", which helped develop a strong sense of identity of the profession. On the other hand, he fought insistently against the insufficiencies of British engineering education and stressed the importance of careful investigation, measurement, and testing (Pacey, 1992, p. 179). Part of the British renewal of engineering education would, however, only happen extensively in the 19th century, influenced by the waves of innovation coming from Germany through the United Sates.

The third tradition emerged in Germany in the early 19th century, inspired by the university reforms proposed by Wilhelm von Humboldt, at the University of Berlin and during his short mandate as Prussian minister of education. Breaking with the medieval tradition of European universities, which emphasised the maintenance and reproduction of knowledge more than the discovery of new knowledge, von Humboldt championed a vision of the university based on the unity between teaching and research. Although, paradoxically, the model did not show much immediate influence in Germany, it was quickly adopted at the major universities in the United States, where it attained enormous success and pioneered the development of outside agency-funded large-scale university research. Only after its triumph in the United States did the model travel back to Europe to be adopted, first in Britain and Germany, and then in the rest of Europe. At a time when engineering education was a topic of heated debate in the United States, this model had naturally a major influence in that debate.

Before 1900, most universities worldwide were still guided by the goal of producing good practical engineers. They had replaced apprenticeships as the central means of education but they still conveyed, to a large extent, the culture of engineering practice (Seely, 1997, p. 345). Sixty years later, the scenario had changed drastically and roughly 50% of undergraduate instruction was based on "scientifically derived theory expressed in the language of mathematics" (Seely, 1997, p. 346). This change happened in a troubled period for the recognition of the status and brand image of engineers in English speaking countries, and particularly in the United States. In France, an engineer was a respectable professional who had graduated from *École Polytechnique* or from a university that replicated its scientific and formal ideals. In the United States, no such unique reference was recognised. In England, Thomas Tredgold (1788–1829) had championed the recognition of engineering as an applied science, borrowing from science the reputation it had accrued since the 17th century (Layton, 1991, p. 60). This interpretation was welcomed by American engineers in their efforts to overcome the lack of respect and low status accorded to their profession. Unlike the members of other professions, such as law and medicine, whose values and professional identities stressed their independence from employers and clients, engineers had been recognised as loyal servants of their paymasters and were actually held responsible, in the public eye, for many of the horrors practiced in the factories, mills, and mines during the Industrial Revolution (Layton, 1991, p. 60; Seely, 2005, p. 122). The affiliation within the science camp was, thus, an appealing one. Indeed, American engineers, who saw themselves as responsible for "designing, operating, and maintaining the large systems on which Americans increasingly depended, ranging from water and power systems in cities to massive bridges and railroad networks" (Seely, 2005, p. 116), could hardly bear the frustration and inferiority complex of their lack of social position (Seely, 2005, p. 116). An enthusiastic defender of this 'applied science' view of engineering was Vannevar Bush (1945), an influential voice in the debate about the nature of the engineering profession in the United States.

Another important contribution to the strong turn to science of American engineering education was the massive funding of university research by the federal government, following World War II, especially after the introduction of the space program in 1957. This turn to science made engineering graduates almost indistinguishable from science graduates and complaints from employers about the excessively analytical approaches of academia, the poor problem-solving abilities of graduates, and the absence of mutual understanding between universities and the corporate world were increasingly voiced (Seely, 2005, p. 117). The debate continues today, but important measures have been taken by agencies such as the Accreditation Board for Engineering and Technology (ABET) to redress the balance between science and design and to incorporate elements of the social sciences and humanities, so as to prepare engineering graduates "for leadership positions in society and business" (Seely, 2005, p. 119).

7 THE FOUR DIMENSIONS OF ENGINEERING

In the previous historical account, I have identified four main dimensions of engineering practice: the crafts, design, the basic sciences, and the social sciences (including business and management). In this section I discuss them individually and as a whole by referring to the holonic diagram in Figure 1, adapted from Figueiredo (2002; 2008) and Adams *et al.* (2011). The labelling of the vertical axis suggests that the disciplinary dimensions in the top half have historically lent themselves to more solid theoretical grounding, while the dimensions in the bottom half have been historically more inspired by the realm of practice. The left half suggests that the corresponding dimensions are more directed to people, while the right half is more concerned with nature (life, matter, energy) and information (with abstract mathematical entities included in the concept of information). My conjecture is that each engineer incorporates the distinctive threads of the above historical tradition in varying combinations, stressing one or the other according to personal style, maturity in the profession, and the pressure of circumstances. Pure practising engineers, if they existed, would be strongly committed doers following the cultural values of the crafts (Fig. 1). Pure design engineers would excel as visionaries and integrators, preoccupied with imagination, synthesis and projective thinking (Fig. 1). Pure engineering scientists would act as thinkers, concerned with analysis and explanation (Fig. 1). Pure business/management engineers would shine as entrepreneurs, managers, negotiators, and communicators (Fig. 1). However, as no pure engineers exist, each engineer has a multiple identity that embodies the cross-fertilisation of the four dimensions.

The concept of a person's multiple identities, which had difficult acceptance some years ago, is well recognised today. Marvin Minsky, a cognitive scientist, criticises,

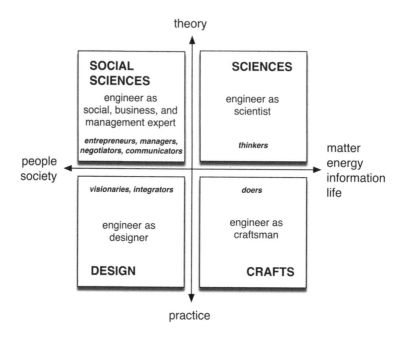

Figure 1 The four dimensions of engineering.

in *The Society of Mind*, the absurdity of the traditional belief in a single self which he claims has gravely impeded the progress of psychology (Minsky, 1988, p. 51). Boltanski & Thévenot, authors of *On Justification: Economies of Worth*, a foundational work of post-Bourdieu sociology, argue that people support the justification of their actions on various conflicting logics that compete to legitimise their views (Boltanski & Thévenot, 1991). The concept of multiple identities is also seen as essential by experts in complexity theory, such as Kurtz & Snowden (2003, p. 464), who remind us that a single person can be simultaneously a parent, a sibling, a spouse, and a professional, and can behave differently, in multiple combinations of these roles, depending on the context. François Laplantine, a French cultural anthropologist, also argues for the irreducibility of the person to a unitary behaviour and describes people as composite beings under the influence of several affiliations (Laplantine, 1994).

7.1 The crafts

As observed earlier, primitive engineering began in the realm of the crafts, originally inspired by utilitarian and artistic reasons. The spirit of the crafts in engineering remained strong until the 18th century when it suddenly started to be replaced by the spirit of the sciences. Still in the 18th century, the reverence for the practices of the craftsman was visible in the widespread acceptance of publications such as the French 35-volume *Encyclopédie, ou Dictionnaire Raisonné des Sciences, des Arts et des Métiers* (Encyclopedia, or Rational Dictionary of the Sciences, the Arts and the Crafts), published between 1751 and 1772 under the guidance of Denis Diderot, which became a worldwide bestseller, finding its way to readers as varied as Catherine the Great of Russia and merchants in New York (Mitcham, 1994, p. 115; Sennett, 2008, p. 90). The French *Encyclopédie* celebrated the daily practices of labouring and the values of those craftsmen who committed themselves to doing work well for its own sake (Sennett, 2008, p. 90). Like the artist-engineers of the Renaissance, the authors of the *Encyclopédie* realised that, as most of the craftsman's knowledge was tacit, beyond what can be expressed in words, its dissemination required not only the abundant use of images but also the immersion of its disseminators in practice (Sennett, 2008, p. 96). This was how Denis Diderot and some of his collaborators, including prominent French scientists and philosophers, came to realise the value of trial and error as a research tool and the relevance of error for the progress of knowledge and practice (Sennett, 2008, pp. 96–97). Other evidences of the importance of craftsmen in the 18th and early 19th centuries are illustrated by the decision of the British government to pass a law prohibiting the emigration of skilled craftsmen and the testimony that in spite of that law, which lasted for over a century, thousands of craftsmen emigrated in the early decades of the 19th century from Britain to the continent, attracted by much higher wages (Landes, 1998, pp. 278–280).

The popularity of the crafts suggested by the publication of the *Encyclopédie* and by the flow of technological talent from Britain to the continent in the 18th and early 19th centuries would not, however, last long and this period would mark the irreversible decline of craftsmanship as the dominant foundation of engineering. Many inventions of the industrial revolution had been produced by craftsmen who were relatively ignorant about science (Pacey, 1992, p. 178). However, industrial-age craftsmen-engineers such as James Watt, George Lee, William Reynolds, and John Banks had already a considerable scientific background and the technical problems posed by the use of iron technology and steam power in the textile industries were definitely getting beyond the limited scientific grasp of craftsmen, whatever their manual and intellectual skills (Pacey, 1992, p. 178). This had also been recognised in France, where, as mentioned earlier, the establishment of the *École des Ponts et Chaussés* marked the recognition that engineering education should be strongly anchored on the principles of science and mathematics. In spite of what can be described today as the extinction of the crafts, various authors (Mintzberg, 1987; Suchmann & Trigg, 1996; Ciborra, 1998; Sennett, 2008) advocate the renewal, in the professions, of the spirit of the crafts as a way to overcome the overly mechanistic approach of our times.

Even though craftsmanship is no longer a central foundation of engineering, much of its spirit and values are deeply engrained in the culture of the profession and contribute distinctly to its richness. It is useful, for this reason, to summarise here its main contributions to the legacy of engineering practice. Craftsmen are doers, in the sense that they have a strong drive to get things done. They are also perfectionists, in their sustained accumulation of skills, deep engagement, pride in their work, and compulsion to do their job well for its own sake (Sennett, 2008). They feel the need to explore worlds of possibility by improvising, using their imagination, and persistently overcoming resistance and ambiguity (Sennett, 2008). Although they naturally like problem solving, they usually excel at linking problem solving with problem finding (Sennett, 2008). The most tangible expressions of the success of good craftsmen are their top-quality accomplishments which stand before the world (Figueiredo, 2008).

7.2 Design

If we consider design in the sense attached to the word *disegno* in the Italian Renaissance, the ability to conceive something new, design lies at the very heart of engineering. *Disegno*, from the Latin verb *designare* (to designate, assign a meaning to, destine, imagine a destiny for), has a broad range of meanings from 'to draw' to 'to conceive' but the most consistent one, persisting to this day, is 'to conceive'. Design thus means the ability to imagine and produce new apparatuses, systems, and solutions. It is in this sense that Edwin Layton (1976, 69) states:

> "From the point of view of modern science, design is nothing, but from the point of view of engineering, design is everything. It represents the purposive adaptation of means to reach a preconceived end, the very essence of engineering".

Fifteen years later, he insists that "the engineer is a systems builder and engineering is the art and science of building systems" (Layton, 1991, p. 73). He also maintains that the 'ability to design' is the defining characteristic of the 'true' professional engineer (Layton, 1991, p. 68). The identification of engineering with design, understood in this sense, became widespread after the Second World War (Layton, 1991, p. 68), with design representing to some authors the distinctive ground for a whole philosophy of engineering (Lifson & Kline, 1968; Lewin, 1983; Lewin, 1979). Addressing specifically engineering research, Douglas Lewin (1979, 114) asserted that, to avoid the degeneration of engineering research into conventional scientific research, the engineering purpose and application had to be kept firmly in mind and research "should originate from the design function itself" (Lewin, 1979, p. 114).

The big revolution of design, which changed radically the practice of engineering, happened in the Renaissance, with the discovery of the ability to master the non-verbal component of technological creation. This is why the historian of engineering Eugene Ferguson declared in the opening of his book *Engineering and the Mind's Eye* (1992, xii):

> "an engineering education that ignores its rich heritage of nonverbal learning will produce graduates who are dangerously ignorant of the myriad of subtle ways in which the real world differs from the mathematical world their professors teach them".

This ability of the designer to 'see' with the mind's eye had already been recognised by Karl Marx in his comparison between the bee and the architect:

> "A spider conducts operations that resemble those of a weaver, and a bee puts to shame many an architect in the construction of her cells. But what distinguishes the worst architect from the best of bees is this, that the architect raises his structure in imagination before he erects it in reality." (Marx, 1906, p. 157)

The recent popularity of graphic design as an artistic and professional discipline that focuses on visual communication has considerably confused the traditional understanding of design, since it stresses an artistic essence that may seem to go against the ability to imagine and produce complex technical apparatuses, systems, and solutions, the core of engineering endeavours. To overcome this confusion, some authors refer to 'engineering design', but this is not the same thing: engineering pertains to design as much as the graphic arts, or information systems, or many other disciplines, including the social disciplines. Design cuts transversally across multiple disciplines, and it is in this general and increasingly rich sense that it should be understood. So, when discussing the nature of engineering practice, it is in this general sense that the word is used.

The nature of design as a pillar of engineering is easier to grasp if we contrast it with the nature of science. As far as method is concerned, the distinction offered by Gregory is illuminating:

> "The scientific method is a pattern of problem-solving behaviour employed in finding out the nature of what exists, whereas the design method is a pattern of behaviour employed in inventing things of value which do not yet exist. Science is analytic; design is constructive" (Gregory, 1996).

Also on method, specifically on its role in the process of creating new things, Nigel Cross observes that while method is vital to the practice of science, where it is essential to validate the results, it may be irrelevant to the practice of design, where the results do not need to (and in many cases must not) be repeated (Cross, 2001, p. 54). He also claims that, besides our traditional science-based approaches, we must concentrate on the designerly ways of knowing, thinking and acting (Cross, 2007).

The argument in favour of the acceptance of other ways of knowing, thinking and acting, beyond those of science, had also been explored extensively by Archer (1992), who noted that the scientific process seeks to isolate a phenomenon and abstract generalizable principles, with no major concern for the immediate usefulness of the results, letting scientists turn their minds to whatever interests them, provided they do it scientifically. On the contrary, design "is directed towards meeting a particular need, producing a practicable result and embodying a set of technological, economic, marketing, aesthetic, ecological, cultural and ethical values determined by its functional, commercial and social context" (Archer, 1992, p. 8). This view of design resonates with an identical view of engineering expressed by Vincenti:

> "For engineers, in contrast to scientists, knowledge is not an end in itself or the central objective of their profession. Rather, it is . . . a means to a utilitarian end . . ." (Vincenti, 1990, p. 6).

The importance of design as a core dimension of engineering justifies the identification of the most distinctive features of engineers when they act as designers. Above all, designers value synthesis much more than analysis and root their practice on holistic, contextual, and integrated visions of the world, rather than on the partial visions that are sought in science. In this sense, designers are strategic: they are led by a vision of the whole, an evolutionary dream where the whole generates the parts and the parts generate the whole. Unlike scientists, designers enjoy complex and ill-defined problems which they like to approach in exploratory ways, tolerating ambiguity, the exploration of alternative and compromise, the practice of improvisation, the probing use of non-scientific thinking and decision making in the face of incomplete knowledge, with the help of intuition, experience, and, very often, teamwork and co-creation (Cross, 2007; Figueiredo & Cunha, 2007; Adams *et al.*, 2011). They favour generative over deductive reasoning, they welcome analogy and metaphor, and they like to progress dialectically, moving between 'seeing that' and 'seeing as' (Wittgenstein, 1953). This is one of the reasons why their practices, sometimes described as design thinking, are frequently seen as better suited than those of science to confront the increasingly complex engineering problems of today's world.

7.3 The sciences

Just as the birth of science in Ancient Greece took a long time to influence the practice of engineering, the rise of modern science in the 17th century also propagated very slowly into the realm of engineering (Kirby *et al.*, 1990, p. 126). In contrast, the experimental attitude of engineering played an important role in stimulating the creators of modern science. Some of them were, in fact, engineers and many others exchanged views with engineers or collaborated in the discussion of the relationship between their theories and the design of their experimental setups. This has been explicitly recognised in the opening of Galileo's *Dialogues Concerning Two New Sciences* (Kirby *et al.*, 1990, p. 126). The influence of science upon the advance of engineering would only start becoming expressive with the French *École des Ponts et Chaussés* in the 18th century, just before it suddenly became overpowering and irreversible in the 19th century. Although some branches of modern engineering, such as civil and mechanical engineering, originated in ancient times, others would never have existed in the absence of advanced scientific contexts. Two major examples are electrical engineering and chemical engineering which developed in close collaboration with physics and chemistry respectively (Auyang, 2004, p. 45). Computer science, which originally developed as a mathematics discipline but has evolved to encompass engineering, is another example.

The dimension of the sciences is today, for good reasons, the strongest component of engineering. It should be seen, however, not as the core of engineering but as an essential aid. As put by Herbert Hoover, 31st president of the United States, himself an engineer, in his explanation of engineering as a profession:

> "It is a great profession. There is the fascination of watching a figment of the imagination emerge through the *aid* of science to a plan on paper. Then it moves to realization in stone or metal or energy. Then it brings jobs and homes to men. Then it elevates the standards of living and adds to the comforts of life. That is the engineer's high privilege" (my italic) (Hoover, 1954).

The sciences, seen as a core dimension of engineering practice, emphasise rigorous methods, mathematical abstraction, logical reasoning, measurable evidence, and the accurate prediction of outcomes. When engineers act in this dimension their preferred activity, for which they are highly regarded, is engineering science and the most tangible expressions of their success are original claims to knowledge that can be published in advanced research journals and presented at top research conferences (Figueiredo, 2008).

7.4 The social sciences

Hardy Cross (1885–1959), a structural engineer and university professor whose valuable reflections on the nature of engineering were posthumously published in the book *Engineers and Ivory Towers*, maintained that engineers are not primarily scientists:

> "If they must be classified, they must be considered more humanists than scientists. Those who devote their life to engineering are likely to find themselves in contact

with almost every phase of human activity. Not only must they make important decisions about the mere mechanical outline of structures and machines, but they are also confronted with the problems of human reactions to environment and are constantly involved in problems of law, economics and sociology" (Cross, 1952, p. 5).

Expressing his views on the nature of engineering, Edwin Layton observed that "engineering involves social change in an essential way" (Layton, 1991, p. 74). When interviewed for *Interactions*, an ACM journal, Austin Henderson stressed that when we design in the technical space we are also designing social activity (Ehrlich, 1998).

Unlike scientists, who often work for their own intellectual enjoyment regardless of immediate utility, engineers always work for patrons and in response to requests. They work for people and with people. What they design and build changes the world, often irreversibly, and it can affect hundreds, thousands, or millions of people. The social nature of their action includes the clarification of technical and business requirements with their clients; the negotiation of design options with the populations affected, such as politicians, environmentalists, and other pressure groups; the mobilisation of their teams; the compromises between partakers in the construction and deployment of their solutions; and even the anticipation of the social risks of those solutions. They must also develop strong insights into the business dimension of their projects, which they must, very often, manage from beginning to deployment. In full opposition to the stereotype of hackers turned inwards, the most successful engineers are, on the contrary, remarkable social experts.

The recognition of the social nature of engineering seems to have been evolving from ignorance to acceptance, with some setbacks. The popular definition of engineering by Thomas Tredgold (1788–1829) as "the art of directing the great sources of power in nature for the use and convenience of man", inscribed in the 1828 charter of the (British) Institution of Civil Engineers (Layton, 1991, 60; Ferguson, 1992, p. 1), has received multiple improvements over time. One such improvement, apparently introduced to overcome the sense of guilt felt by the engineering profession about the negative social effects of technology, was to redefine the core values of engineering to align them with the good of society (Layton, 1991). This redefinition was, however, far from effective in countering the belief in the neutrality of action of the engineers, who, trusting the objectivity of science, saw their activity as value-free (Layton, 1991). By decomposing tasks and isolating them from the overall picture, the bureaucratisation and division of labour of engineers in large organisations also contributed to reinforce this belief (Pacey, 1999, p. 176). An expressive example is given by Pacey: as napalm originally did not stick enough to the bodies of the victims, who could scrape it off to save their lives, the engineers involved in the project were requested to solve the abstract scientific problem of making napalm stick, which they did effectively, without apparently giving a thought to the tragic consequences of their solution (Pacey, 1999, p. 176). Today, a more widespread consensus seems to exist on the recognition that the technological is social (Bijker & Law, 1992, p. 4) and that engineering "is an instrument of social change and social revolution" (Layton, 1991, p. 74).

Another aspect of the social nature of engineering practice is the engagement of engineers in business. Layton (1971) analysed this dimension extensively in his book *The Revolt of the Engineers: Social Responsibility and the American Engineering*

Profession, where he emphasised the critical role of the tensions between science and business in shaping the engineering profession. On one hand, he noted: "The engineer is both a scientist and a businessman. Engineering is a scientific profession, yet the test of the engineer's work lies not in the laboratory, but in the marketplace" (Layton, 1971, p. 1). On the other hand, he stressed that the key problem of the engineer is that "he is expected to be both a scientist and a businessman, but he is neither" (Layton, 1971, p. 2). This strained dilemma is one the multiple aspects of the practice of engineering that the diagram in Figure 1 is an attempt to illustrate. Independent of the multiple solutions to the dilemma which engage the other dimensions of the diagram, management, which is a core component of the business activity, is also one of the core components of the activity of an engineer. It might be useful to recall in this context that Frederick Taylor (1856–1915), the father of modern management, was a mechanical engineer who tried to promote the re-foundation of management on the basis of engineering principles and believed, with his fellow members of the American Society of Mechanical Engineers (ASME), that all social problems could be solved by applying engineering methods (Layton, 1971).

The recognition of the social sciences as a core dimension of engineering lets us see engineering professionals not just as technologists but also as managers, business people and social experts capable of getting to grips with the social nature of the markets they address and act upon and also with the social complexity of the teams they belong to. When seen in this dimension, engineers are particularly concerned with the creation of social and economic value, the satisfaction of end users, and the effectiveness and motivation of their teams (Figueiredo, 2008).

8 THE WHOLE ENGINEER

The framework of Figure 1 puts together the four disciplinary dimensions of engineering discussed in this chapter. If we take into account the differences between dimensions examined in the previous section and if we bear in mind that epistemology is concerned with the inquiry into the nature and possibility of knowledge, the methods of producing knowledge, and the value of knowledge in a given domain, then we may anticipate major epistemological difficulties in combining the four dimensions. Examples of the tensions between the dimensions of business and science in engineering have been discussed by Layton (1971). Similar tensions have been observed in the debate between 'science advocates', 'design advocates', and 'systems advocates' in Williams' (2002) account of the disciplinary wars at MIT in the late 1990s. A more conciliatory and integrative example can be found in Henderson's suggestion that science, design, and engineering represent complementary activities of analysis, imagination, and implementation, respectively (Ehrlich, 1998).

In his reflection on the biological bases of creativity, Martindale (1999, p. 137) reminds us that creative thinking requires the combination of a small set of relatively common traits (intelligence, perseverance, unconventionality, and differing thinking). What makes creativity uncommon, he observes, is that these traits are rarely present in a single person. Something identical happens with engineering. According to our framework, a 'whole engineer' is the combination of a strategist/integrator, a scientist/thinker, a businessman/humanist/negotiator, and a doer. Each one of these dimensions is relatively frequent, but the widespread academic trend to educate engineers as one-dimensional scientists is making the combined occurrence of the four dimensions in a single graduate almost impossible. This transformation of the whole engineer into a formatted engineering scientist is actually accelerating, since most of those who decide on the future of engineering education are already one-dimensional engineering scientists in a process of fast convergence towards self-perpetuation.

In this context, Rosalind Williams could be right when she declares that engineering as a profession is disintegrating. This would happen not only because, as she points out, the popularity of engineering is making the profession widespread and diluted, but also because the four disciplinary components that, over the millennia, made engineering a powerful and coherent whole are now falling apart. In fact, some major engineering achievements of our times have not been sparked by people who graduated in engineering, but by people who, in the light of the four historical dimensions of our framework, can be described as whole engineers.

The approach taken by the Wright brothers, in the early 1900s, in their invention of the controllable fixed-wing powered aeroplane and of the three-axis control system of today's aeroplanes was "that of the engineer – open-minded trial, systematic tabulation of data, the intelligent interpretation of results, together with boldness to follow where the facts led" (Kirby *et al.*, 1990, p. 416). The fact that the Wright brothers were bicycle mechanics rather than engineers stresses the key role of their craftsman's culture of persistence in the search of perfection. The wind tunnel of the Wright brothers, where they tested more than 200 types of winged surfaces and measured lift and drag in monoplane, biplane and triplane models (Kirby *et al.*, 1990, p. 416), is an illustrative example of this culture. Thomas Edison, who never took a university degree but was one of the most celebrated innovators in electrical engineering, was a persistent doer, a keen businessman, and an exceptional designer (Hughes, 1991; Carlson, 1999) who surmounted his lack of scientific knowledge by employing prominent scientists such as Francis Upton, von Helmholtz's former student in Germany (Pacey, 1990, p. 170). Another outstanding contributor to the progress of electrical engineering, Nikola Tesla, never obtained an engineering degree but was considered one of the most brilliant engineers of his time. Not surprisingly, he fulfilled the four dimensions of our model of the whole engineer. In more recent times, entrepreneurs such as Steve Jobs and Bill Gates offer examples of unparalleled engineering innovators who did not obtain an engineering degree but assumed with resounding success the four historical dimensions of the whole engineer.

Between the inevitable disintegration suggested by Williams (2002) and the perfect reconciliation I would like to defend, one could possibly attempt some intermediary compromises. The development of common theoretical understanding in a domain of knowledge such as engineering, made up of various disciplines, would suggest the exploration of the mutual interpenetration between the epistemologies of these disciplines. This would mean adopting a transdisciplinary approach to explore the continuous synergy and interaction between the four disciplines in a wide variety of contexts of application. The disintegration of the engineering profession could thus be avoided by exploring the transdisciplinary nature of the bonds between the four constituent disciplines. It would then be possible to talk about an ideology of engineering (Layton, 1976) and about the occupational identities (Becker & Carper, 1956) of its professionals. Of course, this would not require every engineer to be a full expert in all the four dimensions. On the contrary, the personal style and stage of professional development of each engineer would determine both the emphasis given to each dimension and the approach adopted to manage the tensions between dimensions. In any case, the ability to recognise the existence of the four dimensions would help visualise the opportunities for collaboration across them. It would also facilitate the dialogue and synergy between the professionals matching the myriad combinations possible within the framework. If taken into account in the education of future professionals, it would also guide them in the selection of their profiles and improve their ability to cross borders and maintain the dialogue within multi- and transdisciplinary teams. Although no engineering professional could cover well all the four dimensions, any prospective professional should ideally be aware of the attractiveness and coherence of the whole and of the challenge of being able to collaborate across the borders.

Two major difficulties still stand in the way of a consolidated whole view of engineering. One is the current systematic neglect of the core role of design in engineering. The other is the narrow understanding of engineering practice as mere problem solving. The two difficulties are related and they both seem to result from a failure to contextualise engineering in the complexity of today's world. Globalisation, the increasing power of computers, the exploding number of human and technological systems, the rising dependence between such systems, the escalating speed of communications, the changing competitive environments and organisational needs, and, above all, the human and social dimensions of all these issues are contributing to increase dramatically the complexity of our world. Today, the design of a system often develops from a puzzle of incomplete requirements that keep changing, with design and implementation co-existing. This uncertainty is found in the growing 'wicked problems' (Rittel & Webber, 1973) of our time, in which the interdependence between technology and social and organisational factors requires design strategies where formulation and solution develop in parallel, feeding each other, and where emergence needs to be intelligently accounted for (Figueiredo & Cunha, 2007). This means that the most prominent role of engineering today is not to solve problems but to clarify and formulate them in contexts that require holistic visions, social awareness and projective thinking. Science is essential to enable engineers to work on the parts but it is their overarching design approaches that let them see the whole. The more the world becomes socially and technologically complex and interdependent, the more critically it depends on the ability of engineers to act as designers of whole sociotechnical systems.

9   CONCLUSIONS

In the opening of this chapter, when commenting on Williams' (2002) claim that engineering is suffering a process of "expansive disintegration", I recalled the multiple identity crises engineering professionals have gone through in the last two centuries. I also suggested that those who wish to call themselves engineers and those who are responsible for the education of future engineers would have much to gain by looking into the heritage of thousands of years of engineering practice. In this chapter, I tried to traverse those thousands of years so that I could identify, with the help of a simple framework, the defining characteristics of the engineer who remains true to the historical values of the profession. This 'whole engineer' is, to some extent, the antithesis of the 'disintegrated engineer' described by Williams. I believe, however, that both concepts are useful. It is a well-known fact of history and legend that the most virtuous creations of the new are often built over the ruins of a glorious past. As Williams announces the end of a glorious past, we may be reminded of the phoenix, the mythological bird that burns itself to ashes every few hundred years before it is reborn from its own ashes, combining the energy and creativity of youth with the maturity and wisdom of its valuable past. What I tried to do in this chapter was to contribute to the groundwork for such a rebirth.

REFERENCES

Adams, R., Evangelou, D., English, L., Figueiredo, A. D., Mousoulides, N., Pawley, A. L., Schifellite, C., Stevens, R., Svinicki, M., Trenor, J. M. & Wilson, D. M. (2011) Multiple Perspectives on Engaging Future Engineers. *Journal of Engineering Education*, 100 (1), 48–88.

Archer, B. (1992) The Nature of Research in Design and Design Education. In: Archer, B., Baynes, K. & Roberts, P. (eds.) *The Nature of Research into Design and Technology Education: Design Curriculum Matters*. Loughborough, Department of Design and Technology, Loughborough University.

Auyang, S. Y. (2004) *Engineering – An Endless Frontier*. Cambridge, Mass, Harvard University Press.

Becker, H. S. & Carper, J. W. (1956) The Development of Identification with an Occupation. *The American Journal of Sociology*, 61 (4), 289–298.

Bijker, W. E. & Law, J. (1992) *Shaping Technology/Building Society: Studies in Sociotechnical Change*. Cambridge, Mass, MIT Press.

Bjerklie, D. (1998) The Art of Renaissance Engineering. *MIT's Technology Review Magazine*. [Online] Available from: http://www.technologyreview.com/featured-story/400136/the-art-of-renaissance-engineering/ [Accessed 3rd January 2013].

Boltanski, L. & Thévenot, L. (1991) *De la Justification: Les Économies de la Grandeur*. Paris, Gallimard.

Bush, V. (1945) *Science The Endless Frontier*. United States Government Printing Office. [Online] Available from: http://www.nsf.gov/od/lpa/nsf50/vbush1945.htm [Accessed 3rd January 2013].

Buxton, B. (2007) *Sketching User Experiences: Getting the Design Right and the Right Design*. San Francisco, Focal Press.

Capra, F. (2007) *The Science of Leonardo*. New York, Anchor Books.

Carlson, W. B. (1999) *Banishing Prometheus? Edison, Invention, and Representation*. Critical Studies Workshop: Writing Science. The Stanford Humanities Center. [Online] Available from: www.stanford.edu/dept/HPS/WritingScience/etexts/Carlson/Prometheus.html [Accessed 3rd January 2013].

Chiu, Y. C. (2011) *A History of Ancient Project Management: From Mesopotamia to the Roman Empire*. Delft, Eburon Academic Publishers.

Ciborra, C.U. (1998) Crisis and Foundations: an Inquiry into the Nature and Limits of Models and Methods in the Information Systems Discipline. *Journal of Strategic Information Systems*, 7, 5–16.

Cross, H. (1952) *Engineers and Ivory Towers*. New York, McGraw-Hill, Inc.

Cross, N. (2001) Designerly ways of knowing: design discipline versus design science. *Design Issues*, 17 (3).

Cross, N. (2007) *Designerly Ways of Knowing*. Berlin, Birhäuser.

Cypher, M. & Richardson, I. (2006) An Actor-Network Approach to Games and Virtual Environments. *Proceedings of the 2006 International Conference on Game Research and Development. 4–6 December 2006, Australia*, Murdoch University. pp. 254–259.

Dewey, J. (1930) *Human Nature and Conduct*. New York, Modern Library.

Ehrlich, K. (1998) A conversation with Austin Henderson. *Interactions*, 5 (6), 36–47.

Ferguson, E. S. (1992) *Engineering and the Mind's Eye*. Cambridge, Mass, The MIT Press.

Figueiredo, A. D. (2002) Accreditation and Quality Assessment in a Changing Profession. *Proceedings of the International Conference on Engineering Education, ICEE 2002, July 30–August 3 2012, UK*. Manchester.

Figueiredo, A. D. & Cunha, P. R. (2007) Action Research and Design in Information Systems: Two Faces of a Single Coin. In: Kock, N. (ed.) *Information Systems Action Research: An Applied View of Emerging Concepts and Methods*. New York, Springer. pp. 61–96.

Figueiredo, A. D. (2008) Toward an Epistemology of Engineering. *Proceedings of the Workshop on Philosophy & Engineering (WPE 2008), 10–12 November 2008*. London, Royal Engineering Academy.

Foote, N. N. (1951) Identification as the Basis for a Theory of Motivation. *American Sociological Review*, 16 (1), 14–21.

Galluzzi, P. (1987) The Career of a Technologist. In: Terry, J. (ed.). *Leonardo da Vinci: Engineer and Architect*. Montreal, Montreal Museum of Fine Arts.

Gregory, S. A. (1966) *The Design Method*. London, UK, Butterworth.

Kirby, R. S., Withington, S., Darling, A. B. & Kilgour, F. G. (1990) *Engineering in History*. New York, Dover Publications.

Kurtz, C. F. & Snowden, D. J. (2003) The New Dynamics of Strategy: Sensemaking in a Complex and Complicated World. *IBM Systems Journal*, 42 (3), 462–482.

Hill, D. (1996) *A History of Engineering in Classical and Medieval Times*. London, Routledge.

Hoover, H. (1954) Engineering as a Profession. *Engineer's Week*. [Online] Available from: http://www.hooverassociation.org/hoover/speeches/engineering_as_a_profession.php [Accessed 3rd January 2013].

Hughes, T. P. (1991) The Electrification of America: the System Builders. In: Reynolds, T. S. (ed.) *The Engineer in America: A Historical Anthology from Technology and Culture*. Chicago, The University of Chicago Press. pp. 191–228.

Imhausen, A. (2006) Ancient Egyptian Mathematics: New Perspectives on Old Sources. *The Mathematical Intelligencer*, 28 (1), 19–27.

Kuhn, T. S. (1970) The Structure of Scientific Revolutions. In: Neurath, O., Carnap, R. & Morris, C. (eds.) *Foundations of the Unity of Science*. Vol. 2. Chicago, The University of Chicago Press.

Kuhn, T. S. (1977) Second Thoughts on Paradigms. In Suppe, F. (ed.) *The Essential Tension*. University of Chicago Press.

Kirby, R. S., Withington, S., Darling, A. B. & Kilgour, F. G. (1990) *Engineering in History*. New York, Dover Publications. Inc.

Landes, D. S. (1998) *The Wealth and Poverty of Nations*. London, Abacus.

Laplantine, F. (1994) *Transatlantique. Entre Europe et Amérique Latine*. Paris, Payot.

Layton, E. T. (1971) *The Revolt of the Engineers: Social Responsibility and the American Engineering Profession*. Cleveland, Ohio, The Press of Case Western Reserve University.

Layton, E. T. (1976) American Ideologies of Science and Engineering. *Technology and Culture*, 17 (4), 688–701.

Layton, E. T. (1991) A Historical Definition of Engineering. In: Durbin, P. T. (ed.) *Critical Perspectives on Nonacademic Science and Engineering*. Cranbury, NJ, Lehigh University Press. pp. 60–79.

LeMoigne, J.-L. (1999) *Les Épistémologies Constructivistes*. 2nd edition. Paris, Presses Universitaires de France.

Lewin, D. (1979) On the Place of Design in Engineering. *Design Studies*, 1 (2), 113–117.

Lewin, D. (1983) Engineering Philosophy: The Third Culture. *Leonardo*, 16 (2), 127–132.

Lifson, M. W. & Kline, M. B. (1968) *Design: The Essence of Engineering*. Los Angeles, University of California.

Madsen, D. A. & Madsen, D. P. (2011) *Engineering Drawing and Design*. 5th edition. Delmar Cengage Learning.

Martindale, C. (1999) Biological Bases of Creativity. In: Sternberg, R. (ed.) *Handbook of Creativity*. New York, Cambridge University Press.

Marx, K. (1906) *Capital: A Critical Analysis of Capitalist Production*. Swan Sonnenschein & Co.

Minsky, M. (1988) *The Society of Mind*. New York, Simon & Schuster.

Mintzberg, H. (1987) Crafting Strategy. *Harvard Business Review*, 65 (4), 66–75.

Miroslav, V. (2001) *The Pyramids: The Mystery, Culture, and Science of Egypt's Great Monuments*. New York: Grove Press.

Mitcham, C. (1994) *Thinking Through Technology: the Path Between Engineering and Philosophy*. Chicago, The University of Chicago Press.

O'Leary, D. L. (1949) *How Greek Science Passed to the Arabs* (1st ed.). London, Routledge & Kegan Paul.

Oleson, J. P. (2008) *The Oxford Handbook of Engineering and Technology in the Classical World*. Oxford, UK, Oxford University Press.

Pacey, A. (1990) *Technology in World Civilization: A Thousand Year History*. Cambridge, Mass., The MIT Press.

Pacey, A. (1992) *The Maze of Ingenuity: Ideas and Idealism in the Development of Technology* (2nd ed.). Cambridge, Mass., The MIT Press.

Pacey, A. (1999) *Meaning in Technology*. Cambridge, Mass., The MIT Press.

Pindur, W., Rogers, S. E. & Kim, P. S. (1995). The History of Management: a Global Perspective. *Journal of Management History*. 1(1). pp. 59–77.

Reynolds, T. S. (1991) The Engineer in 19th-Century America. In: Reynolds, T. S. (ed.), *The Engineer in America: A Historical Anthology from Technology and Culture*. Chicago, The University of Chicago Press, pp. 7–26.

Rittel, H. & Webber, M. (1973) Dilemmas in a General Theory of Planning. *Policy Sciences*. 4. pp. 155–169.

Roberts, J. M. (2002) *The New Penguin History of the World* (4th ed.). London, Penguin Books.

Romanyshyn, R. D. (1989) *Technology as Symptom and Dream*. London, Routledge.

Sennett, R. (2008) *The Craftsman*. New Haven, CT, Yale University Press.

Seely, B. E. (1997) Research, Engineering and Science in American Colleges: 1900–1960. In: Cutcliffe, S. H. & Reynolds, T. S. (eds.) *Technology and American History: A Historical Anthology from Technology and Culture*. Chicago, The University of Chicago Press. pp. 345–387.

Seely, B. E. (2005) Patterns in the History of Engineering Education Reform: A Brief Essay. *Educating the Engineer of 2020: Adapting Engineering Education to the New Century*. Committee on Engineering Education, National Academy of Engineering, Washington, National Academic of Sciences.

Simson, O. G. von (1952) The Gothic Cathedral: Design and Meaning. *Journal of the Society of Architectural Historians*, 11 (3), 6–16.

Simson, O. G. von (1956) *The Gothic Cathedral: Origins of Gothic Architecture and the Medieval Concept of Order*. New York, Harper & Row.

Suchman, L. A. & Trigg, R. H. (1996) Artificial Intelligence as Craftwork. In: Chaiklin, S. and Lave, J. (eds.) *Understanding Practice*. Cambridge, UK, Cambridge University Press. pp. 144–178.

Vincenti, W. G. (1990) *What Engineers Know and How They Know It: Analytical Studies from Aeronautical History.* Baltimore, The John Hopkins University Press.

Williams, R. (2002) *Retooling: A Historian Confronts Technological Change.* Cambridge, Mass, MIT Press.

Wilson, C. S. J. (2000) *Architectural Reflections: Studies in the Philosophy and Practice of Architecture.* Manchester, UK, Manchester University Press.

Wittgenstein, L. (1953). *Philosophical Investigations. Translated by G.E.M. Ascombe.* New York, Macmillan Publishing, Co., Inc.

Chapter 2

Towards a theoretical framework for engineering practice

James Trevelyan
School of Mechanical and Chemical Engineering, The University of Western Australia,
Perth, Western Australia

I INTRODUCTION

This book provides, perhaps for the first time, a series of detailed studies of engineers at work, with the aim of providing insight into engineering as it is practiced today in several different settings in different countries. Our project emerged from discussions that highlighted the scarcity of such literature. The small set of extant contributions lie in such diverse media that they are hard to find, especially for engineers and engineering educators. Even though more have appeared in the last decade, the relative lack of such literature is surprising, given the critical importance of engineering in sustaining human life on our planet today and in the future (Barley, 2005; Trevelyan & Tilli, 2007). This lack of readily available literature might explain why contemporary notions of engineering have drifted far from the realities of practice and are in urgent need of revision.

Taken together, the contributions in this book and particularly this introductory chapter, argue that the landscapes of practice traversed by professional engineers are shaped by many factors that seldom emerge in contemporary discourses on engineering education and practice. Each of the chapters reveals particular aspects of practice, many through different "lenses", each lens associated with a particular contemporary research community represented by our contributors who have all studied engineers at work.

Antonio Dias de Figueiredo gives us a historical perspective which helps us understand how we have arrived at this juncture, acknowledging the relative scarcity of both historical and contemporary studies of engineering. He reminds us of historical and sociological perspectives on engineering that emerged in recent decades identifying the elements of craft, physical sciences, social sciences and design. Frédéric Kaplan, Matthieu Hubert, and Dominique Vinck describe the emergence of particular technologies as actor networks. These allow participants to understand the complex social dynamics that shape the constraints within which they perform their technical work.

Jim Borgford-Parnell and his colleagues show how conversation and thinking aloud protocol analysis can reveal insights into social interactions and design thinking in a series of closely observed design team meetings. This technique has been extensively used in the design studies and troubleshooting research communities.

Several chapters rely on different qualitative observation and analysis methods. Robin Adams and Tiago Forin, writing from an education and learning science perspective, use phenomenography to investigate the differences and similarities in the way that people experience cross-disciplinary practice in engineering contexts. Whereas some experience this simply as working together, others find it transformative. Rachel Itabashi-Campbell and Julie Gluesing, taking an organisation science perspective, explore engineering problem solving in the automotive industry which also, inevitably, crosses disciplinary boundaries. Successful problem solving turned out to be dependent on a social performance: negotiating the meaning of the problem and gaining the alignment of diverse stakeholders. Andrew Chilvers and Sarah Bell expose the constraints within which engineers and architects "lock in" future sustainability into our buildings, writing from an environmental sustainability perspective. Donna Rooney and her colleagues draw our attention to workplace learning by engineers, working with theoretical perspectives that have been developed to describe occupational practices in many communities.

Eileen Goold and Frank Devitt's chapter recognises the fundamental significance of mathematical thinking in contemporary engineering practice, finding that the application of tacit mathematical knowledge underpins technical thinking in engineering, rather than the explicit use of particular mathematical techniques.

My fellow editors, Bill Williams and José Figueiredo, compare aspects of engineering practice in Portugal with my own earlier published results from Australia. Their evidence provides further support for the idea that the predominant aspect of engineering practice is informal technical coordination (Trevelyan, 2007).[1] My own studies of engineering practice were motivated by observations of practice in South Asia as described in a later chapter in the book. Perhaps in contrast to other contributors, we found ourselves working in relative isolation from established research communities. This isolation, and the shared challenges we encountered in locating relevant literature that could help us, led to the meeting in Madrid[2] from which most of the chapters in this book emerged.

The aim for this chapter, therefore, is to offer a theoretical framework within which each of the different disciplinary perspectives of the chapters contributes explanations for aspects of engineering practice in a way that can inform practitioners, educators and researchers.[3] This framework has emerged from reflections on several related studies of engineers across most main disciplines in several countries by the author and his students.[4]

Many of the chapters in this book originate from research communities that normally may have only tenuous links with engineering practitioners. The research communities represented by the authors can be regarded within a traditional "engineering" academic community as peripheral to "hard core technical" engineering, at best somehow related to management. The different bodies of work that provide the intellectual foundations for each of the chapters are hardly known in most engineering schools. To the extent that they are known in management academies, most address "niche" issues, beyond the reach of most mainstream interests.

---

[1] Further evidence of this was presented by (Anderson *et al.*, 2010).
[2] Round Table on Engineering Practice Research, Madrid, October 8, 2011.
[3] Some readers might associate this term with a particular school of thought, or a body of theory that informs a particular study. In this paper I have used the term with a broader meaning.
[4] For further details of methods, data collection and analysis see (Trevelyan, 2010, 2012b) Student papers are available at http://www.mech.uwa.edu.au/jpt/pes.html.

The objective of this chapter, therefore, is to argue that the diverse contributions and perspectives in this book hold great value for young engineers, and their educators. Until now, few of them would find it easy to explain their relevance for practice. They could easily dismiss this collective contribution as fluffy psychology and philosophising with little relevance for real engineering.

Why, then, is this book so relevant?

The contemporary theory-driven approach that underpins engineering education can be traced to the 1950s when many argued that traditional practice-based instruction was unable to sufficiently empower students to make use of rapid technical and scientific advances (Jørgensen, 2007). The Grinter report (1955) explained the purpose of engineering in these terms: "Continual maintenance and improvement of man's material environment, within economic bounds, and the substitution of labour-saving devices for human effort" as well as an essential role in maintaining public safety.[5] At that time, intellectual theories that had the most explanatory or predictive power to assist with these objectives were based in physical sciences and expressed in the language of mathematics.

Definitions of engineering naturally coalesced around the espoused body of knowledge taught in engineering schools: the application of science in the context of design and technical problem solving such as optimisation (Layton, 1991). In the almost completely unnoticed absence of detailed evidence from systematic observation of engineers at work, there were few well-founded arguments for challenges (Trevelyan, 2012c). This narrow view of engineering practice influenced many others. For example, Donald Schön focused his analysis of engineering on the technical inventor and problem-solver while at the same time calling for professionals with greater awareness of why they perform their work (Schön, 1983, pp. 172–182, p. 274). Engineers themselves described their work in those terms, regarding the rest, often the greater part of their work, as non-technical and therefore not engineering, perhaps management. This view persists: many of the engineer participants in our studies have indicated that they hardly do any "real engineering" (Trevelyan, 2010), often framing this as a confession, some with expressions of guilt. Engineers often describe themselves as technical problem solvers, even though real industrial problems that can be resolved entirely within a single technical domain are infrequent (Jonassen et al., 2006; Korte et al., 2008). Even so, these particular episodes are valued highly by many engineers, some of whom point to them as landmarks in their careers.

Compared with the 1960s when today's engineering curricula solidified around the seed crystals provided by mathematics and the physical sciences, we face a very different intellectual landscape. There are four main differences.

First, there is now a small but growing body of evidence from diverse sources demonstrating that engineering practice is dominated by intellectually challenging

---

[5]The change in engineering education was strongly advocated by the Grinter Report in which mathematics, physics, and chemistry were recognised as the basis for engineering education because of "great changes" in these sciences in preceding decades and the likelihood of further advances to come. The value of study in the liberal arts and humanities was strongly advocated in response to comments on a draft report that contained much less emphasis on these aspects (the comments came mainly from industry). While the committee advocated spending 20% of study time on these aspects, few if any engineering programmes have achieved this (Grinter, 1955).

socio-technical activity that cannot easily be reconciled with earlier descriptions of engineering based on solitary technical design and technical problem solving activity. For example, Dominique Vinck has demonstrated how particular technical issues need to be 'represented', to have a voice through a human actor, a person who has been assigned special responsibility in order for them to be considered meaningfully (Vinck, 2003, Ch. 1).[6] Most chapters in this book represent further contributions to this evidence which is now hard to refute.

Second, engineers are aware that their traditional education does not prepare them well for practice (e.g. Martin *et al.*, 2005). The traditional model of engineering, based on the application of physical sciences to technical problem solving, optimisation and design, relies mainly on the idea that mathematical expressions derived from physical sciences and mathematics can explain and predict the boundaries of engineering feasibility.[7] As the evidence presented in the chapters of this book demonstrates, we have now moved beyond this relatively simplistic idea. While the strength of materials like steel and concrete impose limitations on the height of tall buildings, there are many other factors that shape the landscapes and boundaries of engineering feasibility, mostly involving human behaviour, even for novices. Bill Williams and José Figueiredo, for example, explain how reputation, regulations and commercial practices by other engineers in a competitive market also influence these landscapes and their boundaries.

Third, the evidence presented in this book rests on intellectual foundations that came from the advances in many fields outside the traditional basic sciences on which most contemporary academic engineering discourse is based. These advances now provide powerful explanatory insights into many more of the human behavioural factors that shape contours of engineering feasibility: the issues that demand most attention from engineers in their workplaces. Systematic observations have revealed many of these issues. The predominance of technical coordination and the human perception and behaviour issues that influence sustainability are examples of these factors.

Fourth, it is now apparent that the results of engineering practice strongly depend on localised social, economic and political factors, even though the underlying principles of the natural sciences are universal, as explained in my later chapter in this book. The influence of these local factors requires us to understand how social, cultural, philosophical, even religious beliefs affect practice, the sharing and distribution of human knowledge, and the financial capital on which it depends.

The fields that have contributed to these advances include education, neurophysiology, psychology, philosophy, linguistics, economics, sociology, anthropology, history, finance, accounting, marketing, management, organisation science, design studies, political science, and the biological sciences.

It is time, therefore, to rethink our conceptions of engineering in the light of these developments, and provide a much stronger intellectual foundation for practice and, ultimately, educators.

---

[6]In chapter 1 of his book, Vinck describes how design flaws in a machine only came to light when an engineer was specifically allocated the task of designing, in this case, a shielding wall for a particle accelerator.
[7]On page 3, we read: "Engineering practice is, in its essence, problem solving". (Sheppard, Macatangay, Colby, & Sullivan, 2009, p. 3)

The need for revision can be illustrated by some readily observable aspects of engineering practice and education today. The Grinter report (1955) emphasised "the inability of engineers to express themselves in clear, concise, effective, and interesting language" and "the importance to engineers of an acquaintance with the humanities and social sciences". These comments continue today, almost unchanged, and the consistency of such comments invites a search for a deeper intellectual explanation.[8]

Even though engineers would readily associate lack of proficiency in a technical skill with inadequate education, there is a widespread view within the engineering community that communication can be learned merely by practice, without formal instruction.[9] Formal instruction, when employed, tends to be resisted by students (e.g. Paretti, 2008). As long as communication retains its label as a non-technical, generic skill, outside the theoretical framework of engineering, this is likely to continue.

Another aspect is the consistency with which the "non-technical" label is assigned to any social interaction, even management activity within engineering.[10] This is an intriguing assumption contradicted by observations from the workplace that reveal that the content of engineers' social interactions, especially in the workplace, is dominated by technical issues.

Further, while there is continuing debate within engineering schools on the extent to which science fundamentals and mathematics should be taught by engineers, there is little if any debate on the extent to which relevant aspects of social sciences should be taught by engineers. To the extent that liberal arts, social sciences and humanities are part of an engineering curriculum (accepted often with considerable reluctance from engineering academics), such teaching is almost invariably sourced outside engineering faculties. This reinforces the notion that they lie outside the accepted framework of engineering knowledge.

The links between engineering practice and the physical sciences are unquestioned yet the equivalently strong links between engineering practice and the human sciences seem to be missing. The language of the Grinter report (1955) reflects this separation. Mathematics, physics and chemistry were labeled as the "foundations of engineering curricula". Inclusion of studies in the humanities, "an acquaintance with the social sciences", on the other hand, was justified on the basis that a graduate[11] "must be not only a competent professional engineer, but also an informed and participating citizen, and a person whose living expresses high cultural values and moral standards". This language remains unchanged in engineering academies today where professional skills are widely regarded as non-technical components of engineering, on the curricular margin,[12] yet this is refuted by many of the observations of engineering practice reported in this book that demonstrate that collaboration based on social interactions

---

[8]Similar comments have appeared in numerous contemporary reports, for example (Dahlgren *et al.*, 2006; Spinks *et al.*, 2006).

[9]See, for example, the frequent reliance on mere practice to develop teamwork skills in engineering education, reported in (Sheppard *et al.*, 2009, p. 67).

[10]Two examples will illustrate this. First, "professional skills, practices beyond the core of technical problem solving", (Downey, 2009, p. 69) Second, (Engineering Education Australia, 2012).

[11]Referred to as "he" in the report. However, a similar report produced today would almost certainly use more inclusive language.

[12]See, for example, similar language in (American Society of Civil Engineers, 2008).

lies at the core of technical practice. As the Grinter report acknowledged, insights gained from the arts, humanities and social sciences are critical for professional competence.

Now, it is time to acknowledge the evidence from the contributions of this book, as well as other published literature, and reconstruct our concepts of engineering in a way that includes human sciences alongside mathematics, physics, chemistry and biology in the scientific foundations of engineering.

2 ENGINEERING PRACTICE – A MUFFLED DISCOURSE?

The research studies that eventually led to this chapter were stimulated by curiosity: seeking an answer to the question "What do engineers do?" The question emerged from observations in Pakistan and India that challenged long-held assumptions about the capabilities of engineers one could take for granted in an industrialised country.[13]

Therefore, the primary focus of our research has been to understand aspects of engineering practice and provide some answers for these questions. Together we have explored a small number of engineering microcosms, shining tiny lights on small features in a largely unexplored cavern of darkness. We have perhaps exposed fragments of the work of only 200 engineers.[14]

Several other researchers have been similarly motivated, including Bechkey, Bailey, Tonso, Horning, and Faulkner.[15] However, they were writing for audiences other than engineers. For us as engineers, unlike many working in the social sciences, the primary objective at the time was a description of practice that would explain the work of our constituency.

Then, just a few months back, one of our rare engineering academics with extensive industry experience asked me what I meant by the term 'engineering practice'. Given that he had worked for a decade as an engineer, it was surprising that the term meant nothing to him. I explained it in terms of the knowledge that engineers need to be effective practitioners, such as how to devise a robust specification document that would help the supplier provide suitable materials or equipment and techniques for technical coordination.

This was just one of many indications that have come to us pointing to evidence of amnesia within the engineering community about practice knowledge. Another came soon afterwards.

With a colleague, I was asked to provide comments on a new and completely revised draft of professional engineering competencies for Engineers Australia, the pre-eminent organisation in Australia that sets the requirements for entry to the profession. The draft that had emerged from extensive consultation in the engineering community omitted many significant aspects of practice. Critical aspects of competence that create economic value were missing, as were competencies important for earning income as

---

[13] See the later chapter in this book.

[14] Some are listed in (Trevelyan, 2010, 2012b; Trevelyan & Tilli, 2010). Other contributions are available on-line at http://www.mech.uwa.edu.au/jpt/pes.html.

[15] (Bailey & Barley, 2010; Bechky, 2003; Faulkner, 2007; Gainsburg *et al.*, 2010; Horning, 2004; Tonso, 2006).

an engineer, such as gaining approval for money to be invested in a project. Creating value was listed as one of the four core competencies, but it was interpreted entirely in the sense of achieving a degree of technical perfection: economic and social values were not included. Some of our research has demonstrated how engineers find it difficult to articulate the economic value of their contributions but it was still surprising to find this aspect completely absent (Trevelyan, 2012b).

In a third development, some of our research participants have explained recent experiences in companies when key organisational processes have been forgotten and the supporting documents misplaced over the last few decades.

On reflection, I realised that my students and I had perhaps taken on the role of Foucault's historians, retelling engineering practice from our observations and experiences and the stories that engineers have related for us. We are, perhaps, reawakening a "muffled discourse that has at least partly disappeared".[16] The realities of engineering practice, even within the profession itself, but particularly within the academy, have become a muffled discourse.

My own experience of engineering practice, the first 20 years of my career, impressed on me that the technical engineering and mathematical sciences on which most engineering careers are built do not explain much of the landscape of practice. While they certainly provide helpful insight into machines, systems, information and materials that populate object worlds (Bucciarelli, 1988), Foucault's planes of objects (1984) that engineers feel comfortable thinking about, technical engineering sciences provide limited insight into many other factors that shape the contours of engineering feasibility.

In other words, engineers have to work within many constraints (Koen, 2009). Some are imposed by the laws of physics, usually described in mathematical terms (e.g. gravity, strength of materials, electrical conductivity, etc.).

However, creation of engineering systems and artefacts relies on human actions; therefore feasibility is limited by human capabilities as well as the laws of physics. Practice is constrained by how much we can remember in our heads, and how reliably and how fast we can learn more, how effectively we can tell this to others, time available to complete the work, how much finance investors are willing to make available, the skills and capabilities of the people at hand, how they are organised, and many other factors.

Therefore, if we conceive of a map that shows all the possibilities that could provide effective solutions for a particular engineering project, we can also imagine contours of difficulty. The low contours include possibilities that are easier to achieve, and the high contours denote more difficult possibilities. The boundaries are partly determined by the laws of physical sciences, but can equally be shaped by finance, timescale, precedent, regulations and the distribution of human know-how.

Contemporary engineering science draws on natural science theories, for example, in physics, chemistry, mathematics, continuum mechanics, computation, geology, biology, physiology and materials science. These theories provide the underpinning for engineers to think about object worlds. However, engineering science cannot take us very far beyond the world of objects. The engineering performance at its most

---

[16]Adapted from (Foucault, 1984) "A muffled discourse that has almost disappeared".

fundamental level involves the actions of people who re-arrange objects and materials to achieve some useful outcome (Bucciarelli, 1994; Trevelyan, 2010; Zussman, 1985). Engineering only happens because people intervene in the world of objects.

Engineering is a human performance: people bring machines, tools, components, materials and information together in some pre-conceived way using engineering knowledge to produce new products or services for people. The propositions from the technical sciences provide guidance for interventions, but this guidance has to be tempered with years of contextual and tacit knowledge developed through experience: the craft of engineering.[17] Therefore, human sensorimotor, learning and cognitive limitations constrain an engineering performance just as much as the laws of engineering science, such as the ultimate strength of materials. To understand human limitations, we need insights from cognitive science, psychology, neurophysiology, physiology, ergonomics, and human factors. Since engineering relies on being able to predict performance, and human limitations constrain engineering performance, it follows that the human sciences can contribute to engineering in the same way as physical sciences and mathematics.

Observations of practice demonstrate that most engineering only happens in large extended networks of people: even solitary engineers need the support of many other people – technicians, suppliers, tradespeople, drafters, production supervisors and planners to name a few. Engineering itself is a social system and can only happen within an extended social context as it relies on the willing and conscientious cooperation of all these people and many others (Bucciarelli, 1988; Trevelyan, 2007). Together they construct knowledge: the knowledge of practice is created by practice.[18] Further, the technical expertise on which engineering depends is itself distributed in the minds of the participants (Larsson, 2007; Mukerji, 2009; Trevelyan, 2010). Therefore, the social interactions between people impose further constraints on the results. In the same way that cognitive science, neurophysiology and psychology can contribute to our understanding of individual human performances, the social sciences, organisation science, the learning sciences, anthropology, linguistics, philosophy, history, science and technology studies all provide insights and theoretical explanations that can help engineers understand constraints arising from social performances. These insights will be critical in 'localising' engineering practices, identifying local and regional factors that mediate the results of practice, and devising both local and global strategies that promote effective practice in different cultural contexts.

Given that many people are involved in engineering, not much can happen in engineering without large amounts of money (and human effort) and often there are ill-defined resource limits. The reputations of engineers, particularly their ability to keep the consequences of uncertainty within acceptable limits, builds the required confidence for investors to provide capital funding up front, long before they can expect a financial return.[19] It demands considerable ingenuity from engineers to achieve predicted results within a predetermined time schedule and budget. Insights from theories

---

[17]The chapter on the history of engineering practice explores this in more detail. See also (S. Barley & Orr, 1997; Horning, 2004; Meiksins & Smith, 1996; Vincenti, 1990).

[18](Gherardi, 2009; Nonaka, 1994) Also the chapter in this volume by Rachel Itabashi-Campbell and her colleagues and the chapter by Donna Rooney and her colleagues.

[19]The chapter by Bill Williams and José Figueiredo presents evidence on these aspects.

of business, economics, finance, accounting, marketing, decision theory and politics, provide further help and understanding on these constraints.

Engineering demands highly educated and skilled people: the investment in skills and education demands long-term investment of human effort from a society in raising and educating young people. Engineering, in turn, extends human capabilities, potentially freeing up human resources for education, healthcare and governance that provide the social environment in which engineering itself can flourish. Human societies only provide extensive investment for engineering because of the confidence that comes from experiencing the benefits arising from earlier engineering investments. Social theories of ethics, justice and equity can inform practice and, in doing so, help to explain the critical role of engineering in a civilised society (Lucena, 2010; Lucena, *et al.*, 2010).

It is startling that, within engineering discourse, there is mostly a muted silence on the social significance of engineering, and both engineers and their professional associations have difficulties explaining their roles.[20]

It is not surprising, therefore, that students find it difficult to explain the value of engineering in a human context. While they have no difficulty explaining the value of being a lawyer (defending human rights, providing justice) or a doctor (curing sickness, prolonging life), the value of being an engineer is often difficult for them to explain. A typical response is "That's a difficult question!" After some thought, students will often respond with "Developing new products, solving technical problems ..." often with a questioning shrug of the shoulders.[21] It was only when I was confronted by the absence of effective engineering in the back streets of South Asian mega-cities that I started to appreciate the immense social significance of engineering for myself.

There are now sufficient research studies based on observations of engineering in the workplace to demonstrate that there is a body of knowledge that constitutes engineering practice, much of it unwritten, built through practice.[22] It is this implicit knowledge that equips engineers to plan and organise predictable socio-technical human performances that rely on a high level of technical knowledge when every individual performance is, to a certain extent, unpredictable, even when many of the people involved are forgetful, bored, tired, prone to making mistakes, disinterested and anxious to get home. Yet most of this knowledge, even knowledge of its existence, lies outside engineering academies today.

3 NOTIONS OF PRACTICE IN THE ENGINEERING ACADEMY

Setting aside, for a moment, the practical difficulties involved in setting a boundary for any occupational group, it might be useful to assert first that:

engineers are people for whom their primary occupational identity is based on knowledge associated with engineering schools and allied communities of practice.

---

[20](Trevelyan, 2012b) Recent discussions on social justice (referred to later) have helped to re-awaken this discourse.

[21]The chapter on observations of engineering practice in Australia and South Asia provides some comparisons that highlight the social and economic value created by more effective engineering practices.

[22]Rooney and her colleagues expand on this in a later chapter.

Even though this causes difficulties for many practicing engineers who report that they hardly do any 'real engineering' in their work,[23] it is still their primary link to the engineering profession. Engineering science knowledge seems to be the one thread that ties together most of the disparate groups that identify themselves as engineers.

In leading contemporary engineering schools that set the education agenda there is often an overwhelming representation of engineering technology and science researchers among the faculty; the result of university recruitment, research funding, and promotion practices. For engineering educators, their engineering identity is important because they use it to distinguish themselves from the other physical and mathematical science disciplines. Quinlan's observation that many faculty see engineering in terms of "scientific process of developing new theories from which the viability of new designs can be tested" reflects the research identity that characterises these schools (Quinlan, 2002). In other words, engineering faculty subscribe to a generalised view of engineering expressed in terms of the engineering that they practise themselves. She also described how 'design division' faculty saw engineering as a creative discipline through which new products are developed. These different views shaped their teaching, disputes on education priorities, and hence the experiences of students in their classes.

Sheppard and her colleagues (2006) provided further insights in a study of perceptions of about 300 faculty and students based on semi-structured interviews and focus groups in seven major American universities. These perceptions centered on problem solving based on expert theoretical and contextual knowledge, supported by a combination of formal processes and creativity.

Pawley (2009) found that calls to reshape the discipline were unlikely to influence teaching. She perceived three 'universalized disciplinary narratives': engineering as applied science and mathematics, engineering as solving problems, and engineering as making things. Williams distinguished three diverging movements within academies: engineering science, design, and management systems, the latter two nourished from pragmatic commercial interests (Williams, 2002, 2003). She argued that historical and technological developments have led to an identity crisis in engineering and, as a result, education has become a 'contested domain'.

In the ethnographies by Stevens and his colleagues (2008) looking at engineering educators, and by Tonso (2006) looking at student teams, we can see how engineering education shapes the 'accountable disciplinary knowledge', skills, values, attitudes and identities as students 'grow into engineering'. Educators assume the responsibility for appropriately shaping this developmental process and their notions of engineering practice can have a profound effect on their students' beliefs.

Difficulties arise, however, for the majority of their graduates who emerge from university or college and practice in a different setting. When graduates experience engineering as practiced in most industries other than research and education, they can feel disoriented. "*When I started, I felt completely unable to do anything useful*", one graduate reported to the author recently. Martin and her colleagues described how graduates found they were not well prepared to work with other people and lacked practical skills, factors widely reported in many other similar studies (Dahlgren *et al.*,

---

[23]'Real engineering' being what they learned in engineering schools, as explained above.

2006; Martin *et al.*, 2005; Spinks *et al.*, 2006). In Australia, most companies assert that it takes 3–5 years for a novice engineer to become reasonably productive in a commercial context. The transition into industrial practice can be discomforting for many novices and employers alike. Many Australian employers have expressed their dissatisfaction with the capabilities of graduate engineers and prefer to recruit engineers with five or more years of experience if they can find them (Trevelyan, 2012a). Our research has helped clarify the many aspects of the work of young engineers on which formal education provides little if any enlightenment.

Medical educators have embraced extensive clinical practice and situate themselves in, or close to, teaching hospitals to promote the successful transfer of academic learning to practice. Most, if not all, of the instructors in the later stages of medical education have extensive and continuing exposure to medical practice. Engineering educators, on the other hand, have to prepare their students for a much greater diversity of career settings, and real engineering settings often require secrecy or are too large, too expensive or too hazardous to accommodate within a teaching institution. In addition, universities have a contemporary research agenda that excludes engineering practitioners from all but the margins of the academy.[24] Perhaps because of this, the combination of impracticality and policy barriers that keep practice outside the academy, notions of practice held by the academy have diverged far from the reality encountered by graduates, as the research in this book helps to demonstrate.

Texts, mostly written from within the academy, further reinforce this separation. Analysis reveals numerous aspects of practice omitted from contemporary texts (Trevelyan, 2011, 2012c). It is possible that the authors knew about practice, but decided to focus on engineering science. However, the consistency of omission in all the texts suggests, once again, the onset of amnesia within the engineering community.

In Australia and many other countries, at least until the 1980s, most engineers gained their first employment in a government engineering enterprise (or firms depending on government contracts) such as public works, water or electricity distribution, defence or telecommunications. These organisations continued to enact the practice-based approach to engineering inherited from education in the first half of the twentieth century. While many graduates found them bureaucratic and governments complained about their inefficiency, these organisations provided valuable years of practice training to complement university education. In retrospect, much of the inefficiency reflected their role as schools of engineering practice. Private firms used them as recruiting 'pools', taking on engineers as and when commercial opportunities provided the need. In the last few decades, these organisations have been privatised as governments decided that engineering was "not a core business activity", reflecting a world-wide drive to rationalise government business. Once exposed to the imperative for commercial efficiency, the implied emphasis on training and skill development has been eliminated, and replaced by a few short courses on generic communication and leadership skills (Tilli & Trevelyan, 2010; Trevelyan, 2012a). By the end of the coming decade, most of these practice-trained engineers will have retired.

---

[24]In Australia, there are few academics left with significant experience of practice outside the academy, and those that have usually experienced highly technical and specialised practice (Cameron, Reidsema, & Hadgraft, 2011).

The lack of interest in practice training is not surprising: given the scarcity of research on practice and lack of detailed written knowledge of practice, companies interested in training would find it hard to locate learning materials.

The picture that emerges is that our understanding of practice in the academy is all but non-existent. Government engineering organisations that stepped in to rectify that deficit have largely been disbanded. This book represents a small step towards rebuilding engineering practice knowledge that might otherwise disappear completely.

4   A THEORETICAL FRAMEWORK FOR ENGINEERING PRACTICE

Any attempt to rebuild practice knowledge in engineering schools would be easier with a theoretical framework. Such a framework would encompass different theoretical approaches and offer explanations, validated by research studies, that can help graduates understand more of the landscape of engineering practice. A framework would not only label knowledge as relevant for engineering, but also provide a scaffold on which students can build their practice knowledge.

Such a framework would help to ensure that issues that are relevant for engineering practice cannot be dismissed as "mere practical issues" or "non-technical" and therefore, by implication, not relevant for engineers. The following quote, adapted for this chapter, seems appropriate:

> "Unless we admit that rules of thumb, the limited experience of engineers in a particular setting, and the general sentiment of the street are the sole possible guides for engineering decisions with major consequences, it is pertinent to enquire how the best representative practices of engineers generally may be made available as a broader foundation for such decisions, and how proper theories of engineering practice are to be obtained."[25]

Such a framework would need to be built on a sound understanding of 'engineering' and, in doing so, help to define what we mean by engineering and the purpose it serves.

At a recent gathering of eminent engineers, confronting the challenge of devising a philosophical approach to engineering and engineering education,[26] I listened as participants discussed what they thought the purpose of engineering was: innovation, development of new technology? Designing new-products? Improving sustainability through new technology? I was one of a few participants with extensive experience of engineering outside the academy. I observed that the provision of affordable energy, water, transport, sanitation, communications, defense and security, food processing and storage were not mentioned, yet these are activities in which engineers are

---

[25] Adapted from an excerpt from the first editorial of the first issue of the Harvard Business Review issued in September 1922, quoted by Adi, "Towards a Proper Theory of Business". 2012.

[26] Construction of a philosophy of engineering, driven by the desire for improvements in education, has occupied several engineers, educators and philosophers. (Heywood, 2011; Heywood *et al.*, 2011)

pre-eminent actors, activities that are essential to sustain civilised human existence. Gradually I realised that there is a school of thought that sees engineers predominantly as change agents: technical specialists whose aim is to achieve change by innovation, mainly in the form of new technologies. To some in this school, engineers are distinctly different from managers: the activities of engineers are circumscribed by managers' agendas.[27]

These perspectives echo Layton's (1991) essay on engineering. However, research studies reveal the need for a different view of engineering. Engineers also organise and supervise production work, construction, operations and maintenance in large and small engineering enterprises. Even the most junior engineers find themselves exercising informal leadership, coordinating technical work performed by others, with only a tenuous comprehension of social and organisational factors that impose constraints on their actions. Engineering is much closer to the notion of productive activity suggested by Carl Mitcham's (1991) essay in the same volume as Layton.

Earlier, I proposed to assert that engineers are people for whom their primary occupational identity is based on knowledge associated with engineering schools and allied communities of practice.

In addition I propose to assert that:

engineering is more than what engineers do.

In doing so I have chosen to depart from a conventional analysis of practice where one would read that "science is what scientists do in their laboratories": conventional studies of practice focus on a community of like practitioners, such as scientists.[28] Here I am arguing that engineering practice, what engineers do, needs to be appreciated in the context of an enterprise so we can understand how engineers influence outcomes produced by a larger community of people.

Engineers are people who cast their own identity in terms of engineering, or who are recognised by others as engineers. One such category is professional engineers, formally recognised by government statutes in some jurisdictions, in others formally recognised by one or more professional organisations. Professional engineers often complain that other people (not recognised by their profession) call themselves or are referred to by others as engineers. Aircraft engineers, highly skilled technicians responsible for maintenance and repairs for aircraft, particularly in Australia, take a high public profile to argue for better employment conditions, playing on public fears in relation to safety of air travel.[29] In some countries, railway locomotives are driven by "engineers". In some countries, automotive engineers are skilled technical experts who repair or enhance the performance of motor vehicles, particularly with

---

[27]Engineers are not only seen by others, but also see themselves as applied scientists who solve technical problems for management (American Society of Civil Engineers, 2008; Goldman, 2004, p. 164, p. 166).

[28](Gherardi, 2009, p. 120). The chapter by Donna Rooney and her colleagues presents a selection of literature on theories of practice.

[29]Examples of media coverage include: FWA upholds Qantas safety check cuts. (2012, June 14). ABC News. Qantas set to shed another 30 engineers. (2012, June 14). AAP; Lannen, D. (2012, November 9). 260 Avalon Jobs Axed: Qantas cuts even more maintenance workers, Geelong Advertiser.

knowledge of engines and power transmission elements. One of the most famous cases in constitutional law in Australia is the "Engineers Case" arising from arguments relating to the pay of boiler-makers and tube fitters.[30]

Who is and who is not recognised as an engineer is as much a social and political issue as a philosophical one. The reality is that a large part of both the identities and work roles of many engineers in the categories described above are concerned with mediation between machines and systems on the one hand and people on the other hand. These engineers have specialised technical, cognitive and sensorimotor skills to interact with machines and systems. Most professional engineers, on the other hand, have neither the time nor inclination to develop these special skills. Instead, their work is more concerned with abstract ideas and coordination of technical work performed by other people. It is difficult, therefore, to decide whether a particular action performed by an engineer might be 'legitimately' recognised as engineering. Another difficulty is that engineers perform work in many occupational areas outside engineering where their special skills with quantitative analysis can be productively employed. Examples include providing advice for financial investment decisions and even real estate. Engineers are found here because of their quantitative analytical skills, which they have in common with many mathematicians, physicists, computer scientists and others. It might be misleading to include these performances in 'engineering'.

If we look at the work of an individual engineer in isolation, it can be difficult to appreciate the significance or value arising from that activity. Take, for example, an engineer who checks specifications and incoming components to ensure they comply with safety standards. Taken in isolation, the work seems to have contributed no real value. The engineer has not contributed any design or performance improvement or cost reduction, indeed his work has increased the overall cost and possibly the time needed. However, in the context of an enterprise, the engineer has reduced the perceived risk of failure just because the checks have been made, even if no non-compliance was found. In doing so, he has increased the perceived economic value of the enterprise.[31]

Therefore, rather than answer the questions "What is an engineer?" or "What is engineering?" I would propose that it is more useful to ask a sequence of questions related to 'an engineering enterprise'. Referring to an enterprise as the 'unit' or 'locus' of engineering action is helpful because, in nearly all instances where engineers are engaged, many different people make contributions, not all of them with special engineering expertise.[32] Further, there is little difficulty in arguing that all those who call themselves engineers can be shown to have special knowledge on which their enterprises depend.

---

[30]Gavan, D. J. (1920, August 31). High Court of Australia: Engineers Case. Melbourne.

[31]This relies on economic theories of utility in which risk perceptions strongly influence perceived value. (Kahneman, 2011; Kahneman & Lovallo, 1993)

[32]Scholars of science, technology and society (STS) have shown how the traditional objects at the core of engineering thinking can be conceived in terms of socio-technical networks of influence. Chapters by Hubert, Kaplan and Vinck in this volume illustrate how this approach reveals the heterogeneity of engineering and that its course of development needs explanations that go beyond technical rationality. Even though engineers and what they do are the focus of our interest, we need to develop an understanding of the social and technical context in which they perform their practice. For further examples, refer to Bijker, 1995; Latour, 2005; Law, 1987; Law & Callon, 1988; Suchman, 2000.

An engineering enterprise is only very seldom contained within a single individual or even a single firm. More commonly, an enterprise will involve people and engineers who work in sections of a number of different organisations, including government. The following attributes could help to identify the characteristics of such an enterprise in a social context.

An engineering enterprise requires contributions from engineers, people who have special technical expertise that is based in, or originates from an engineering discipline or recognised body of knowledge, who contribute to the work of the enterprise. Engineers perform a critical role because their technical knowledge, when applied, greatly influences the performance of the entire enterprise.

The people involved can come from several different firms, all collaborating with the same purpose – delivering a product, a service or information. Even the customers, clients and end-users are part of the enterprise: without their knowledge of how to use the products and services and their willingness to pay, there would be no enterprise. The end-users may be directly involved in design decisions. Local communities may participate in risk assessment reviews. The investors who provide finance are also part of it. The people in the enterprise are not all engineers, in fact there may be very few engineers and they may be a very small minority of the people involved. Engineering relies on many different people collaborating, including the financiers, clients, end users, consultants, even government regulators.

Engineering enterprises use tools, machines, components, materials, energy and information to deliver products, information or services using special knowledge contributed by engineers, underwritten by financial capital from investors. The aim is predictable delivery of reliable products, services or information, all provided with the most economic use of financial capital, human effort and material resources possible, within acceptable health, safety, environmental and sustainability performance criteria. The degree of reliability and predictability, and other performance criteria, depends on the quality of the skilled performances by people involved in the enterprise. Technical design contributes to reliability and economy, and analysis with appropriate optimisation improves predictability, economy and reliability. Greater predictability and reliability helps to build the confidence needed for a society (or business) to commit large amounts of human effort and resources for an engineering enterprise, long before the benefits can be experienced.

The products, information and services are often provided as inputs to other engineering enterprises in a complex web of connections linking companies and people in a network that can extend across the planet.

Some of the necessities of human existence that engineering enterprises support are:

- Provision of water, sanitation, energy supplies, raw materials, food production and processing (and associated activities such as fertiliser production, irrigation etc.), transport (infrastructure and vehicles), healthcare facilities, and communications,
- Construction of shelter and other infrastructure to protect people from natural exposure, natural disasters etc.,
- Manufacture of products such as clothing, food preparation and storage equipment, communication devices, vehicles, roads, entertainment devices,

- Defence installations, law enforcement facilities and associated equipment that help to protect people and social institutions from potentially destructive or violent human behaviour, and
- Information systems and associated equipment that facilitate productive human effort and efficient use of resources, disseminate explicit knowledge, and facilitate social justice and fair access to opportunities.

I would also argue that it is not possible for an engineering enterprise to remain independent of the social context in which it operates. An engineering enterprise requires resources from other members of the society in which it operates, at least to provide space and sustenance for those engaged in the work of the enterprise.

Therefore there is a mutually dependent symbiotic relationship:

- The enterprise depends on continuing resources and support, and
- The society in which it operates has no alternative but to place its trust in the engineering enterprise to deliver the economic productivity benefits, the promise of which motivated the decision to provide resources in the first place.
- The people involved in the work of the enterprise are also necessarily members of the community in which it operates, and therefore are influenced by community values and expectations.
- The expertise on which the engineering enterprise depends is distributed among the people involved in its work, even end-users (not just engineering expertise, but also the expertise contributed by all the other people).
- Engineers have an ethical duty to ensure that decisions by other people are made in the light of their special knowledge and insight (Davis, 1998).

Engineers, as we shall see in later chapters of this book, are engaged in solitary performances, as well as performances with numerous other people.[33] The term 'engineers' here includes all those who identify themselves, or are identified by others, as engineers. While this might cause some discomfort to those who identify themselves as 'professional engineers' and who carry some formal recognition from a professional society of engineers, it is best to avoid unnecessary restrictions of scope as long as possible.

Research using data derived from observations of engineers can help to reveal the nature of their human performances.

Although there are a limited number of well documented studies so far, researchers have observed engineers performing 6 principal groups of activities: (Trevelyan, 2010, online appendix).

1   Educating clients and others on engineering possibilities, eliciting and negotiating client requirements,
2   Conceiving solutions to meet client requirements, design,

---

[33] Adapting the notion of engineering as a human performance explained in (Trevelyan, 2010), also using the performance metaphor as a means of describing engineering communication in (Evans & Gabriel, 2007). The view of practice is from the 'inside': "to know is to be able to participate with the requisite competence in the complex web of relationships among people, material artefacts and activities". (Gherardi, 2009)

3   Predicting performance (using simulation, experiments, analysis etc.), predicting risks, etc. and preparing information to assess the benefits and uncertainties from a proposed investment,

4   Organising, mobilising resources, planning, detailed design,

5   Monitoring production and service delivery, diagnosing and correcting performance deficiencies, providing maintenance and support, sustainment, constructing, operating systems and machines, and,

6   Organising, mobilising resources for recycling, remediation, decommissioning, re-use.

Any one engineer seldom performs all of these but may well perform all of them through the course of a long career. Further research is likely to contribute evidence that the list needs to be extended or changed. One can also expect engineering practice to evolve over time. Further, as more researchers take an interest in this field, the completeness of the list, and knowledge of regional, setting and disciplinary variations will improve.

A theoretical framework for engineering practice could be advanced as a description and exposition of the body of knowledge and the capabilities that enable engineers to perform, to practice all these aspects of engineering. Since part of the body of knowledge comprises tacit, procedural, contextual and implicit knowledge, and phronesis,[34] none of which can be easily explained, the framework could also explain how engineers acquire this knowledge and associated capabilities, particularly in the context of workplace learning (Eraut, 2004, 2007; Korte, *et al.*, 2008).

Such a framework might also contribute insights that would enable engineers to systematically assess the quality of performances, even providing criteria that would help to assess the level of 'effectiveness', 'expertness' or 'mastery' demonstrated.

It is desirable, therefore, to distinguish those components of practice that are more or less common to all engineering disciplines and settings. With limited resources at our disposal, it would be prudent to focus on this first, and then later focus on aspects of engineering practice that are specific to certain disciplines, settings, geographic regions, and so on.

5   SCOPE OF A THEORETICAL FRAMEWORK

Engineering already has theoretical frameworks: they just need to be expanded beyond the current conventional theories based on a restricted set of natural sciences. However, whereas existing frameworks are restricted almost entirely to inanimate physical objects and materials, the inclusion of human sciences, particularly the social sciences, brings with it some fundamental and necessary intellectual transformations.

Theories that have been 'imported' into engineering have often taken new meanings and interpretations. For example, the fundamental engineering relationship that relates work and energy in thermodynamics portrays 'work' as a useful output of a thermodynamic system. In the physical sciences where thermodynamics originated, however, 'work' takes on the opposite meaning: it is the work done on a system by an outside agency.

---

[34]Expert judgement, see theory of expert performance (Ericsson, 2003; Ericsson *et al.*, 1993).

The choice of theories in existing engineering frameworks is at least partly discipline-specific. Quantum theory, for example, is essential for understanding electronics and optics in electronic and computer engineering. However, it is of marginal, if any, utility in mechanical and civil engineering, with the possible exception of certain chemical interactions.

As explained earlier, contemporary engineering science provides little guidance for engineering practice beyond the world of objects. As explained earlier, a theoretical framework for engineering practice must address the major components of practice: it must provide useful insights on most of the challenging issues that engineers work with, to the extent that they can be described by natural, physical, human and social sciences.

Admitting theories from the social sciences into engineering requires at least three transformative intellectual changes in outlook from us, as engineers.

The first change is to realise that social science theorising is different from the natural sciences in that it mostly aims to clarify and explain.[35] While natural science theories also aim at understanding, their utility is as much in making predictions about the future as in offering compelling explanations. Indeed the measure of a successful natural science theory is its predictive power. Engineering science is distinctly different from the natural sciences because of the concern with developing methods for accurately predicting the performance of artefacts, systems and materials that do not yet exist. Hence the predictive power of a theory is an important attribute for engineering. While some theories in the social sciences have a degree of predictive power, the intrinsic complexity and variability of people means that prediction, in the sense that engineers are accustomed to, is usually impractical. Even statistics, the preferred engineering tool for handling uncertainty, may be of little help.

This limitation in the social sciences does not mean that theories have less value, however. Social science theories offer powerful explanations for human behaviour that help anticipate obstacles and spaces where solutions might be difficult or impossible. For example, value expectancy theory (Eccles, 2005) and value economics (Akerlof & Kranton, 2005) help explain why relatively autonomous workers such as engineers favour certain tasks above others, given a choice.

The second change is to realise that any given situation can often be explained satisfactorily with entirely different theories and concepts. There is no 'right' or 'wrong' theory to use. Sometimes there may be a choice from ten or more 'schools of thought', particularly in the social sciences. While the proponents of some schools might write strident critiques of other schools, it does not mean that one school is better than another. Equally informative explanations may emerge from several different theoretical approaches. One school may become 'fashionable' in a particular research community in a particular country for a while, and then may fade away to be displaced by another. Each research community will have developed different yet rigorous observation methods that need to be applied carefully with deep appreciation of the implications. Choosing any one particular method may preclude the possibility of reaching certain findings.

---

[35] With the possible exception of critical theory and pragmatism that provide a theoretical basis for social action to support change.

The final change, and perhaps the most difficult for engineers, is to be able to step outside our traditional positivist empirical mental framework in which we can conceive of 'objective' observations of a phenomenon with identifiable causes and effects. We also need to step outside our traditional idea that words have fixed, standardised usage and meanings and recognise the need to seek different understandings and interpretations through conversation and dialogue.

These are significant intellectual challenges and for many engineers they do not come easily. These challenges reflect difficulties that have been identified when engineering faculty engage in education research (Borrego, 2007).

One potential distraction that needs to be addressed is the distinction between a theoretical framework for engineering practice and theories of management. Given that the aim is to seek theories that explain aspects of engineering practice that are common to all disciplines and settings, thereby omitting (for the time being) discipline-specific aspects, some may argue that such a framework would substantially overlap theories of management, to the extent that a separate framework for engineering practice would be redundant.

A common reaction from students on learning about engineering practice research is "That's all management and psychology stuff, isn't it? It's nothing to do with engineering".

Our engineering faculty colleagues may feel more comfortable to imagine that management schools can offer a comprehensive theory of engineering practice, encompassing everything beyond the terrain covered by engineering science.

Unfortunately, management science has not provided many substantial intellectual contributions that can explain engineering practice (Barley, 2005). For a time, during the mid-20th century, the discipline of industrial engineering did encompass some of this terrain, but this discipline has since retreated largely to mathematical analysis (Bailey & Barley, 2005).

Some of the difficulties faced by social researchers from organisation research in engineering enterprises have been noted earlier (e.g. Whalley & Barley, 1997; Zussman, 1985). One is the abstract intellectual activity associated with the technical dimension: such activity is inevitably invisible: we cannot directly observe what is happening inside people's heads. Manual work study at least offers the possibility of directly observing the results of thinking. However, another obstacle is the language used by engineers. Engineers often describe immensely complex intellectual processes with just a few words (Tang, 2012). They also impose significant barriers that can deflect investigations of the social aspects of their work because they often describe these aspects using technical terminology (Trevelyan, 2010).

Management or, to be more precise, organisation science is based on social psychology, and to a lesser extent anthropology. Organisation science has much to offer in the context of engineering practice because it provides explanatory models of specialised communities, enterprises and firms.[36] Related disciplines in marketing, accounting and finance also have rich perspectives that help engineers explain practice, particularly

---

[36] Being based mostly in social sciences, organisation science principally aims to be explanatory, though management researchers claim elements of predictive power from their theoretical analysis, mostly through prescriptive comments that commence "Managers should ...".

factors affecting financial constraints.[37] Apart from management academics, management and organisation science journals claim to address "managers". It is difficult, however, to imagine many managers having the time to read organisation science literature: perhaps the primary impact comes from exposure of carefully selected journal articles to students in management schools.

For example, within organisation science, the study of "practice" has emerged, illuminating a degree of tension between studies of workplace activity without evaluation of practice, and normative accountability, what counts or could be evaluated as competent or exemplary practice (Gherardi, 2009). However, the study of practice has, as one its main aims, a theoretical framework within which to understand practice in a general sense, rather than to inform practitioners in one specific discipline.

In contrast to organisation science, therefore, I would argue that the primary audience for engineering practice theorising must be practicing engineers and their educators, particularly novices navigating the troubling liminal spaces between formal education and established professional competence.[38] It is this difference in audience that distinguishes the agenda of management schools from the engineering practice agenda.[39] Some might discern in some of the management literature an agenda to exercise power and control over others, including engineers. To the extent that engineers coordinate technical work performed by many other people, the exercise of social power may have to be part of practice and, with that, the necessity to act with justice and restraint.

There are several different research communities that could contribute to a theoretical framework for engineering practice. One of the most obvious, engineering studies, has emerged with the journal of the same name as a field of study largely based in social sciences, with interests in history, sociology, literature, gender studies and education. It is a developing field of academic enquiry, and there are not yet academic departments or even large research teams focused on this endeavour. For now, it provides spaces for individuals who are often sitting in a precarious marginal position, neither fully in engineering nor the humanities.[40] Alongside Science and Technology Studies (STS), engineering studies can offer insights into the heterogeneity of engineering practice, and local cultural and historical factors that shape the landscapes of practice.[41] Within STS, engineering still occupies only a marginal niche, however.

Engineering education is another research community with the potential to contribute. At the time of writing, most in this community are firmly focused on undergraduate engineering studies, with at least the utilitarian promise (yet to be

---

[37] The most popular choices for a second major to engineering in Australia are provided by business schools.

[38] Liminal space is a term borrowed from my engineering education colleagues studying threshold concepts – it describes the intellectual confusion experienced by students as they attempt to accommodate concepts that are transformative and, once mastered, enable them to achieve a new level of intellectual capability.

[39] As Downey has argued, the audience shapes the research agenda from the earliest stages. (Downey, 2009)

[40] The primary audience largely consists of people within INES, the community from which the journal emerged. SHOT and SSSS seem to be favoured annual meetings.

[41] See, for example, Bijker, 1995; Faulkner, 2007; Law, 1987; Law & Callon, 1988.

fully realised) of reforming curricula and teaching capabilities.[42] It is interesting to observe a degree of reluctance among people in this field to come to terms with the notion that engineering practice might differ from accepted orthodoxy based on design and problem solving. This is a rich multi-disciplinary field that brings all engineering disciplines together with education researchers and teachers. Many of the research methods used in engineering education can be readily adapted for studies of engineering practice.[43]

There is limited overlap between engineering studies and engineering education. Some engineering education academics see engineering studies as a receptive community in which to discuss work on social justice, sustainability, and philosophical arguments that would not be seen by engineering education journal editors to be of immediate interest to their constituencies, front-line engineering academics. Social justice is another of the constraints that engineers have to confront, particularly as the engineering enterprise encompasses end users. If benefits are not shared with sufficient attention to distributive justice, social disruptions can impede or even stop an engineering enterprise. This issue is often encountered in engineering practice in the context of social sustainability and ethics.

Design studies is another related field, largely separate from engineering studies and engineering education. It was established for many years before engineering education with journals such as Design Studies[44] and Journal of Design Research. Intentionally multidisciplinary, it embraces architecture, engineering, computer graphics, art, and others. It is strongly focused on visual appreciation and imagery, but in recent years has embraced linguistic techniques as well. Historically there have been strong interests in developing computer software, and there are close links with the computer aided design software field. Their audience comprises mainly design academics in architecture and some engineering faculties. There are limited interests in engineering education as it applies to design and visualisation skills. While design studies writers claim that design is central to engineering (and architecture)[45] recent research has shown that it sits alongside several other significant aspects of engineering practice. Design studies offer valuable insights; engineering practice, however, embraces a much wider range of human activity than design.

There are many other disciplines such as linguistics, neurophysiology, philosophy, economics, history, political science and psychology that have developed rich theories that could explain many aspects of practice. The chapters of this book all serve as examples from research communities that can contribute to an engineering practice theoretical framework.

---

[42]The journal of engineering education sees the broad mass of engineering faculty as their primary audience. Regular meetings include ASEE, REES, FIE, ICEE, INEER, SEFI and several other regional meetings like AAEE in Australia and New Zealand.

[43]See for example, the chapter by Robin Adams and Florin Tiago.

[44]Regular meetings include ICED and many others. The most prominent issue on which contributions have recently focused is design cognition – the psychological aspects of design thinking. The principal remaining issues include product design methods, design theory and design practice including social studies of design (Analysis contributed by Dominique Vinck).

[45]For example, "Design is widely considered to be the central or distinguishing activity of engineering". (Dym et al., 2005, p. 103)

6 EDUCATION PROSPECTS AND CHALLENGES

Being realistic, it is important to write a few remarks on some of the difficulties inherent in rebuilding practice knowledge in engineering schools.

Very few engineering faculty, most of whom today are engineering scientists, are interested in engineering practice at the moment, even within the small minority of faculty who engage in engineering education research. If there is any awareness, it is apprehensive, even defensive, because the notion of engineering practice might seem to undermine the idea that engineering science, coupled with design, offers a complete body of knowledge that defines engineering. Maybe it is because engineering practice might be perceived to be a threat to the pre-eminent position of engineering science to attract research funding from government. Engineering studies may not constitute a real threat, but certainly I (and others) have encountered strong resistance to the idea that engineering practice offers a body of knowledge with utilitarian value to engineers.

The most immediate challenge, apart from enlarging the current small research base for the discipline, is to work towards viable education programs. The challenge for educators is to bring together many disparate aspects from the different disciplines that offer explanatory power relevant for engineering practice. In particular, the challenge is to construct a coherent theoretical 'scaffold' that provides a starting point for young engineers to learn more when they need to. Just as an engineering science curriculum at the undergraduate level no longer aims to be comprehensive, it would not be possible for an engineering practice curriculum to be comprehensive. Instead, it must allow students to develop the necessary intellectual skills for them to take responsibility for their own learning as they progress through life. Some of the evidence coming from our research studies suggests that the absence of any consideration that extends beyond the world of objects in engineering education significantly inhibits the ability of young engineers to learn about engineering practice.

For the time being, notwithstanding some well-intentioned attempts by progressive engineering educators to reform undergraduate curricula, engineering practice needs a home in a postgraduate environment, preferably with students already exposed to the world of work and sensitised to the need to understand social interactions.

It is interesting to speculate whether, with further development, a theoretical framework for engineering practice would need to be subdivided into different disciplines. We know that engineering science subdivides into separate engineering disciplines like civil engineering, mechanical engineering, software engineering, chemical engineering, electrical and electronic engineering. These subdivisions reflect practice: engineers practice in a number of distinct disciplines, each with its own base in theory. One of the contested aspects of engineering education is the extent to which there should be a common intellectual foundation that supports all engineering disciplines. While many institutions have adopted such a common foundation in their curriculum, many others continue with specialised curricula for each different discipline. At least in our own research, we have not yet seen sufficient differences between settings or technical discipline practice to warrant a separate body of theory and analysis. There is more in common among engineering disciplines than differences between them. However, this may well reflect the very early stage of theory development in which we find ourselves today. If there are significant subdivisions, they are more likely to be between different forms of practice rather than following existing technical discipline boundaries. For example, there are very significant differences between engineering practice within a consulting firm, compared with a contracting company or a manufacturing enterprise. It is more likely that divisions would follow these distinctions in practice.

An education curriculum in engineering practice could benefit from a theoretical framework in many different ways.

First, informative theories can also reduce the complexity arising from descriptions by enabling us to see patterns with simple explanations.

At an earlier stage of our research, it seemed like a significant achievement to reduce hundreds of thousands of words to a list of 82 aspects of practice, 37 categories of technical knowledge, and 27 aspects of organisational knowledge. It seemed like progress to identify between 40 and 50 aspects of practice in which students develop fundamental misunderstandings. However, this level of complexity can only be relegated to the appendix of even a research publication. While we know that human behaviour is not simple, and we can only understand engineering fully as a human performance, we should not be satisfied unless we can reduce this complexity to something that students can comprehend in an undergraduate education. That is work in progress.

Second, even in a research context, theories can be useful as a means of standardising terminology, albeit within certain research communities or schools of thought. A theoretical framework could help us understand each other and help us learn from a much wider set of research communities.

Third, a framework could help standardise research methods and reporting. Social science theories such as Actor Network Theory provide guidance on data collection, analysis and writing the results. There is some similarity here with the physical sciences. A typical paper giving the structure of complex organic molecule derived from a complex series of X-ray diffraction studies is typically only one or two pages long. It is short because the complex methods employed in the analysis are described only by references to an extensive body of prior research and descriptions of experimental and analysis techniques. Anyone with the appropriate equipment could reproduce the analysis in a different setting. This enables the experiments described in the short paper to be reproduced elsewhere, gradually building greater confidence in reported scientific results.

Fourth, a framework for engineering practice could help students learn more quickly. Theories provide ways of thinking about phenomena that enable students to learn more quickly from their mistakes. This is well known in education (Bransford *et al.*, 2000, Ch. 1). Graduates from courses that emphasise theory learn faster when they encounter the world of practice because they can interpret observations in terms of theories and learn how to improve performance. The explanatory power of engineering science theories developed in the 20th century underpinned the extraordinary industrial progress that has been achieved in the last few decades, for at least part of the world's population.[46]

---

[46] We could side-step inevitable differences of opinion on this for now by agreeing that we have reduced the human effort needed to sustain many essential services.

In summary, therefore, developing a theoretical framework for engineering practice offers many practical advantages. It could:

- help engineers make better decisions,
- help to define 'engineering' and 'engineering practice',
- help us understand why we do engineering,
- help explain practice in more simple terms, making it easier to see patterns in complex descriptions,
- help understand how local cultural, social and other factors mediate the results of engineering practice,
- help researchers with shared terminology and ideas so that we can learn more from each other and from the wider scientific community,
- help researchers with standardised data collection and analysis methods, and ways of presenting results,
- address the substantial issues that preoccupy engineers, helping them make better decisions,
- enable students to learn about engineering practice and improve their own performance, and,
- establish engineering practice as knowledge that needs to be built and transmitted.

Eventually, it may also enable us to make predictions about the effectiveness of certain approaches to engineering practice in a given setting.

If this undertaking, the development of a theoretical framework for engineering practice, progresses beyond the preparation of this book and perhaps one or two successors, there will still be a long and hard road ahead. However, there are many promising indications from the papers in this volume that many of the frustrations experienced by engineers in all disciplines could be alleviated, or at least satisfactorily explained, by such a framework.

ACKNOWLEDGEMENTS

Thanks are due to all the engineers who participated in these studies, my students and colleagues who contributed ideas through shared research enquiries and discussions, and the grant agencies, companies and donors who contributed to individual studies. Thank you also to the reviewers and other contributors to this volume, without whose efforts and informed comments this chapter could not have reached this stage.

REFERENCES

Akerlof, G. A. & Kranton, R. E. (2005) Identity and the Economics of Organizations. *The Journal of Economic Perspectives*, 19 (1), 9–32. doi: 10.1257/0895330053147930.

American Society of Civil Engineers. (2008) Civil Engineering Body of Knowledge for the 21st Century: *Preparing the Civil Engineer for the Future*. 2nd edition, Vol. 2008. Virginia, USA, ASCE Committee on Academic Prerequisites for Professional Practice.

Anderson, K. J. B., Courter, S. S., McGlamery, T., Nathans-Kelly, T. M. & Nicometo, C. G. (2010) Understanding engineering work and identity: a cross-case analysis of engineers within six firms. *Engineering Studies*, 2 (3), 153–174. doi: 10.1080/19378629.2010.519772.

Bailey, D. E. & Barley, S. (2005) Return to work: Toward post-industrial engineering. *IIE Transactions*, 37, 737–752. doi: 10.1080/07408170590918308.

Bailey, D. E. & Barley, S. (2010) Teaching-Learning Ecologies: Mapping the Environment to Structure Through Action. *Organization Science Articles in Advance*, 1–25.

Barley, S. & Orr, J. (eds.) (1997) Between Craft and Science: *Technical Work in US Settings*. Ithaca, New York, Cornell University Press.

Barley, S. R. (2005) What we know (and mostly don't know) about technical work. In: S. Ackroyd, R. Batt, P. Thompson & P. S. Tolbert (eds.) *The Oxford Handbook of Work and Organization*. pp. 376–403. Oxford, Oxford University Press.

Bechky, B. A. (2003) Sharing Meaning Across Occupational Communities: The Transformation of Understanding on a Production Floor. *Organization Science*, 14 (3), 312–330. doi: 1047-7039/03/1403/0312$05.00.

Bijker, W. E. (1995) *Of Bicycles, Bakelite's, and Bulbs: Toward a Theory of Sociotechnical Change*. Cambridge, Massachusetts, MIT Press.

Borrego, M. (2007) Conceptual Difficulties Experienced by Trained Engineers Learning Educational Research Methods. *Journal of Engineering Education*, 96 (2), 91–102.

Bransford, J. D., Brown, A. L., Cocking, R. R., Donovan, M. S. & Pellegrino, J. W. (eds.) (2000) *How People Learn: Brain, Mind, Experience and School*. Washington DC, USA, National Academy Press.

Bucciarelli, L. L. (1988) An ethnographic perspective on engineering design. *Design Studies*, 9, 159–168.

Bucciarelli, L. L. (1994) *Designing Engineers*. Cambridge, Massachusetts, MIT Press.

Cameron, I., Reidsema, C. & Hadgraft, R. G. (2011, December 5–7) Australian engineering academe: a snapshot of demographics and attitudes. *Proceedings of the Australasian Conference on Engineering Education, 2011 AAEE, 5–7 Dcember 2011, Fremantle, Western Australia*.

Dahlgren, M. A., Hult, H., Dahlgren, L. O., Segerstad, H. & Johansson, K. (2006) From Senior Student to Novice Worker: Learning Trajectories in Political Science, Psychology and Mechanical Engineering. *Studies in Higher Education*, 31 (5), 569–586. doi: 10.1080/03075070600923400.

Davis, M. (1998) *Thinking Like an Engineer*. New York, USA, Oxford University Press.

Downey, G. L. (2009) What is engineering studies for? Dominant practices and scalable scholarship. *Engineering Studies*, 1 (1), 55–76. doi: 10.1080/1937 86 209 0278 6499.

Dym, C. L., Agogino, A. M., Eris, O., Frey, D. & Leifer, L. J. (2005) Engineering Design Thinking, Teaching, Learning. *Journal of Engineering Education*, 94 (1), 103–120.

Eccles, J. S. (2005) Subjective task value and the Eccles et al Model of Achievement Related Choices. In: A. J. Elliot & C. S. Dweck (eds.) *Handbook of competence and motivation*. New York, The Guildford Press.

Engineering Education Australia. (2012) Graduate Program in Engineering: A program of non technical skills for the real world of work Retrieved. Available from: http://www.eeaust.com.au/graduateprogram.html [acessed 23th February 2013].

Eraut, M. (2004) Transfer of knowledge between education and workplace settings. In: H. Rainbird, A. Fuller & A. Munro (eds.) *Workplace Learning in Context*. pp. 210–221. London, Routledge.

Eraut, M. (2007) Learning from other people in the workplace. *Oxford Review of Education*, 33 (4), 403–422. doi: 10.1080/03054980701425706.

Ericsson, K. A. (2003) The Acquisition of Expert Performance as Problem Solving: Construction and Modification of Mediating Mechanisms through Deliberative Practice.

In: J. E. Davidson & R. J. Sternberg (eds.) *The Psychology of Problem Solving*. pp. 31–83. New York, Cambridge University Press.

Ericsson, K. A., Krampe, R. T. & Tesch-Römer, C. (1993) The Role of Deliberate Practice in the Acquisition of Expert Performance. *Psychological Review*, 100 (3), 363–406.

Evans, R. & Gabriel, J. (2007, October 10–13) Performing Engineering: How the Performance Metaphor for Engineering Can Transform Communications Learning and Teaching. *Proceedings of the 37th ASEE/IEEE Frontiers in Engineering Education Conference*, 10–13 October 2007, Milwaukee, Wisconsin, USA.

Faulkner, W. (2007) Nuts and Bolts and People. *Social Studies of Science*, 37 (3), 331–356. doi: 10.1177/0306312706072175.

Foucault, M. (1984) The Order of Discourse: Inaugural lecture at College de France December 1970. In M. Shapiro (ed.) *Language and Politics*. pp. 108–138. Oxford, Blackwell.

Gainsburg, J., Rodriguez-Lluesma, C. & Bailey, D. E. (2010) A "knowledge profile" of an engineering occupation: temporal patterns in the use of engineering knowledge. *Engineering Studies*, 2 (3), 197–219. doi: 10.1080/19378629.2010.519773.

Gherardi, S. (2009) The Critical Power of the 'Practice Lens'. *Management Learning*, 40 (2), 115–128. doi: 10.1177/1350507608101225.

Goldman, S. L. (2004) Why we need a philosophy of engineering: a work in progress. *Interdisciplinary Science Reviews*, 29 (2), 163–178. doi: 10.1179/030801804225012572.

Grinter, L. E. (1955) Report of the Committee On Evaluation of Engineering Education (Grinter Report). *Journal of Engineering Education*, 44 (3), 25–60.

Heywood, J. (2011) A Historical Overview of Recent Developments in the Search for a Philosophy of Engineering Education. *Proceedings of the Philosophy of Engineering and Engineering Education*, 12–15 October 2011, Rapid City, SD, USA.

Heywood, J., Carberry, A. & Grimson, W. (2011) A Selected and Annotated Bibliography of Philosophy In Engineering Education. *Proceedings of the Philosophy of Engineering and Engineering Education*, 12–15 October 2011, Rapid City, SD, USA.

Horning, S. S. (2004) Engineering the Performance: Recording Engineers, Tacit Knowledge and the Art of Controlling Sound. *Social Studies of Science*, 34 (5), 703–731. doi: 10.1177/0306312704047536.

Jonassen, D., Strobel, J. & Lee, C. B. (2006) Everyday Problem Solving in Engineering: Lessons for Engineering Educators. *Journal of Engineering Education*, 95 (2), 139–151.

Jørgensen, U. (2007) History of Engineering Education. In: E. F. Crawley, J. Malmqvist, S. Östlund & D. R. Brodeur (eds.) *Rethinking Engineering Education: The CDIO Approach*. pp. 216–240. New York, Springer.

Kahneman, D. (2011) *Thinking, Fast and Slow*. London, Penguin, Allen Lane.

Kahneman, D. & Lovallo, D. (1993) Timid Choices and Bold Forecasts: A Cognitive Perspective on Risk Taking. *Management Science*, 39 (1). doi: 0025-1909/93/3901/0017$01.25.

Koen, B. V. (2009) The Engineering Method and its Implications for Scientific, Philosophical and Universal Methods. *The Monist*, 92 (3), 357–386.

Korte, R. F., Sheppard, S. D. & Jordan, W. (2008) A Qualitative Study of the Early Work Experiences of Recent Graduates in Engineering. *Proceedings of the American Society for Engineering Education (ASEE) Annual Conference & Expositio*. 22–26 June 2008, Pittsburgh, PA.

Larsson, A. (2007) Banking on social capital: toward social connectedness in distributed engineering design teams. *Design Studies*, 28 (6), 605–622. doi: 10.1016/j.destud.2007.06.001.

Latour, B. (2005) *Reassembling the Social: an Introduction to Actor Network Theory*. Oxford, Oxford University Press.

Law, J. (1987) Technology and Heterogeneous Engineering: the Case of the Portuguese Expansion. In: W. Bijker, T. Hughes, P. & T. J. Pinch (eds.) *The Social Construction of Technological*

Systems: New Directions in the Sociology and History of Technology. pp. 111–134. Cambridge, Massachusetts, USA, MIT Press.

Law, J. & Callon, M. (1988) Engineering and Sociology in a Military Aircraft Project: A Network Analysis of Technological Change. *Social Problems*, 35 (3), 284–297.

Layton, E. T. (1991) A Historical Definition of Engineering. In: P. T. Durbin (ed.) *Critical Perspectives on Nonacademic Science and Engineering*. pp. 60–79. Cranbury, New Jersey, USA, Associated University Presses.

Lucena, J. C. (2010) What is Engineering for? A Search for Engineering beyond Militarism and Free-markets. In: G. L. Downey & K. Beddoes (eds.) *What is Global Engineering Education For? The Making of International Educators, Part 1*, Vol. 1, pp. 361–384. London, Morgan and Claypool.

Lucena, J. C., Schneider, J. & Leydens, J. A. (2010) *Engineering and Sustainable Community Development*. London, Morgan and Claypool.

Martin, R., Maytham, B., Case, J. & Fraser, D. (2005) Engineering graduates' perceptions of how well they were prepared for work in industry. *European Journal of Engineering Education*, 30 (2), 167–180.

Meiksins, P. & Smith, C. (1996) *Engineering Labour: Technical Workers in Comparative Perspective*. London, Verso.

Mitcham, C. (1991) Engineering as Productive Activity: Philosophical Remarks. In: P. T. Durbin (ed.) *Critical Perspectives on Nonacademic Science and Engineering*. pp. 80–117. Cranbury, New Jersey, USA, Associated University Presses.

Mukerji, C. (2009) *Impossible Engineering: Technology and Territoriality on the Canal du Midi*. Princeton, Princeton University Press.

Nonaka, I. (1994) A Dynamic Theory of Organizational Knowledge Creation. *Organization Science*, 5 (1), 14–37. doi: 10.1287/orsc.5.1.14.

Paretti, M. (2008) Teaching Communication in Capstone Design: The Role of the Instructor in Situated Learning. *Journal of Engineering Education*, 97 (5), 491–503.

Pawley, A. L. (2009) Universalized Narratives: Patterns in How Faculty Members Define "Engineering". *Journal of Engineering Education*, 98 (4), 309–319.

Quinlan, K. M. (2002) Scholarly Dimensions of Academics' Beliefs about Engineering Education. *Teachers & Teaching: Theory and Practice*, 8 (1), 41–64. doi: 10.1080/13540600120110565.

Schön, D. A. (1983) *The Reflective Practitioner: How Professionals Think in Action*. Basic Books Inc., Harper Collins.

Sheppard, S. D., Colby, A., Macatangay, K. & Sullivan, W. (2006) What is Engineering Practice? *International Journal of Engineering Education*, 22 (3), 429–438.

Sheppard, S. D., Macatangay, K., Colby, A. & Sullivan, W. (2009) *Educating Engineers*. Stanford, California, Jossey-Bass (Wiley).

Spinks, N., Silburn, N. & Birchall, D. (2006) *Educating Engineers for the 21st Century: The Industry View*. Henley, England, Henley Management College.

Stevens, R., O'Connor, K., Garrison, L., Jocuns, A. & Amos, D. M. (2008) Becoming an Engineer: Toward a Three Dimensional View of Engineering Learning. *Journal of Engineering Education*, 97 (3), 355–368.

Suchman, L. (2000) Organising Alignment: a Case of Bridge-Building. *Organization*, 7 (2), 311–327. doi: 10.1177/135050840072007.

Tang, S. S. (2012) *An Empirical Investigation of Telecommunication Engineering in Brunei Darussalam*. (PhD). The University of Western Australia, Perth.

Tilli, S. & Trevelyan, J. P. (2010, May) Batter graduate development programs could fix skills shortages. *Engineers Australia*, 82, 40.

Tonso, K. L. (2006) Teams that Work: Campus Culture, Engineer Identity, and Social Interactions. *Journal of Engineering Education*, 95 (1), 25–37.

Trevelyan, J. P. (2007) Technical Coordination in Engineering Practice. *Journal of Engineering Education*, 96 (3), 191–204.

Trevelyan, J. P. (2010) Reconstructing Engineering from Practice. *Engineering Studies*, 2 (3), 175–195.

Trevelyan, J. P. (2011) Are we accidentally misleading students about engineering practice? *Proceedings of the Research in Engineering Education Symposium*, 4–7 October 2011, Madrid, Spain.

Trevelyan, J. P. (2012a) Submission to Senate Enquiry on the Shortage of Engineering and Related Employment Skills. Canberra, Australian Parliament.

Trevelyan, J. P. (2012b) Understandings of Value in Engineering Practice. *Proceedings of the Frontiers in Education 2012*, 3–6 October 2012, Seattle, WA, USA.

Trevelyan, J. P. (2012c, December 3–5) Why Do Attempts at Engineering Education Reform Consistently Fall Short? *Proceedings of the Australasian Association for Engineering Education (AAEE) Conference*, 5–8 December 2010, Sydney, Australia.

Trevelyan, J. P. & Tilli, S. (2007) Published Research on Engineering Work. *Journal of Professional Issues in Engineering Education and Practice*, 133 (4), 300–307.

Trevelyan, J. P. & Tilli, S. (2010) Labour Force Outcomes for Engineering Graduates in Australia. *Australasian Journal of Engineering Education*, 16 (2), 101–122.

Vincenti, W. G. (1990) *What Engineers Know and How They Know It: Analytical Studies from Aeronautical History*. Baltimore, The Johns Hopkins University Press.

Vinck, D. (ed.) (2003) *Everyday Engineering: An Ethnography of Design and Innovation*. Boston, MIT Press.

Whalley, P. & Barley, S. (1997) Technical work in the division of labour: stalking the wily anomaly. In: S. Barley & J. Orr (eds.) *Between Craft and Science: Technical Work in US Settings*, pp. 23–36. Ithaca, New York, Cornell University Press.

Williams, R. (2002) *Retooling: a historian confronts technological change*. Cambridge, Mass, MIT Press.

Williams, R. (2003, January 24) Education for the Profession Formerly Known as Engineering, *The Chronicle of Higher Education*, p. B12.

Zussman, R. (1985) *Mechanics of the Middle Class: Work and Politics Among American Engineers*. Berkeley, University of California Press.

Chapter 3

The practical confrontation of engineers with a new design endeavour: The case of digital humanities

Frédéric Kaplan[1] *& Dominique Vinck*[2]

[1]*DHLAB, Ecole polytechnique fédérale de Lausanne (EPFL), Lausanne, Switzerland*
[2]*LaDHUL, Institut des Sciences Sociales, University of Lausanne, Lausanne, Switzerland*

I INTRODUCTION

This chapter shows some of the practices engineers use when they are confronted with completely new situations, such as when they enter an emerging field where methods and paradigms are not yet stabilised. This is the case for engineers and computer scientists who engage with human and social sciences to imagine, design, develop and implement digital humanities (DH) using specific hardware, software and infrastructure. Observing engineers in such a situation helps to shed light on their practices whenever they are confronted with new fields and new interlocutors.

Before introducing the DH case studies, we will position our understanding of engineering as sociotechnical work.

The classical representation of engineering as design and problem solving (the systematic process used to resolve problems with canonical examples like building a bridge, but also the process to define some ill-defined problems), which is grounded in mathematical modelling and solitary technical work, has been challenged for a long time. Whatever the sector of activity, engineering involves uncertainty; problem solving lacks clear structure and outcomes remain unpredictable (Vincenti, 1990). Thus, engineering is far from simply following a clear logic or applying design rules and scientific knowledge.

A completely different picture of engineering emerges when researchers study the work engineers really do. In this perspective, Youngman *et al.* (1978) revealed engineers spent 60% of their time interacting with other people. Not only senior engineers with managerial responsibility but also novice engineers working on 'technical' things devote an average 60% of their time to direct social interactions (Trevelyan, 2010). A qualitative study of the design work performed by a young mechanical engineer illustrates very well how such social interactions contribute to the real technical work (Vinck, 2003, chap. 1). It shows how engineers collaboratively work through the design process. It emphasises its social nature, with constant negotiation with many people on technical aspects (constraints of all kinds, good practices, possible solutions, validity) and on the purposes of the client and the strategies of his/her organisation. In that sense, engineering is coordinating the work of others (Trevelyan, 2007, 2008). Anderson *et al.* (2010, p. 155) wrote: "Engineers need competence in coordinating technical teams, negotiating a complex, nonlinear system, and solving problems" while Downey (2009, p. 60) calls to mind the

> "contingencies of practice, such as the myriad negotiations among forms of engineering knowledge, strategic activities of heterogeneous agents, specificities of territorial formations, distributions of money, etc."

This messy part is constitutive of engineering practices and justifies field and shop floor experience, and the learning and development of know-how and social skills (Bailey & Gainsburg, 2003). Engineers need to develop know-how and practical skills to produce working results in and for specific contexts, learning also through failures (Petroski, 1985). Engineering needs to be understood as a social performance, in which expertise is distributed among the participants and emerges from their social interactions (Trevelyan, 2010). The foundation of engineering practice is the distributed expertise enacted through social interactions between people harnessing the formal and tacit knowledge and expertise carried by a variety of people (see the Adams & Forin chapter in this book).

The specificity of the settings in which engineers work and of the tacit part of their work explain the variety of engineering cultures. Inside the same professional group, organisation, industrial sector, country, experienced working collaboration, design office, etc., common shared knowledge and tacit rules develop a variety of cultures. Their specificities are grounded into organisational routines, working places, mutual knowledge of people and material cultures (Vinck, 2003). It is not a surprise then to discover that engineering practices differ from one kind of sector or activity to another (Mer, 2003) and from one design office to another (Blanco, 2003), even for the same kind of design activity (Ravaille & Vinck, 2003). Different styles of practice, material cultures, organisational routines and language are developed in different settings and engineering disciplines.

Furthermore, engineers are neither isolated nor separated from the rest of society. In the shaping of their work, executive mandate, marketplace needs and the norms of the organisation play a dominant role (Bucciarelli, 1996) which engineers try to understand, translate into their work and, sometimes, negotiate. Engineering is more of a social process than it is the application of scientific principles. Engineers devote so much time to social interactions because they need to understand what the society needs (the client, the markets, the whole society), to convince other people to give them the resources they need to work, to convince them of the quality of the technical solutions they propose and to shape the perceptions of other people (Trevelyan, 2010; Vinck, 2003). Their work relates to various parts of society and to a broad range of actors. They are engaged in many inter-dependencies and/or aligning work with a variety of social actors they need to convince (Suchman, 2000). They are connected to trends and structures in which they are involved. They also actively connect their work to the structures and the transformations of societies (Meiksins & Smith, 1996; Lam, 1996, 1997). In this perspective, we position engineering as a social process of negotiation across object worlds (market, policy, organisational and social norms). One of the consequences is the fact that society also influences engineering cultures. For instance, American engineers put the emphasis on creating low-cost goods for mass consumption; this is their primary 'code of meaning' (Downey & Lucena, 2004).

Thus these social interactions, for the engineer, are less concerned with communicating technical or management information than with negotiation between heterogeneous actors (Law, 1989). So we can understand why expert engineers gather more heterogeneous information and more carefully consider the context and constraints than do young engineers (Anderson *et al.*, 2010). They know how important this is to enable them to succeed in technical work. They report that most of the failures they have encountered relate to social interactions, rather than errors in terms of scientific and technical knowledge and work. Of course, these negotiations influence the definition of the problems and the technical results. Downey (2009) mentions frictions with non-engineers, who might define problems differently. To be careful on the way to cooperate, to circulate information, to negotiate, etc. appears to be a crucial component of engineering work.

The last major aspect of our understanding enforces the idea of engineering as sociotechnical work, which means social and technical aspects are completely intertwined and constitute one unique reality. Social interactions relate not only to the definition of ill-defined problems (Sheppard *et al.*, 2006) and problem setting,[1] but also to the core of engineering and technical practice (problem solving, integration of knowledge, validation and implementation of the solutions) which involves feeling, judgment and creativity. Social researchers in science and technology studies have long recognised these technical and social aspects are intertwined and cannot be separated (Bijker & Law, 1992; Hughes, 1983). The distinction between social and technical aspects is more a social construction, locally enacted (Lagesen & Sorensen, 2009) than a repertoire of explaining factors (Latour, 1987). Technical knowledge is distributed between participants (Trevelyan, 2010); the exchange of data and information between engineers, technicians and others occupies a central place. Observing design practices, Vinck (2011) shows that technicians and engineers continually search for information by questioning colleagues, walking around their desks and looking inside shared databases. But the information is embedded into people, artifacts, practices and tacit knowledge. Technical workers also have to work to recover associated information: they strive to know not only where information or ideas come from and the history of their construction (e.g. the identity of the people who produced the information), but also the degree of trust they should have in it and what they can do with it. They do not trust isolated and de-contextualised information because they do not know where it has come from or who has shaped it. The example of exchanging technical information reveals we cannot separate technical content and social interaction: mobilising technical information is mobilising people; qualifying technical data is qualifying people; shaping the technique is shaping social relations. Engineers' work involves improving connections between information, people and things (Davis, 1998). This

---

[1] In engineering terms, problem setting is a quest to resolve the right problem (a question of effectiveness). To do that, the engineer needs to be able to formulate it well (Schön, 1979). In ANT terms, problem setting is conceptualised 'problematisation' which means the process by which the engineer determines a set of entities or actors, and build a network of relationships by defining the problems and identities of these entities in such a way as to establish him/herself as an obligatory passage point in the network. Thus 'problematisation' is more than the simple formulation of a problem; it touches on elements which are parts of both the social and the technical worlds, and involves establishing hypotheses on their identities and on the links between them.

applies throughout the production and circulation of intermediary objects and their equipping (Vinck, 2011). Designing involves using, interacting, shaping and circulating many intermediary objects of all kinds (industrial drawings, sketches, screen views) (Vinck, 2003). Through the process of supplying intermediary objects, engineers confer new properties on them (e.g. a status like 'draft' or 'validated' design) and shape the design space (a space of exchange) and collective work (Vinck, 2011). Intermediary objects and their equipping are a central concern of engineers involved in design and production processes. This emphasises the sociotechnical nature of engineering, which means that technical problem setting and problem solving are completely intertwined with social interactions.

2 ENGINEERS CONFRONTED BY NEW FIELDS

In this chapter, we would like to explore what happens when engineers engage in a completely new field. Studies on engineering practices have focused on (mechanical and software) design and innovation (Bucciarelli, 1996; Leonardi & Bailey, 2008; Sonnentag *et al.*, 2006; Vincenti, 1990; Vinck, 2003), technical support (Barley & Bechky, 1994; Barley & Orr, 1997; Orr, 1996) or production (Bechky, 2003; Mason, 2000). Even when the authors study engineers involved in innovation processes, these engineers appear to be working in their own engineering discipline with a lot of validated technical knowledge and methods. Of course, in many situations, they have to collaborate with other engineering disciplines or even with non-engineers, but the working experience is mainly framed by a relatively stabilised sector (car industry, software industry, aeronautics, etc.) where practices relate to established norms, standards, organisations and technical solutions. They refer to embedded sets of practices that have become routine and are not often openly questioned. In this chapter, we set up case studies as a departure from previous engineering practice studies.

Can we identify some different practices or a new style of practices? To generate exploratory data, we began our investigation with the case of a newly created research lab inside an engineering faculty. The research lab is dedicated to digital humanities.[2] Initially, it included one professor, himself an engineer trained in artificial intelligence, and a few young engineers. Before being named professor, the director of the DH lab was the director of a small company (with 20 employees) active in the field of digital publishing. He is co-author of this chapter and provider of most of the field information we use to qualify the engineering practices occurring in the new setting. Thus, the investigation is a kind of self-study, orientated by a social scientist in order to qualify the practices used. The constitution of the data and their interpretation is thus also, for the DH lab director, a reflective practice. The main methodological problem encountered by the authors relates to the ex post reconstitution of the project's journey and associated practices and to the difficulty in dissociating them from the intentions of

---

[2]Digital humanities is a relatively new field, the name being coined in 2001. It is the result of a lexical shift from humanities computing (automated textual analysis) (Kirschenbaum, 2010). It is a research and teaching field dedicated to the use of information technologies for the humanities, related to a shift from the texts towards a broad variety of corpuses (images, sounds, artifacts) and to the problems of heterogeneous and big data.

the DH lab director. We came to the story we explore in these cases through the external perception of the social scientist that there could be a different dynamic occurring in this lab than in the emerging field of DH.

Our analysis concentrates on two case studies representative of the diversity of practices occurring in the new DH lab. The first one concerns the production of an electronic edition of the complete works of Jean-Jacques Rousseau, initially conducted by a publishing company in collaboration with an academic editor and a design and programming company, and then integrated as a research project in the academic DH lab. The second case study concerns the on-going creation of a large historical database dealing with data from different epochs. This project involves collaboration with a Swiss newspaper and several other academic partners.

2.1   Case Study 1: the electronic edition of the complete works of Jean-Jacques Rousseau and the 'Facebook of the 18th century'

A Swiss publisher introduced, in June 2012, a new version of Jean-Jacques Rousseau's *Complete Works*. This new edition had been scientifically directed by two famous scholars in the humanities, and involved 21 Rousseau specialists. Published in paper format, it comprises 24 volumes and more than 15000 pages.

The publisher decided to produce an electronic (web-based) edition working with a local design and programming company specialising in electronic publishing. This company was known to have worked on other publishing projects, including related adaptations, for instance a digital 'enhanced' version of Voltaire's Candide. The director of the company assigned a designer and an engineer to work collaboratively on the project. He took the role of project leader. The designer had already worked on similar previous editions, in particular on the Voltaire project. The engineer was a specialist in large-scale web-based projects. At the beginning of the project, they met with the publisher to understand how he imagined this digital adaptation of the Rousseau edition. In order to orient the design of a technical solution, the project leader tried to imagine the expectation of the potential reader of this electronic edition, first by discussing with the publishers and then with one of the scientific editors of the *Complete Works*. In some cases, the two views diverged. The scientific editor considered a copy-and-paste function to be mandatory. For him the main relevance of the tool was the capability to search for particular occurrences (the word 'friendship' for instance) in a very large corpus of text in order to quote the corresponding sentences in academic articles. In his context of academic scholar, the copy-and-paste function was the most interesting feature of the digital edition. In contrast, the publishers were worried by potential abuses of such a function, imagining readers using the tool to export large parts of the text, beyond the fair use of academic quotation. During the implementation of the first prototype, the design and engineering team resolved this tension by limiting the range of the copy-and-paste function.

Thanks to these initial discussions, the engineer and the designer started to have a clearer image of the typical academic user of the future tool and adapted the design accordingly with an efficient and work-oriented interface rather than a seductive and didactic layout (as was the case for the Voltaire project).

Thus, the starting point of the search for a solution was not only the expression of a demand coming from the client (the publisher) but also the personal identification, made by the engineer and the designer, with a potential user. Based upon his personal (as reader) and professional (as digital publisher) experience, the project leader suggested to the designer and the engineer that the potential reader should be envisaged as probably facing a navigation problem in using such a large set of documents. This way of thinking also reflects a position he took in a book he authored (Kaplan, 2009). He identified navigation as the challenge and was then confronted by the necessity to find an efficient way for the user to navigate within a text of 15000 pages. The argument he articulated was that because a volume of the paper edition was a concrete three dimensional object, the reader could easily understand where he/she was. On a two dimensional screen, it seemed necessary to introduce additional navigational elements to compensate for the missing third dimension. The way of framing this development in the context of a dimensionality problem was actually linked with the notion of document dimension developed by Pascal Robert, an author the project leader often cites in his academic papers (Robert, 2010). The designer working on the project was familiar with this issue and had successfully designed similar solutions during the digital adaptation of smaller books.

The designer worked alone for a couple of days on this specific issue and imagined several solutions such as: menus proportional to the amount of content they gave access to, navigation bars taking the form of histograms indicating the textual density of each page of a book. These solutions were meant to give additional information to the reader, which would facilitate navigation in the large corpus of text. They were modelled using a vector graphics software package the designer was familiar with. This led to a skeletal framework known as a website wireframe which is a common method used by Interaction Design professionals. It allows the arrangement of elements taking into account knowledge and issues regarding visual perception, information architecture and user interaction concerns such as interface conventions and navigation. The project leader decided to use all these elements combined as their structuring effects reinforced one another.

The resulting representation lacked aesthetics (no typographic style, no colour...), since the objective was to represent and to explore the content and the functionality of what the screen would do. The small design and engineering team usually preferred to work this way, regarding it as more reliable, but in this case, with a client not used to working with wireframes, the project leader decided to start the graphical design process at this early stage and to present the design in the form of screens to the publisher. The idea was that such a prototype required less effort for the publisher to understand because he would perceive directly what the final presentation of the results could be. Furthermore, the small team decided to present only one solution, the solution which appeared the most coherent to them (Figure 1).

The meeting took place at the publisher's office. The project leader and the designer were present. The publisher approved it without much discussion, raising only questions aimed at achieving understanding. He did not discuss the design decisions and was very appreciative of the graphical appearance. He just asked for a second meeting with the scientific director of the Rousseau edition. This second meeting took place in town and only the project leader attended. He presented the same demo but the

Figure 1   Navigation flux presenter on screen to the client.

discussion this time focused on the ergonomics of the tool, in particular the ease of use of the copy-and-paste function.

Immediately after the first meeting, the programmer started programming coding the interface according to the wireframe and early graphical design presented to the client. The designer stopped being involved at this stage of the project; he expected only to check the coherence between the design he produced and its actual implementation at the end of the project.

The programmer built a search engine with the objective of navigating easily in the corpus. This search engine included a full-text search permitting a query against all the pages containing a particular word, which is a standard expected feature in this kind of digital edition. Although this was not an initial request of the client, the project leader suggested that this search engine could also use part of the work the Rousseau scholars had done for the paper edition, notably the establishment of a large index of persons' names. Indeed, a given person might appear under different alternate names in Rousseau's works. The programmer then built a feature enabling the automatic creation of a web page for each person appearing in the corpus. This page pointed to all the passages in which this person appeared. The programmer immediately designed a first version of the search engine. The project leader corrected a couple of ergonomic elements.

All the functional and design features mentioned so far could be considered state-of-the-art. This was what the publisher expected from the design and programming company. But an unforeseen innovation step occurred when all this was in place. The project leader liked the idea of having a personal web page for each person appearing in Rousseau's work. Furthermore, this inspired him with the idea of telling more about this person then just what could be read on the page where he or she appeared. Indeed, there was an analogy between this web page and a social network personal page (e.g. a Facebook page, which was a popular web service at the time they were developing the technical solution for the electronic edition, see Figure 2). The project leader argued to the programmer and the designer that the user might have thus expected to find in this context the standard extended metaphor of social network features, such as a list of names of other persons linked with the person featured on the web page. The question then was: is it possible to represent the network of the person by harvesting from the *Complete Works* the links of this person with other persons in the corpus? This involved complex text-mining techniques to find the ways people were connected and to find a way to display each personal network. Both the programmer and the designer found the challenge interesting and fun.

The project leader and the programmer brainstormed the best technical way to tackle this challenge. After considering several complex text mining techniques to recreate the network of persons inside Rousseau's *Complete Works*, a simple algorithm was tested by the programmer. The algorithm would assume there is a link between two persons of the corpus if they appear conjointly in a significant number of pages. The project leader decided the fact that two persons appear in the same page might not be significant but if this co-occurrence is found in at least five pages then these two persons can be considered to be linked in some way or another. The nature of this link is unknown but a link probably exists. In consultation with the project leader, the programmer tested different limit values and then included this feature in a prototype of the electronic edition. It was now possible to display for each person in Rousseau's

Figure 2 A personal web page (Facebook type) for a person appearing in Rousseau's work.

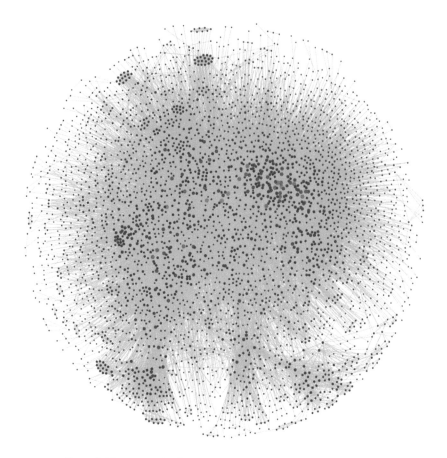

Figure 3 The network of persons inside Rousseau's *Complete Works.*

works their own 'personal network' showing all links with other persons (see Figure 3), and to navigate within the corpus using this new feature. To make explicit the nature of the link found by the algorithm, the user could just display the list of pages where the two names co-occurred and read them to infer the underlying relationship to explain their statistical co-occurrence.

The publisher found this feature exciting but, as it constituted an unexpected innovation step, he sought approval from at least one of the scientific directors of the edition to check its relevance. They organised a meeting with one of the scientific directors of the edition. The scholar found this new feature exciting and it was decided to keep it in the final edition. A few weeks later, it was presented to a larger group of Rousseau scholars at the occasion of Rousseau's 300th anniversary. The fact that Rousseau's work implicitly contained a 'Facebook of the 18th century' was perceived as an unexpected innovative feature of the new edition.

Interestingly, this single feature also transformed the 'standard' publishing project (digital publishing with functionality like a browser) into an academic Digital Humanities project (which includes exploring ways to generate new information and to interact with a corpus). When the project leader was appointed as a professor, he continued the project inside his newly created DH lab. One reason for this is that the set of the links calculated by the algorithm constitutes a well-studied object in computer science: a network. Networks have been studied extensively in a variety of contexts (biological, economical, ecological and social networks) and are now associated with sound mathematical theory. One consequence of this is that as soon as one encounters a network in a study, a collection of measures and visualisations can be computed in a straightforward manner. For instance, one may compute the centrality of each node of the network using a standard measure, giving a way to evaluate its social importance.

As a consequence, a new phase of the project started in the academic setting of the DH Lab, in which a large number of measures could be tried. This triggered a new phase of computational analyses around the Rousseau network. A mathematician, a specialist in network theory, was hired to start a PhD work on this subject at the DH lab. The PhD student expected that only a subset of network measures would be relevant for bringing an innovative view on Rousseau's work. This meant that, after an almost-blind generation phase where many state-of-the-art computations were tried, a discussion with a Rousseau scholar was planned in order to 'prune' and interpret the interesting findings among all the various possible measures and representations generated. The mathematician of the DH lab worked directly with the engineer of the company to produce the network measures. He also attended a meeting with the publisher to fully understand the context of the project. This collaboration resulted in a joint submission to the Digital Humanities conference, co-authored by the mathematician, the engineer and the project leader.

A couple of months later, when the first results were produced, the project leader organised the meeting with a Rousseau scholar. The mathematician presented the work in development and the first graph that had been produced. In order to assess the relevance of the results, the scholar suggested applying the same methods to other autobiographical works (the *Memoires* of Saint Simon as an example earlier than Rousseau and Chateaubriand as an example after Rousseau's period). This research project is still on-going.

This case study illustrates at least two processes relatively common in the formation of a Digital Humanities project and in many engineering design projects. The first is the use of an analogy (a personal page on a social networks site, a well-established contemporary service in 2010), in a different context (the 18th century). This metaphor provided a bridge between the perspectives of the different people involved in the project (scholars in the humanities, publisher, interface designer, programmer, project leader, scholars in network theory). Its narrative force helped in translating the complexities of project. It also opened new perspectives and opportunities, which led to new goals and challenges. The creation of a webpage for each person mentioned in Rousseau led 'naturally' (due to shared knowledge relating to successful web practices at the time) to the idea of computing an associated social network. The idea of creating a 'Facebook of the past' crystallised in a swift and efficient way a new sub-goal for the project. Most participants could understand the idea easily because it corresponds to a well-known popular use of the Internet and, as a consequence, made appropriate design and programming decisions. It provides a common place for understanding all the design decisions and helped the transition from designing a navigation tool to designing a social network. Such kinds of analogy[3] permit the association of a strong narrative force for explaining the project and results with the potential for high visibility in the popular press. A technically complex project is thus summarised by a short and efficient narrative process.

The second process at work in this case study is a 'fast-forward' move occurring when a standard academic object (the network and its theory) is connected to the project. This makes the project move very quickly, opening the possibility to take a well-defined scientific 'superhighway' with a lot of associated concepts, methods and tools. Many results can then be quickly generated, reusing state-of-the-art measures, visualisation and reasoning. They must then be mobilised and introduced into the collaboration with humanities scholars (e.g. those who are involved into digital publishing) to open new approaches for human sciences investigation on textual corpuses.

In general, it is interesting to note the opportunistic nature of these moves, as the most innovative feature of the project was not part of the standard job planned in the beginning. The actors who chose to introduce these new features (a new way to navigate within an enormous text) were dynamically inspired by the discovery of common patterns of contemporary digital culture (Facebook) or by spotting a well-mastered concept from their scientific toolbox (network analysis).

2.2   Case Study 2: Creation of a large historical database based on uncertain sources and the generalisation dilemma

Before the DH lab was officially created, its director initiated collaboration with a Swiss newspaper focusing on their database of 200 years of digitalised articles (4 million articles). This base is currently accessible in image and text mode. Users can access it freely and select articles using a search-engine. The project is about transforming this large number of disparate, incomplete and (to a certain extent) unreliable archives into a coherent and searchable reconstruction of the past. The idea is to extract from this textual and visual database semantic information that could permit users to browse and visualise differently the information contained in the corpus.

At the beginning of this project, only the DH lab director was involved. The DH lab director anticipated that text mining algorithms could be used to automatically extract spatiotemporal data (information about people, places and events) from the base of a billion items. For him, one major challenge of this project was that these spatiotemporal data were intrinsically uncertain both in time and space (e.g. if an article mentions an event that happened 'last year' or 'in Argentina', it is not precise enough to place it in a precise historical geospatial time system). The information can also be inconsistent from one dataset to another, e.g. about a single event. For the director of the DH lab, a research challenge identified early on was to invent new ways to deal with such intrinsic uncertainty, in particular during the process in which large sets of information are merged. For the time being, he considered using fuzzy logic and probabilistic approaches to integrate these data in a coherent framework (a typical fast-forward move based on well understood methods) but he was not sure

---

[3]Other Digital Humanities projects have used a similar kind of approach. For instance, a team in Stanford built Orbis, a travel service mimicking the functionalities of standard online travel-route services but for the Roman Empire (http://orbis.stanford.edu).

about the relevance of these methods in this particular context. As in the previous case study, the director of the lab therefore sought collaboration with a history scholar.

This project, of reconstructing a historical probabilistic, geographic information system out of billions of uncertain data of the last 200 years, is quite ambitious in terms of technological, epistemological and historical science challenges. It could reasonably occupy researchers from computer sciences and the humanities of the lab for several years. However, it aggregates other projects which present similar challenges. For instance, a professor of the University of Lausanne consulted with the DH lab director because he wished to start a collaborative project about the diffusion of prayer texts in the 16th and 18th centuries, as part of a national research project on Digital Humanities. Its objective was to understand the spatiotemporal diffusion of prayer literature based on a corpus of documents, framing the data in a historical geographical information system. The system should model the diffusion route of the printed material from European printers to various publics (e.g. peddlers' networks, book fairs). Because of the intrinsic uncertainty of the information available, these diffusion models would be probabilistic. From the point of view of the DH lab director, this project about early modernity shared many common technological features and challenges (modelling of uncertainty, building of a probabilistic historical geographical information system) with the newspaper project that focuses on sources from the last 200 years. Important differences concerned the amount of data available and the level of uncertainty of the reconstructed models. Nevertheless, the DH lab director considered the possibility of creating a common infrastructure that would be capable of dealing with the two kinds of data.

Another research team was informed about the creation of the DH lab and contacted the director of the lab with the opportunity of collaboration centred around a famous prehistoric cave. Again, the underlying ambition of the project was to create a common geographical framework serving as a base for integrating various data about prehistoric times. Uncertainty in this case was very high and information sparse. Nevertheless, the director of the lab considered the relevance of a common framework could again be argued.

The director of the lab discussed with his team the relevance of building a common approach for all these projects but did not get clear feedback. He decided it was preferable, for the time being, to build specific systems for each of the projects in order to understand their specificity, keeping in mind that at some stage a technical convergence could take place. Part of this convergence would also be linked to the speed of each project, depending on how fast they would be financed.

In these projects, the director and his team were confronted by many questions and difficult strategic decisions: should they try to build a system capable of dealing with epochs ranging from the present period to prehistoric times, with the risk of not making any relevant contributions to any of these domains? Would integrating this constraint from the beginning of the project enrich the whole system or weaken it? In some sense, if a satisfactory way of dealing with uncertain data (e.g. solutions for coding uncertainty, merging together various kinds of uncertain information, representing the level of uncertainty of reconstructed models) is found for an epoch it could perhaps be translated to be relevant to another one. For the early modern period, and obviously for prehistoric times, many elements have to be reconstructed and cannot be directly associated with primary sources. At some point, the director suggested, generative approaches like procedural methods, agent-based simulations or differential equation modelling could be used in such conditions (e.g. Weber *et al.*, 2009; Turchin, 2003). Would such methods also be appropriate for more recent periods, even if more reliable data are available?

The engineers of the DH lab are thus engaged in questioning and thinking about issues which are both fundamental (epistemic) and strategic (being ambitious in order to be innovative). Some of those involved feel it to be too ambitious and, when the DH lab director puts forward his ideas, very few people give interesting feedback. The project seems to lack maturity and the lab director remains relatively isolated. The project is facing what he feels to be a common issue with various digital humanities projects: the generalisation dilemma. The more general a project is, the more it would attract opportunity for collaboration, he supposes. This encouraging feedback would push the project to generalise more in making a computing infrastructure capable of dealing with a variety of research questions and time periods. As an engineer, he feels very tempted to make a system as general as possible. But he expresses a dilemma. On the one hand, if generality is too high, relevance may be lost. On the other hand, putting several research studies that were traditionally academically separated into a common framework could be one of the major contributions of the Digital Humanities approach.

This second case study illustrates the major concern for engineers confronted by a new emerging field. Lacking established references and standards, they have to tackle epistemological questions and strategic decisions. How do they do that? They identify research challenges defined as inventing new ways to deal with the major problems they have identified: the intrinsic uncertainty of the data and the difficulty of their integration into a coherent framework. As we can observe in other engineering settings, they consider these using different established approaches and well-understood methods. By doing so, the engineers expect to engage a 'fast-forward move'. The major hesitation concerns the relevance of these methods to these particular contexts; they raise epistemological questions. As engineers, they define themselves as interested in general, simple and robust solutions; they are tempted to design a common framework capable of dealing with a variety of data and to build a computing infrastructure capable of dealing with a variety of research questions and time periods. Being ambitious in terms of innovation, they willingly accept and aggregate a variety of DH projects across which they look for common technological features and challenges (such as the modelling of uncertainty and the need to build probabilistic historical geographical information systems). They anticipate the more general a project is, the more it will attract opportunity for collaboration. They expect to make a major contribution by putting disparate research studies within a common framework.

Following engineers in action (Latour, 1987) helps to characterise their *de facto* strategy: a search for the features capable of generalisation and for generative approaches. Along the way, they identify the risk of losing relevance and of weakening their contribution but they act as if they believe this is not a major risk. They consider the relevance of a common framework can be convincingly argued and that they will, in the main, enrich the domain. Through a series of practices (aggregating heterogeneous projects, posing a general problem, re-casting specific problems into a general one, looking for simplified general models and for 'fast-forward moving' methods) which the engineers highlight as part of their identity, we can also perceive the force of some unquestioned convictions (better to go fast, better to have a high level of generality, better to have simple solutions, better not to lose oneself in the details of specific contexts) associated with this engineer's identity.

3 CONCLUSION: ENGINEERING IN ACTION – ENGINEERING PRACTICES IN EMERGING DYNAMIC DOMAINS

In this endeavour engineers were involved in the design, development and implementation of two types of digital tool for the humanities and social sciences; social networks within the works of famous authors and a probabilistic historical geographic information system. We observe, firstly, that both cases confirm classical results in the social study of engineering practices: social interactions between heterogeneous actors, negotiation on the purposes of the client and on technical aspects (possible solutions and their validity), distributed expertise (not so evident in the second case study), work on ill-defined problems, influence of the various perspectives of the actors involved, creativity and uncertainty, the opportunistic and contingent nature of all the moves.

Secondly, and more interestingly, these cases show three working processes that could shed some light on engineering practices.

The first one was previously observed in design offices in the car industry where designers have to work under time pressure; they reuse pre-existing solutions they have to hand. In the digital humanities projects the engineers reused off-the-shelf solutions, such as the search engine. They also designed new solutions, using as their starting point the analogy of an existing solution made for a different problem. In the case of the electronic edition of the complete works of Jean-Jacques Rousseau, they began with such well-established and popular products as the personal web page on social networks sites (e.g. Facebook). In the practices of these engineers getting involved in the digital humanities, 'captivating analogies' and 'generative metaphors' seem to frame the problem formulation and problem solving when practitioners are confronted with new situations, as Donald Schön (1979) demonstrated when practitioners are confronted with new situations. Thus, the design is heavily connected to on-hand products, local ecology and the social and market distribution of products and services. Far from starting from scratch or with the application of systematic design methods, they think and work with contemporary, popular (at the moment), and even trivial, solutions. Through the practices of these engineers, the design of tools for digital humanities is related to the trajectory and the dissemination of successful tools. The material and informational ecology surrounding them leads them to analogically connect their everyday contemporary tools (those which are popular for their on-line social networks) to past periods and different contexts. For them, this path is natural, evident; they are not stopped by the risk of anachronism. They apply to the past a contemporary social practice and the tools supporting this practice. They create a webpage for each person who, in the 18th century, was mentioned in the Rousseau work: a 'Facebook of the past'. The idea seemed to be luminous and evident and crystallized the effort. Even if the device could only track only names and words, and not ideas, it functioned as a new paradigm or a design framework for the programming. Inside this framework, well-known and well-established solutions are thus recycled. Furthermore, these solutions have the advantage of being known both by the designers and by the client and humanities scholar's users. Most of them understand the idea 'easily' because it is based on relatively common knowledge of the ICT ecology and associated meanings.

The second practice observed at work is the alignment to and the mobilisation of a whole academic world through a standard academic package (Fujimura, 1992); using state-of-the-art established theory and associated measures, protocols, etc. In the case of the 'Facebook of the 18th century', the concept of 'network' and the associated toolkit allowed the engineers to translate (Callon, 1986) contemporary popular practices and tools, social theory and mathematical modelling. These connections opened up for them a well-defined scientific 'superhighway' in which they could easily progress; their project could engage in a 'fast-forward' move instead of a risky exploration. They could then 'do the same as usual' in terms of network analysis, modelling and programming and so generate many original solutions for humanities scholars. Thanks to the connection to this academic package, they shifted from exploration mode to exploitation activities (March, 1991).

The third practice refers to the definition of the relevant strategy made of heterogeneous project aggregation; of posing a general problem and looking for general and simplified models and fast-forward moving methods. From the multiplication of the various demands (regarding geographical data and historical data of different periods), they made connections that led to the search for a universal solution relevant to all these cases. This required them to re-define and to re-cast the problems of each academic client as a specific instance of the general problem. Constructing a kind of generalisation, they deduced consequences in terms of the attractions of others demands, clients, collaborations, cases, corpuses and sets of data, but also in terms of designing a computing infrastructure capable of dealing with a variety of research questions and corpuses. Confronted by an emerging domain, their practices reveal the temptation to make a system as general as possible; to induce connections between the fragmented divergent academic disciplines (between disciplines like sociology, geography and history, but also between historians specialised in different periods) and to make a major contribution to the (digital) humanities.

Engineering in action, confronted by a new field, engages in two kinds of practices: the first refers to the use of existing solutions coming from another domain (either circulating through an analogy between domains, situations or periods, or connecting to an established domain whose results and tools are mobilised through a mediating concept); the second is the engagement of in-depth theoretical and strategic thinking above any established knowledge or rule of work. If we consider the knowledge ecology, underlined by Bailey and Gainsburg (2003), we could note that this situation reflects the novelty of the field to which engineers are entering. It contrasts with a traditional firm in civil engineering, for instance, in which established ideas and 'helping' done by senior engineers could guide the design. In the digital humanities, the younger and newer technological setting had to be done peer-to-peer, through experimentation, exploration and connection to heterogeneous domains. Thus we can no longer idealise the engineer as a rational man applying the state of the art, nor as a genius craftsman or negotiator sealing compromise into the materiality. Instead, he becomes an investigator and strategist looking for a major orientation that could guide the shaping of the technologies. This chapter illustrates the switch engineers make from 'scientist' mode (where rigorous methods, mathematical abstraction, measurable evidence, and accurate prediction of outcomes can be done) to 'designer' mode (Figueiredo, 2008; see also his chapter in this book) when they engage in a new domain.

REFERENCES

Anderson, K., Courter, S., McGlamery, T., Nathans-Kelly, T. & Nicometo, C. (2010) Understanding engineering work and identity: a cross-case analysis of engineers within six firms, *Engineering Studies*. 2 (3), 153–174.

Bailey, D. & Gainsburg, J. (2003) *Knowledge at Work*. Stanford Institute for Economic Policy Research. [Online] Available from: http://www-siepr.stanford.edu/programs/SST_Seminars/Bailey_Gainsburg_-_SECOND_DRAFT.pdf [Accessed 25th August 2010].

Barley, S. & Bechky, B. (1994) In the Backrooms of Science: The Work of Technicians in Science Labs. *Work and Occupations*. 21 (1), 85–126.

Barley, S. & Orr, J. (eds) (1997) *Between Craft and Science: Technical Work in US Settings*. Ithaca, NY, Cornell University Press.

Bechky, B. (2003) Sharing Meaning across Occupational Communities: The Transformation of Understanding on a Production Floor, *Organization Science*. 14 (3), 312–330.

Bijker, W. & Law, J. (1992) *Shaping technology/building society: studies in sociotechnical change*. Cambridge, MA, London, MIT Press.

Blanco, E. (2003) A prototype culture. Designing a paint atomizer. In: Vinck, D. (ed.) *Everyday engineering. Ethnography of design and innovation*, MIT Press, Cambridge, USA. pp. 119–133.

Bucciarelli, L. (1996) *Designing Engineers*. Cambridge, MA: The MIT Press.

Callon, M. (1986) Some Elements for a Sociology of Translation. Domestication of the Scallops and the Fishermen of St-Brieuc Bay. In Law, J. (ed.) *Power, Action and Belief. A New Sociology of Knowledge?* London, Routledge and Kegan Paul. pp. 196–233.

Davis, M. (1998) *Thinking Like an Engineer: Studies in the Ethics of a Profession*. New York, Oxford University Press.

Downey, G. & Lucena, J. (2004) Knowledge and Professional Identity in Engineering: Code-Switching and the Metrics of Progress. *History and Technology*. 20 (4), 393–420.

Downey, G. (2009) What Is Engineering Studies For? Dominant Practices and Scalable Scholarship. *Engineering Studies*. 1 (1), 55–76.

Figueiredo, A. D. (2008) Toward and Epistemology of Engineering. *Proceedings of the Workshop on Philosophy & Engineering. WPE 2008, 10–12 November 2008*, London, Royal Engineering Academy.

Fujimura, J. H. (1992) Crafting Science. Standardized Packages, Boundary Objects, and 'Translation'. In Pickering, A. (ed.) *Science As Practice and Culture*. Chicago, University of Chicago Press. pp. 169–211.

Hughes, T. P. (1983) *Networks of power: Electrification in Western society, 1880–1930*. Baltimore, Johns Hopkins University Press.

Kaplan, F. (2009) *La metamorphose des objets*. Limoges, FYP Editions.

Kirschenbaum, M. G. (2010) What Is Digital Humanities and What's It Doing in English Departments?. *ADE Bulletin*. [Online] (150). 55–61. Available from: http://mkirschenbaum.files.wordpress.com/2011/03/ade-final.pdf. [Accessed on 26th December 2012].

Lagesen, V. & Sorensen K. (2009) Walking the Line? The Enactment of the Social/Technical Binary in Software Engineering. *Engineering Studies*. 1 (2), 129–149.

Lam, A. (1996) Engineers, Management and Work Organization: A Comparative Analysis of Engineers' Work Roles in British and Japanese Electronics Firms. *Journal of Management Studies*. 33 (2), 183–212.

Lam, A. (1997) Embedded Firms, Embedded Knowledge: Problems of Collaboration and Knowledge Transfer in Global Cooperative Ventures. *Organization Studies*, 18 (6), 973–996.

Latour, B. (1987) *Science in action: how to follow scientists and engineers through society*. Cambridge, MA, Harvard University Press.

Law, J. (1989) Technology and heterogeneous engineering: the case of Portuguese expansion, In: Bijker, W., Hughes, T. & Pinch, T. (eds) *The social construction of technological systems*.

New directions in the sociology and history of technology. Massachusetts, MIT Press. pp. 111–134.

Leonardi, P. & Bailey, D. (2008) Transformational Technologies and the Creation of New Work Practices: Making Implicit Knowledge Explicit in Task-Based Offshoring. *MIS Quarterly.* 32 (2), 159–176.

March, J. (1991) Exploration and exploitation in organizational learning. *Organization science.* 2 (1), 71–87.

Mason, G. (2000) Production Supervisors in Britain, Germany and the United States: Back from the Dead Again? *Work, Employment and Society.* 14 (4), 625–646.

Meiksins, P. & Smith, C. (eds) (1996) *Engineering Labour: Technical Workers in Comparative Perspective.* London, Verso.

Mer, S. (2003) The Structural Engineer in the Design office. A world, its objects and working practices. In: Vinck, D. (ed.) (2003) *Everyday engineering. Ethnography of design and innovation,* MIT Press, Cambridge, USA. pp. 79–91.

Orr, J. (1996) *Talking About Machines: An Ethnography of a Modern Job.* Ithaca, NY, Cornell University Press.

Petroski, H. (1985) *To Engineer Is Human: The Role of Failure in Successful Design.* New York, St Martin's Press.

Ravaille, N. & Vinck, D. (2003) Contrasting Design Cultures: Designing Dies for Drawing Aluminium. In: Vinck, D. (ed.) (2003) *Everyday engineering. Ethnography of design and innovation,* MIT Press, Cambridge, USA. pp. 93–117.

Robert, P. (2010) *Mnémotechnologies: Une théorie générale critique des technologies intellectuelles,* Paris, Hermès-Lavoisier.

Schön, D. (1979) Generative Metaphor: A Perspective on Problem-Setting in Social Policy. In: Ortony, A. (ed.) *Metaphor and Thought.* Cambridge, Cambridge University Press. pp. 254–284.

Sheppard, S., Colby, A., Macatangay, K. & Sullivan, W. (2006) What Is Engineering Practice?. *Int. J. of Engineering Education.* 22 (3), 429–438.

Sonnentag, S., Niessen, C. & Volmer, J. (2006) Expertise in Software Design. In: Ericsson K., Charness, N., Hoffman, R. & Feltovich, P. (eds) *The Cambridge Handbook of Expertise and Expert Performance.* Cambridge, UK, Cambridge University Press. pp. 373–387.

Suchman, L. (2000) Organising Alignment: A Case of Bridge-Building. *Organization.* 7 (2), 311–327.

Trevelyan, J. (2007) Technical Coordination in Engineering Practice. *Journal of Engineering Education.* 96 (3), 191–204.

Trevelyan, J. (2008) The Intertwined Threads of Work. *Engineers Australia.* 80 (2), 38–39.

Trevelyan, J. (2010) Reconstructing engineering from practice, *Engineering Studies.* 2 (3), 175–195.

Turchin, P. (2003) *Historical Dynamics: Why States Rise and Fall.* Princeton University Press.

Vincenti, W. (1990) *What Engineers Know and How They Know It: Analytical Studies from Aeronautical History.* Baltimore, MD, The Johns Hopkins University Press.

Vinck, D. (ed.) (2003) *Everyday Engineering: An Ethnography of Design and Innovation.* Cambridge, MA, The MIT Press.

Vinck, D. (2011) Taking intermediary objects and equipping work into account when studying engineering practices. *Engineering Studies.* 3 (1), 25–44.

Weber, B., Müller, P., Wonka, P. & Gross, M. (2009) *Interactive Geometric Simulation of 4D Cities. Proceedings of the Eurographics 2009 / Computer Graphics Forum, 30 May–3 April 2009,* 28 (2), 481–492.

Youngman, M., Oxtoby, R., Monk, J. & Heywood, J. (1978) *Analysing Jobs.* Farnborough, Hampshire, UK, Gower Press.

Chapter 4

Engineering design teams: Considering the forests and the trees

Jim Borgford-Parnell[1], *Katherine Deibel*[1] *& Cynthia J. Atman*[1,2]
[1]*Center for Engineering Learning & Teaching, University of Washington, Seattle, Washington, US*
[2]*Human-Centered Design & Engineering, University of Washington, Seattle, Washington, US*

I INTRODUCTION

In this chapter, we discuss research on individuals with three levels of engineering design expertise engaged in design tasks in some of our design process and design context research. That research laid the foundation for a subsequent study, wherein we analysed data on an engineering design team of practitioners in action. We expand our data analysis by looking specifically at how and when those design team members paid attention to group processes and contextual issues. Results of our earlier research have been used to foster students' awareness of design processes. Our research results may also help to illuminate the connection between design processes, broad thinking, and teamwork. Additionally, we will present several representations of design processes that we believe may prove helpful in fostering understanding of important aspects of team design.

2 BACKGROUND

What do engineering designers do? They work with others. They collaborate on project teams that often include people from different fields and backgrounds. One of the primary characteristics of effective design teams is having the right mix of knowledge and skill resources represented by their members. Team practices that make best use of those resources focus on how design tasks are divided, how team interdependence is built, and how ideas are shared and weighed.

Much has been written over the past decade about the need for engineers who have the ability to work competently in design teams, to share ideas, and to consider a wide range of issues in the process of designing. Certainly, society needs engineers to develop the ability to communicate well, to understand design processes, and to respect a full range of contextual issues, however, it is well accepted in the field that competent design teams can produce better results than do individual engineers. The creative synergy that takes place when the right mix of team members share and consider each other's ideas is valued by industry and drives important educational efforts, such as the problem- and project-based capstone coursework that often culminates an engineering undergraduate degree. Those courses frequently present novice engineers with their first experience of the complexity of engineering design.

The interaction of design process tasks and group development tasks is an increasingly important area of concern in engineering. The development of clear, timely, and focused dialogue underlies effective collaboration on design teams. This places social processes in a foundational role in a team design process. Obviously this is not a recent revelation. However, it does pose an on-going challenge for engineering educators who are endeavouring to instill in their students a broad conceptual understanding of the myriad pieces and workings that constitute team design. As was made evident by Waite *et al.* from their study of student group work in computer science, if we are to help our students gain an ability to collaborate we will first have to overcome their strong bias against collaboration (Waite *et al.*, 2004). The advent of capstone design and other team-based project design courses is an attempt to address that challenge: to demonstrate to students the types of design problems and design conditions found in engineering practice, and to foster that kind of synthetic understanding. It is also evidence of an educational acknowledgement of the inseparable interaction of skills and knowledge that characterise effective team design processes.

The importance of team design has also been well articulated by engineering industry leaders and educational program accreditors. For example, 'an ability to function on multidisciplinary teams' is one of the student learning outcomes promulgated by the influential ABET engineering accreditation organisation (ABET, 2012). The notion of team design has been incorporated into most engineering capstone design courses (Dym *et al.*, 2005). However, integrating team design experiences at other levels of a typical engineering curriculum has been a slow process. Sheppard *et al.* proposed that the application of technical and scientific knowledge has traditionally defined engineering design, and therefore, those types of knowledge have been emphasised in engineering education (Sheppard *et al.*, 2009). Regardless of how well accepted the importance of team design is to the field of engineering, and despite the efforts that educators make to infuse engineering curriculum with team design experiences, we still have a long way to go. This is especially true if our goal is for engineering students to graduate with more than a proficiency with each of their program's individual learning outcomes (ABET, 2012); to additionally possess a conceptual understanding of how those individual outcomes interconnect in engineering design practice.

3   LITERATURE REVIEW

3.1   Design as a social process

In Donald Schön's influential book The Reflective Practitioner (1983) he proposed a theory of reflection in action that depicted designing as a process of framing, moving, reflecting, and reframing a design problem. Later, Schön and Wiggins (1992) focused on designers' use of sketches as a catalyst for reflective activity. They described the design process as a reflective conversation wherein a designer produces sketches, drawings, or other materials that represent their ideas, and then makes judgments about those ideas, which then sparks new ideas and a continuing development of a design. They called this a reflective process of seeing, doing, and seeing again.

Although they were describing the processes of individual designers, the reflective moves they described fit well with other scholars' descriptions of group design processes (Brereton *et al.*, 1996; Cross *et al.*, 1996; Dym *et al.*, 2003; Hennessy & Murphy, 1999; Minneman, 1991; Paulus & Yang, 2000; Reid & Reed, 2000; Valkenburg & Dorst,

1998). Schön's theory of reflection in action was later used by Valkenburg and Dorst (1998) to analyse and describe the nature of team designing. Similarly, Brereton *et al.* (1996) characterised the social interactions that take place as a design team works as a series of negotiations on topics: topics to focus on, and topics to transition to. Drawing on Goldschmidt's earlier work (1991, 1994), Reid and Reid proposed that a group design process is cyclical, with "figural reasoning alternated unevenly with periods of predominantly conceptual reasoning over a band of periods ranging from 5 to 10 minutes" (2000).

Whether the team design process is described as a series of negotiations, a cycle of figural and conceptual reasoning, or a reflective conversation; it is well-accepted that the individual members of a design team interact with and influence each other's thinking. As Paulus and Yang (2000) proposed, in group brainstorming sessions an exchange of ideas produces additional ideas and therefore each member of a group may be "exposed to more ideas during their session than solitary idea generators" and therefore there is greater potential for "cognitive stimulation" (p. 77). In their influential book, Nonaka and Takeuchi (1995) helped to illuminate how the exchange and building of knowledge during a new product design process is potentially the most important outcome for a company. In Chapter Six in this book, Itabashi-Campbell and Gluesing discuss how the potential in engineering teams is not always realised, and oftentimes problems can be traced to the social aspect of a team. The confluence of social process knowledge and design process knowledge was also captured by Gunther *et al.* (1996) in their investigation of individual and team design processes. They proposed that a designer's work "is part of a complex technical and social process" and thus the performance of a team depends on the "prerequisites of the group" or the combined knowledge and skill the team members represent (p. 117). The ways that group members interact are as much a product of the team's progress through a design process as they are a reflection of their collective social skills.

3.2 What should a team know?

The social interactions discussed above are the primary tools that engineering designers use to leverage the strengths of a design team. In their discussion of everyday problem solving in engineering, Jonassen *et al.* (2006) made the point that not only should engineering students be prepared to tackle ill-structured problems, but students also need to be prepared to work collaboratively in distributed knowledge environments.

"Engineers rarely work alone, they rely on the knowledge of many people to solve workplace problems... different team members contribute their skills and knowledge to the solutions of engineering problems" (p. 144).

Clearly, the knowledge and skill resources represented on a design team are critical to its success. However, as Trevelyan notes in Chapter Two in this book, there may also be social, economic, and political factors in play that require more than technical engineering expertise. Those factors, and others, lend importance to the thoughtful makeup of design teams.

In a study of students' conceptual knowledge Atman and Nair (1996), found that beginning engineering students were both motivated by and interested in the societal context of the challenges that engineers face. They proposed that if engineering educators included contextual issues in their curriculum, they "would find a receptive student audience whose existing conceptual frameworks are well suited to learn more about these issues" (p. 324). In a more recent study that looked at six engineering programs that explicitly provide their students with experiences that blend context and design and promote contextual competence, Palmer *et al.* (2011) found a wide variety of different and effective models. What those researchers also concluded from their study was that contextual competence, as a learning goal, can no longer be viewed as optional, but is instead "a necessity in today's increasingly interconnected society" (p. 24). Contextual competence, they posited, not only affects a design solution, but it impacts an engineer's ability to work effectively on a diverse team. In essence, the diversity that Antonio Dias Figueiredo describes, in Chapter One of this book, as being an essential strength of engineering practice, may not be fully realised without a cadre of 21st century engineers who understand the bigger picture of their design work.

4   OUR EARLIER RESEARCH

4.1   Design process studies

Researchers at the Center for Engineering Learning & Teaching (CELT) at the University of Washington have engaged in the study of how engineers with differing levels of expertise approach design (Atman *et al.*, 2007; Atman *et al.*, 2005; Atman *et al.*, 1999). In this research, data was gathered and analysed from freshman engineering students, senior engineering students, and practicing engineering designers on how they worked on a common design task. This research utilised verbal protocol methods (Atman & Bursic, 1998) in which the research participants frequently spoke about the things they were thinking and doing while they designed. The task was to design a community playground, and the participants were given up to three hours to finish. This research resulted in a large body of rich qualitative data, that was then subjected to rigorous segmentation and coding processes.

The data was initially coded using a coding scheme that represented a generic set of activities in a design process. This was a prescriptive model that was synthesised from seven engineering design texts (Moore *et al.*, 1995). Researchers also counted and categorised the types of information each participant requested during the design process. Additionally, the quality of each participant's final solution was analysed. Table 1 shows the individual design activities that comprise the synthesised prescriptive model. Coding the data segments by design activity provided a unique opportunity to develop design process representations for each individual participant (Atman *et al.*, 2009). Those representations were originally used for analysis and communicating findings; however, they were later shown to be useful pedagogical tools in engineering design courses (Borgford-Parnell *et al.*, 2010).

The representations that have been explored most thoroughly are design process timelines. These timelines represent individual participants' design processes as they

Table I Definitions for the design activities. Code abbreviations are in parentheses.

Design Activities

Activity	Definition
Problem Definition (PD)	Defining the details of the problem
Gathering Information (GATH)	Collecting information needed to solve the problem
Generating Ideas (GEN)	Thinking up potential solutions (or partial solutions)
Modelling (MOD)	Detailing how to build solution or parts of a solution
Feasibility Analysis (FEAS)	Assessing possible or planned solutions (or partial solutions)
Evaluation (EVAL)	Comparing two or more solutions within constraints
Decision (DEC)	Selecting one idea or solution
Communication (COM)	Revealing and explaining design elements to others

Typical Engineering Freshman Student

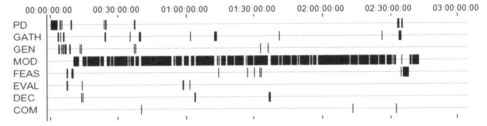

Typical Engineering Senior Student

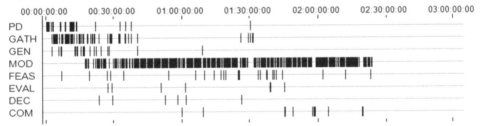

Typical Engineering Expert

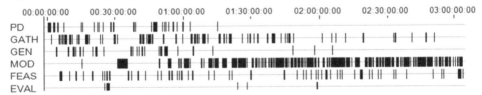

Figure I Design timelines of participants with different levels of experience.

moved from problem scoping activities to project realisation activities. Figure 1 shows the design process timelines of three research participants: an engineering freshman student, an engineering senior student, and a professional engineering designer. Each of these participants had design quality scores in the average range for their group.

By analysing the design timelines of the 69 participants in this study (26 freshmen, 24 seniors, and 19 professionals), researchers developed several comparative findings, including the following:

- Freshmen versus Seniors: (a) Seniors achieved higher quality scores than did freshman; (b) freshmen and seniors spent similar amounts of time in problem scoping and designing alternative solutions stages, however, seniors spent more time than did freshmen in the project realisation stage; (c) freshmen and seniors made similar numbers of information requests, although seniors focused on a wider range of information categories; and (d) freshmen and seniors considered similar numbers of objects (alternative solutions).
- Students versus Professionals: (a) Professionals scoped the design problem more effectively than did students; (b) professionals gathered more information and more categories of information than did students; (c) professionals spent significantly more time solving the problem than did students; (d) professionals spent more time in all design stages, particularly in the problem scoping stage than did students; (e) professionals considered more alternative solutions than did students; and (f) professionals' and seniors' design quality scores were similar.

4.2   Looking at context

Researchers at CELT have also been interested in how narrowly or expansively students and others contemplated the contexts of design problems. As an extension of the information gathering analysis done in the design activity studies (described previously) these researchers, along with colleagues at other institutions, began to examine the types of information considered by designers as a way to characterise the contextual issues they were focused on (Atman et al., 2008; Atman et al., 2008b; Atman et al., 2007; Kilgore et al., 2007; Kilgore et al., 2010; Yasuhara et al., 2009). In those studies researchers developed design tasks and survey instruments to gather data from engineering students on the types of information they gathered or thought were relevant to specific design problems.

In one such study, researchers gathered two sets of data from first year engineering undergraduates (Kilgore et al., 2007). Those students were first asked to describe the factors they would take into account if they were designing a retaining wall system for the Mississippi River in the American Midwest. The researchers were focused on the extent to which students considered broad context issues during the problem scoping stage. In the second part of the study, the students were asked to choose the five kinds of information most likely needed to design a playground. They were given a list of 16 categories of information that were drawn from analysis of transcripts in CELT's design process research, described previously. Findings from both the retaining wall problem and the playground questionnaire were positively correlated and showed that overall these first year students were only moderately concerned about the broader contextual issues for a design problem. However, the female participants were significantly more likely to consider broad contextual issues than were the males.

In further studies that used similar instruments, (Atman et al., 2010; Atman et al., 2008; Kilgore et al., 2010) the researchers found that even though engineering students learned to consider more factors in design, their consideration of broad contextual factors improved little over four years of education. The research suggested that engineering students were not prepared to consider broad contextual issues as practicing engineers.

The research described in this and the previous section focused on design processes and design context considerations of individuals. In the following section we discuss our analysis of a team design dataset.

5 TEAM DESIGN RESEARCH

In 2008, CELT researchers joined with design researchers from across the globe in analysing a common team-design dataset. Researchers were given video recordings and written transcripts from four design meetings that took place in England. Two separate tasks were undertaken, one an architecture task and the other an engineering task. Two team meetings were recorded for each task. Researchers were asked to focus their studies on any one or more of these design meetings, depending on their particular research questions and analytic expertise. The resulting 21 separate studies were the basis for a three-day workshop in the seventh Design Thinking Research Symposium (DTRS7, 2007), and a book (McDonnell & Lloyd, 2009).

5.1 Methods

As our part of this effort, CELT researchers applied analytic methods developed in previous research to the second multidisciplinary engineering design team meeting where engineers were in the process of designing a thermal pen (Atman *et al.*, 2009). We chose the second meeting because the first meeting was limited to brainstorming and the second meeting, although still considered a brainstorming session, also contained a wider range of design activities. For more details about the task, the team members, and additional analyses of the engineering meetings please see About: Designing, Analysing Design Meetings (McDonnell & Lloyd, 2009). Our primary interest in this effort was to explore how contextual issues were handled by the design team. In the following we summarise one analysis we presented in the book and then extend the analysis of the group and context aspects of the process.

We segmented the meeting's transcript as in our earlier research, and then coded the segments using four separate coding schemes (see Table 2). The coding schemes were: (a) Agenda Item, which was a short list of design goals set out by the design team and included in the design brief; (b) Conversation Topic, which were the topics the team members actually discussed during the meeting. Those topics were identified as Pen, Media, Pen/Media, User/Usage, and Group. Group segments were generally utterances by one member of the team that indicated encouragement or agreement with another's statement; (c) Design Activity, which was the same set of codes used in our earlier research (Table 1); and (d) Context Focus, which was a two-category coding scheme – Broad and Close.

We defined Context Focus codes as follows: 1) Close: In the team design process, when Close is a focus of conversation, team members are considering information and ideas about technical components, parts, or features; budgeting; or manufacturability of items; with an intent to further refine an agreed upon design concept

Table 2 Code definitions.

Scheme	Code	Definition
Agenda Item	Features	Basic functions and additions to the base product
	Interface	Interaction between the user and product
	Architecture	Electronics and manufacturing of product
	Applications	Potential uses for the product
Conversation Topic	Pen	Form, abilities, and engineering of the thermal pen
	Media	Various media the pen interacts with
	Pen/Media	How the pen and media work together
	User/Usage	Persons who will use the product, applications of the product, usability issues
	Group	Implicit: indicating understanding and support
		Explicit: regarding the state of the group process
Design Activity	Prob. Definition	Defining the problem
	Gathering Info.	Collecting information
	Generating Ideas	Thinking up potential solutions
	Modelling	Detailing how to build solution or parts of a solution
	Feasibility	Assessing possible or planned solutions
	Evaluation	Comparing solutions within constraints
	Decision	Selecting one idea or solution
	Communication	Revealing and explaining design to others
Context Focus	Broad	End users, marketing, usage concerns, safety
	Close	Design details, technical aspects

(even if the concept is later discarded). 2) Broad: In the team design process, when Broad is a focus of conversation, team members are considering any of the following: (a) people/persons/users, either in general or specific characteristics, dimensions, or aspects of the same; (b) cost and/or information regarding the availability of resources/parts to the end-users, or marketing to the users; (c) health or safety; and (d) other environmental, social, or political concerns. Importantly, these codes only apply to the product being designed. As such, segments that had been previously coded as Group with the Conversation Topic coding scheme were not coded for Context Focus. Figure 2 displays a juxtaposition of the original four design timelines that resulted from this coding scheme.

By juxtaposing the timelines, we were able to identify four sections of the meeting that represented distinct aspects of the design process. The first short section was labeled the Preamble and the following three longer sections were labeled Episodes (Fig. 2). The following are some of the findings that were related to the episodic structure of the design meeting:

– The Preamble discussion was dominated by the team leader (discussed later) and was characterised by a reiteration of the design problem, listing design decisions from a prior meeting and providing information. Details of the thermal pen Interface were discussed, so the primary Context Focus was Close. The Conversation Topic Group also occurred throughout the preamble.
– In Episode 1, the team concentrated on the potential Features and Applications of the Pen. This turned a lot of the conversation towards the Users and Usage

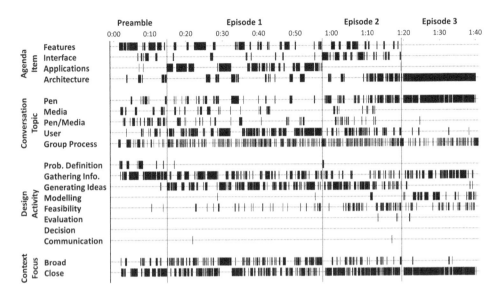

Figure 2 Timelines of four coding schemes with preamble and episode boundaries shown. Reprinted with permission, McDonnell & Lloyd (2009).

of the Pen, and the team's activities were primarily Gathering Information and Generating Ideas. The Context Focus transitioned equally between Broad and Close issues, and a recurring discussion of Group concerns continued.

– In Episode 2, the team completed the discussion of Features and Interface and did not discuss Applications at all. By the end of this episode the Architecture of the Pen dominated the discussion. The team continued Gathering Information and Generating Ideas, and much more Feasibility testing took place, but no additional Problem Definition occurred. The team continued to consider both Broad and Close context issues, however, the Close issues became much more of a focus. Figure 3 is a representation of how time spent in each of the Context Focus codes is accumulated over the course of the meeting. This figure demonstrates how discussion of Broad issues diminished in the final part of Episode 2, and how Close issues dominated the context discussion throughout the remainder of the meeting. As in the earlier sections of the meeting, the team members continued to pay attention to Group issues throughout.

– In Episode 3, the team's Context Focus was on Close technical details (Figs. 2 and 3) of the Architecture of the Pen. Most of the Modelling activity happened in this final episode, and Gathering Information and Feasibility testing continued. Again, discussion of Group issues continued right to the end of the meeting.

When we examined the design meeting as a whole, we found that the highest frequency of Broad Context Focus segments occurred in Episode 1, at the same time that Generating Ideas activities began. Both of those codes diminished in Episode 2. The team identified who Users of the product would be in Episode 1, and once that

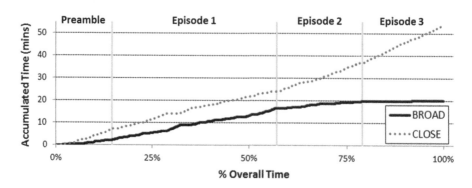

Figure 3 Cumulative Time Plot (CTP) showing extent of context discussion as meeting progressed.

was accomplished discussion of Interface began in earnest in Episode 2. Researchers also found that User/Usage segments were virtually nonexistent when Architecture discussions occurred, and Pen was also consistently associated with Architecture. The importance of group processes is evident in the timelines, as Group was the only Conversation Topic that occurred consistently throughout the meeting. Researchers also identified a Context Focus pattern, in which the discussion switched from Broad to Close over time. This paralleled the change in Conversation Topic from Features and Interface to Architecture.

5.2 Review of previous results

Results of this study demonstrate the importance of contextual issues to a design process, in that everyone on the team considered context to some degree. Additionally, it was shown that more Broad context issues were considered early in the meeting and, as the meeting progressed, a greater emphasis on Close context issues developed. Also, on-going attention was paid to the group process throughout the meeting. Those results support the notion that group process issues and contextual issues are important knowledge and skill areas for practicing engineers and should be integrated firmly in engineering design education. In the following section we report on an examination of the engineering design team data that further explicates the context issues and group processes that occurred.

6 FOCUSING ON INDIVIDUALS, CONTEXT, AND THE TEAM

In this section we discuss the results of another coding of the data that helped us discern the contributions of individual team members. We examine further the complex interplay of design context and group processes that occurred in the team design meeting, and show representations of the data that enable comparisons across team members and within individual's actions. Table 3 provides a brief description of each of the seven design team members, their areas of expertise, and their role in the meeting.

Table 3 Team members.

Name	Profession/Expertise	Role on Team	Present in 1st Meeting
Tommy	Electronics and Business Development	Meeting Facilitator	Yes
Jack	Mechanical Engineer	Technical Specialist	Yes
Patrick	Electronics and Software systems	Technical Specialist	No
Sandra	Ergonomics and Usability	Technical Specialist	Yes
Stuart	Electronics and Software systems	Technical Specialist	No
Rodney	Industrial Design Intern	Project Manager Intern	Yes
Roman	Electronics and Software systems	Technical Specialist	No

As is shown in Table 3, a range of technical backgrounds were present with a majority of team members having a background in electronics and software systems. Note that the meeting we analysed was preceded by an earlier meeting and four of the seven engineers had attended that previous meeting.

6.1 Everyone considers context, everyone considers the team

In the following we display several representations that depict aspects of the design meeting. Figure 4 is a bubble plot that illustrates the relative amounts of time that each of the group members spent on Broad, Close, and Group issues.

To better understand individual contributions, the bubble plot shown in Figure 4 was created to illustrate the relative emphasis that each team member spent talking about Broad, Close, and Group issues. In the plot, each circle's area is proportional to the amount of time spent by each individual on those topics. A quick inspection

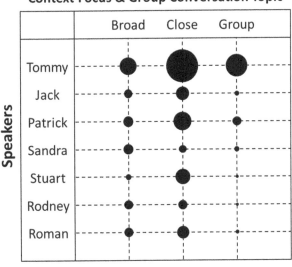

Figure 4 Dealing with context and paying attention to group process.

reveals that Tommy spent more time than any of the other team members discussing contextual and group issues. It also shows that – with the exception of Sandra – the technical specialists (Jack, Patrick, Stuart, and Roman) were more focused on the Close details in the design process than on the Broad context of the design problem. However, it should be noted that they were not limited to just Close details and did contribute to varying degrees in Broad context discussions. The overall Team's ratio of Broad to Close segments was 2:5. This plot shows that two members of the team who focused notably on contextual issues (Tommy and Patrick) also spent a significant amount of time on Group issues.

Figure 5 shows timelines of the original four coding schemes (discussed earlier) and an additional set of Speaker timelines that shows when, and to what extent, the team members individually contributed to the meeting. Again we see that Tommy's facilitator role resulted in his domination of the team discussion. However, looking at the other's individual contributions we begin to get a sense of the importance of the collective expertise and input of the team members. For example, as the frequency of Tommy's contributions diminished slightly over the latter half of the meeting, Patrick's contributions and those of the other electronics and software systems specialists increased. That increase is especially evident during the latter part of Episode 2 and Episode 3,

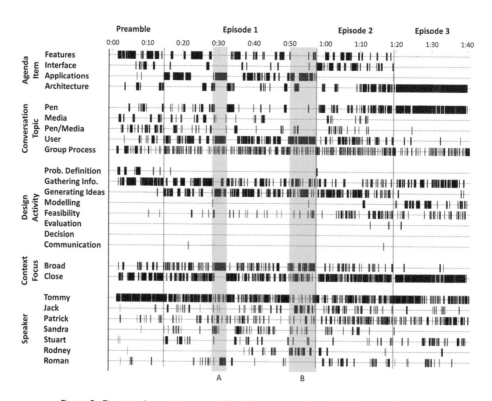

Figure 5 Five timelines juxtaposed. Specific conversations A & B are emphasised.

wherein the design discussion focused primarily on the Architecture of the Pen, a discussion that called for their expertise. It was also at the end of Episode 2 that Sandra chose to leave the meeting.

6.2   Conversation slices

In this section we will look specifically at two interesting slices of the team conversation. The slices were selected after examining the timelines in Figure 5 and identifying the interplay across the codes.

6.2.1   Conversation slice A

Slice A was chosen because of the obvious alignment of dense segments of Application, User, Generating Ideas, and Broad; along with the contributions of several team members. This slice was initiated out of a conversation about the Architecture of the Pen and the Media given two different ideas as to how those two objects would interact. One idea was to store patterns and text in the Pen that could be transferred to the Media, and the other idea was to use the Pen to thermally activate images, patterns, or text that were embedded in the media. The Architecture conversation was on the storage and capacity of a battery. However, Sandra pointed out that they were actually discussing two separate Applications and shortly thereafter the conversation focused on variations of the second idea. Here is what the conversation was like at the beginning of Slice A:

29:02:57–30:00:00

Sandra: "...things like an advent calendar where you uncover a window each..."
Patrick: "Yeah"
Sandra: "...day with the heat, or stories where you could have a different story depending on which word you uncovered. You could have a choice and make your own stories."
Tommy: "Yeah you could, yeah"
Sandra: "The media is then what you're buying."
Tommy: "erm. Are they kind of adventure books or something where you determine the route that you take through the book?"
Sandra: "Yeah, so you could have a story about a teddy or a snowman..."
Tommy: "Mmm."
Sandra: "...or a whatever depending on which line you've decided to uncover so you end up with a different story each time"
Patrick: "Yeah, I quite like the exam, th- practice test idea because you could take that right through and have it so the whole answer..."
Sandra: "Mmm."
Patrick: "...all laid out there just cos kids don't know how to do it at all they could just uncover it step by step instead until they suddenly got it, yeah"
Jack: "Clue one, clue two"
Sandra: "Mmm, the bits that they need"
Tommy: "Yeah."

This 57-second bit of the conversation shows how Sandra (the ergonomics and usability specialist) focused the conversation on how a person might use the Pen. While Sandra revealed her idea, Patrick and Tommy encouraged her with brief "Yeah" and "Mmm" utterances that tended to encourage the idea, and implicitly demonstrated that they understood or agreed. Patrick then turned the conversation to an education application idea, and Sandra, Tommy, and Jack provided encouraging utterances and supporting ideas.

The conversation continued then with a focus on educational applications. Roman entered the conversation with an idea about using the Pen to help children learn calligraphy, at which point Sandra dropped out of the conversation. Patrick and Tommy encouraged Roman to elaborate on his idea and they continued to make encouraging utterances such as 'yeah' and 'OK'. Slice A came to an end shortly after Tommy interjected this Group process question, "OK, alright, OK, erm ... what should we talk about?" Tommy's interjection helped remind the team that they were currently brainstorming potential applications for the pen and could move on to other ideas. This alert prompted the team to pursue other topics which then led to a question regarding the pen's hardware and functionality (Architecture).

6.2.2 Conversation slice B

The Slice B conversation was chosen because it was the final Applications conversation, which drew Episode 1 to a close. As can be seen in Figure 5, this slice is characterised by the alignment of dense segments of Application and User, interwoven Gathering Information and Generating Ideas for the Design Activity, and both Broad and Close for the Context Focus. Slice B also involved most of the participants, with Jack and Rodney substantially increasing their input. This slice was preceded by a short conversation about using the pen to print in Braille. It was focused on Close context Architecture details of the Pen and Media. The Braille idea ended abruptly when Rodney (the project manager) said "I don't know if this one is suited to Braille". Rodney then began Slice B with a new Application idea and a Broad focus, as shown here:

52:32:31–52:52:26

Rodney: "So Simon tells me there's a big demand for being able to initially initialise the sign language. People who use sign language don't consider it to be an extension of English. It doesn't translate directly it's actually basically a language in itself."
Tommy: "Yeah"
Rodney: "So if you were able to print maybe even hand signals ... there might be a big demand for that"

The sign language conversation lasted about 30 seconds and then the conversation switched to a new idea about using the Pen to date-stamp or security mark things, such as postage or train tickets. The new Application discussions got into considerations of Features and Interface of the Pen with a computer. Below is the final segment of

Slice B, where Tommy brings the team back to the main thrust of the design, which is a children's toy or learning tool.

56:43:35–57:27:14

Tommy: "Erm, it would be good to have applications like things that kids will play with in the back of a car, stuff like that."

Patrick: "Yeah but you'd get lots of burn holes in your car seats with this."

All: [laugh]

Patrick: "Great fun you'd get patterns in the back of the seat"

Tommy: "Yeah, well if you were a school thug you could pin people to the ground..."

All: [laugh]

Tommy: "Well you could do that with a pencil you could just stab them in the hand and all sorts can't you?"

Jack: "Yeah"

Patrick: "Yeah, we probably ought to go through this list and look at which ones you can actually do with a pencil."

Tommy: "Erm."

All: [laugh]

Tommy: "That's the screening stage, yeah. Err, right, let's move on to a bit of implementation stuff then."

It is worth noting that when the tone of the conversation turned to a bit of levity Tommy didn't curtail it immediately, but instead he joined in and then deftly brought the conversation back to a more purposeful focus. His suggestion that they move on to implementation stuff marked the end of Episode 1. The next episode began with Tommy revisiting some of the parameters of the design problem. Those segments stand out, on the Design Activity timeline in Figure 5, as the only Problem Definition segments in Episodes 2 and 3. By reiterating design parameters at that point, the team is reminded that there is a meeting agenda they should attempt to complete.

6.3 Three interesting team members

In this section is a brief discussion of three of the team members, Tommy, Sandra, and Patrick. We include figures showing individual timelines for each of the three.

6.3.1 Tommy

Tommy was the meeting facilitator, and therefore primarily responsible for getting through the meeting agenda. Looking at the Agenda Item timelines (Fig. 6) you can see that each item was discussed in the meeting, and at about 1 hour and 20 minutes into the meeting (the beginning of Episode 3) the Architecture of the Pen became the sole agenda item being discussed. Tommy initiated Episode 3 with the following,

"Right, erm OK we've done the other ones. Erm, I'd like to talk a bit about, erm, the (choice of) architecture, and I don't know whether everyone wants to sit through that part so?"

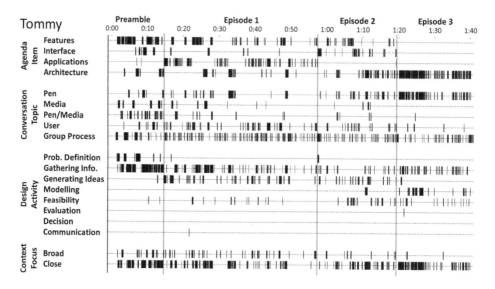

Figure 6 Tommy's timelines (Meeting Facilitator, Electronics and Business Development expertise).

Tommy's timelines shows that the frequency of his contributions to the discussions of Architecture of the Pen increased at that point. The density of the segments on those timelines attests to his deep involvement in the final conversations. Additionally, the Group, Gathering Information, Modelling, and Feasibility timelines (Fig. 6) provide clues as to the specifics of his contributions. Also, the Context Focus timelines clearly convey that Tommy paid attention to both Broad and Close Context issues throughout much of the meeting, and it wasn't until Episode 3 that he focused almost exclusively on the Close details of the Architecture. Figures 4 and 6 show that Tommy spent considerably more time on Close issues than on Broad issues. Tommy's specific ratio of Close to Broad was 7:2.

6.3.2   Sandra

Since Sandra was an ergonomics and usability specialist it makes sense that she would bring more attention to user groups, safety, and comfort issues. This is clearly shown in Figure 7. Many of Sandra's contributions in episodes 1 and 2 were coded as Applications, User, Gathering information, Generating Ideas, and Broad. It is evident that Sandra saw her role as helping the team to keep Broad contextual issues (especially regarding User) in mind. As was mentioned earlier (in section 5.2.1) it was due to Sandra's prompting that the conversation in Episode 1 was focused on how a person might use the Pen.

There were several other instances in the meeting where Sandra deliberately brought the User back into the conversation. For example, at about 33 minutes into the meeting, Tommy, Stuart, Patrick, and Sandra were discussing detailed Architecture and Features of the Pen and Media. They talked for several minutes about battery life,

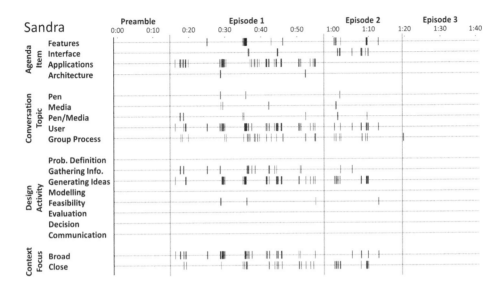

Figure 7 Sandra's timelines (Technical Specialist, Ergonomics and Usability expertise).

patterns and animations on the Media, and whether the Pen had built-in intelligence before Sandra interjected,

> "And the fact that the speed at which you drag it or the angle at which you drag it can make it look different but in a useful way. I mean if it's just a pattern rather than little faces or something then you could actually change the pattern."

And then again, a short time later she stated,

> "Cos of, my initial reaction is why would kids do this? But then I'm thinking of all the sort of fairly mundane things I used to spend hours doing."

The bubble plot in Figure 4 showed that Sandra considered more Broad contextual issues than Close issues. The specific ratio of Broad to Close segments for Sandra was 2:1. The next closest team member's ratio was Rodney's which was 1:1. Obviously, the density of the segments marked on Sandra's timelines pale in comparison to those on Tommy's timelines (Fig. 7) or even to Patrick's in the next section. However, her importance to the team was not only the technical competence she possessed, but appears to have a lot to do with her ability to broaden the focus of the team's conversation and to represent important stakeholders in the design process.

6.3.3   Patrick

Although Patrick, Stuart, and Roman (the Electronics and Software Systems specialists) were added to the design team for this second engineering meeting, Patrick's contributions were far greater than the other two, as could be seen earlier in Figure 5.

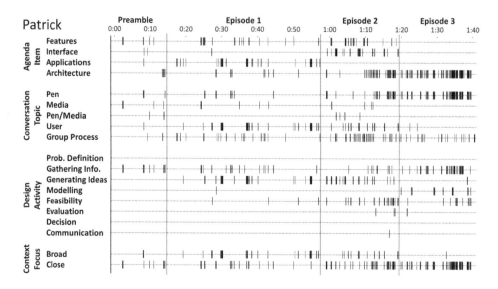

Figure 8   Patrick's timelines (Technical Specialist, Electronics and Software Systems expertise).

As can be seen in Patrick's timelines (Fig. 8) he was involved in the conversations throughout the meeting, although the frequency of his contributions intensified in the final 30 or 40 minutes of the meeting when the conversation became much more focused on Close details of the Pen's Architecture. Patrick paid attention to Group Process throughout the meeting and to both Broad and Close context issues until the beginning of Episode 3, when Tommy explicitly turned the conversation to Architecture. Patrick's specific ratio of Broad to Close segments was 1:3.

7   CONCLUSION

Palmer *et al.* (2011) suggested that engineers who are contextually competent will be better prepared to work with a diverse team. If time spent on contextual issues can be considered an indicator of the contextual competence of a team, then our data representations and findings clearly show a connection between the attention that is paid to context issues and the attention that is paid to group process issues. The data representations show that everyone on the team paid attention to contextual issues and to group processes, however, only Tommy and Patrick did more of both, and they clearly demonstrated the kind of contextual competence needed on design teams. Interestingly, it is not just broad context discussion in the beginning and close context further on in the process – there is a substantial amount of time where the two are interleaved, and take on timely importance.

It is obvious that the type of communication that took place in this engineering design meeting not only kept the design process moving forward, but also kept the social aspects of the team on track. The implicit utterances (e.g. "mmm"", "yeah" and "OK") and the explicit statements and questions (e.g. "Let's move on", "So how do you do that?" or "What should we talk about?") simultaneously accomplished both design and group purposes. Communication and attention paid to the team is important throughout the design process – not just to motivate a problem at the start and sum it up at the end.

Our concern, and an important goal of our research, is to develop a deeper comprehension of engineering design practice and to develop better ways to convey that knowledge to engineering students and practitioners. We think that findings and representations, such as those included in this chapter, can make teaching professional skills more straightforward. Instead of trying to convince engineering students that teamwork, communication, and thinking broadly are important, or telling them that everyone in industry says these professional skills are important, we can demonstrate by example how attention to team processes and the context of design problems are not just important, they are pervasive in engineering team design practice. The representations we presented here have potential as pedagogical tools in the education of competent engineering designers. Additionally, they may enable project managers and other practitioners who are tasked with selecting and building design teams to broaden their conceptions of the skills and knowledge areas that are needed.

Engineering work teams are often made up of individuals with various areas of expertise. While it would be expected that someone with ergonomics expertise would frequently bring up issues of broad context (and indeed we did see that Sandra had the highest ratio of broad to close context comments), it is notable that all the members of the team brought up broad context in addition to close context. This is another area where it could be very useful to show students that as a practicing professional one cannot claim 'I'm only an electrical engineer, I don't need to know context' – everybody needs to do context – both kinds! Just as importantly, everybody needs to know how to work competently as a member of a team.

ACKNOWLEDGEMENTS

This work was supported by National Science Foundation grant EEC-0639895, the Boeing Company, the Mitchell T. and Lella Blanche Bowie Fund, and the Centre for Engineering Learning & Teaching at the University of Washington (CELT). We would like to express our gratitude to the editors, reviewers and proof-readers of this manuscript for their thoughtful insights and great questions. We would also like to acknowledge the people of the larger CELT community who helped to launch our initial analysis with their brainstorming ideas and encouragement, especially Ken Yasuhara, Jennifer Turns, and Julie Provenson. Lastly, we would like to thank Janet McDonnell and Peter Lloyd, who collected the original data, convened the DTRS 2007 Symposium, and edited the resulting book.

REFERENCES

ABET (2012) Criteria for Accrediting Engineering Programs [Online] Available from: http://www.abet.org/accreditation-criteria-policies-documents/ [Acessed 7th April 2013].
Atman, C. J., Adams, R. S., Cardella, M. E., Turns, J., Mosborg, S. & Saleem, J. (2007) Engineering design processes: A comparison of students and expert practitioners. Journal of Engineering Education, 96 (4), 359–379.

Atman, C. J., & Bursic, K. M. (1998) Verbal Protocol Analysis as a Method to Document Engineering Student Design Processes. Journal of Engineering Education, 87 (2), 121–132.

Atman, C. J., Cardella, M. E., Turns, J. & Adams, R. (2005) Comparing freshman and senior engineering design processes: an in-depth follow-up study. Design Studies, 26 (4), 325–357.

Atman, C. J., Chimka, J. R., Bursic, K. M. & Nachtmann, H. L. (1999) A Comparison of Freshman and Senior Engineering Design Processes. Design Studies, 20 (2), 131–152.

Atman, C. J., Deibel, K. & Borgford-Parnell, J. (2009) The Process of Engineering Design: A Comparison of Three Representations. Proceedings of the International Conference on engineering Design, 24–27 August 2009, Stanford, CA, USA.

Atman, C. J., Kilgore, D. & McKenna, A. (2008) Characterizing Design Learning: A Mixed Methods Study of Engineering Designers' Use of Language. Journal of Engineering Education, 97 (3), 309–326.

Atman, C. J., Kilgore, D., Yasuhara, K. & Morozov, A. (2008) Considering Context Over Time: Emerging Findings from a Longitudinal Study of Engineering Students. Proceedings of the Research in Engineering Education Symposium, 7–10 July 2008, Davos, Switzerland.

Atman, C. J. & Nair, I. (1996) Engineering in Context: An Empirical Study of Freshmen Students' Conceptual Frameworks. Journal of Engineering Education, 85 (4), 317–326.

Atman, C. J., Sheppard, S. D., Turns, J., Adams, R. S., Fleming, L. N., Stevens, R. & Lund, D. (2010) Enabling Engineering Student Success: The Final Report for the Center for the Advancement of Engineering Education. San Rafael, CA, Morgan & Claypool Publishers.

Borgford-Parnell, J., Deibel, K. & Atman, C. J. (2010) From Engineering Design Research to Engineering Pedagogy: Bringing Research Results Directly to the Students. International Journal of Engineering Education, 26 (4), 748–759.

Brereton, M. F., Cannon, D. M., Mabogunje, A. & Leifer, L. J. (1996) Collaboration in Team Designs: How Social Interaction Shapes the Product. In: N. Cross, H. Christiaans & K. Dorst (eds.) Analysing Design Activity Chichester. New York, Wiley. pp. 319–341.

Cross, N., Christiaans, H. & Dorst, K. (1996) Analysing Design Activity. Chichester, United Kingdom; New York, New York, Wiley.

DTRS7. (2007) Design Thinking Research Symposium. [Online] Available from: http://design.open.ac.uk/dtrs7/index.htm [Acessed 7th April 2013].

Dym, C. L., Agogino, A. M., Eris, O., Frey, D. D. & Leifer, L. J. (2005) Engineering design thinking, teaching, and learning. Journal of Engineering Education, 94 (1), 103–120.

Dym, C. L., Wesner, J. W. & Winner, L. (2003) Social Dimensions of Engineering Design: Observations from Mudd Design Workshop III. Journal of Engineering Education, 92 (1), 105–107.

Goldschmidt, G. (1991) The Dialectics of Sketching. Creativity Research Journal, 4, 123–143.

Goldschmidt, G. (1994) On visual design thinking: the vis kids of architecture. Design Studies, 15 (2), 158–174.

Gunther, J., Frankenberger, E. & Auer, P. (1996) Investigation of Individual and Team Design Processes. In: N. Cross, H. Christiaans & K. Dorst (eds.) Analysing Design Activity Chichester; New York, John Wiley & Sons. pp. 117–132.

Hennessy, S. & Murphy, P. (1999) The Potential for Collaborative Problem Solving in Design and Technology. International Journal of Technology & Design Education, 9 (1).

Jonassen, D., Strobel, J. & Lee, C. B. (2006) Everyday Problem Solving in Engineering: Lessons for Engineering Educators. Journal of Engineering Education, 95 (2), 139.

Kilgore, D., Atman, C. J., Yasuhara, K., Barker, T. J. & Morozov, A. (2007) Considering Context: A Study of First-Year Engineering Students. Journal of Engineering Education, 96 (4), 321–334.

Kilgore, D., Jocuns, A., Yasuhara, K. & Atman, C. J. (2010) From Beginning to End: How Engineering Students Think and Talk about Sustainability Across the Life Cycle. International Journal of Engineering Education, 26 (2), 305–315.

McDonnell, J. & Lloyd, P. (eds.) (2009) About: Designing, Analysing Design Meetings. London, CRC Press/Balkema Taylor & Francis Group.

Minneman, S. L. (1991) The Social Construction of a Technical Reality: Empirical Studies of Group Engineering Design Practice. (Ph.D.), Stanford University.

Moore, P. L., Atman, C. J., Bursic, K. M., Shuman, L. J. & Gottfried, B. S. (1995) Do Freshman Design Texts Adequately Define the Engineering Design Process? Proceedings of the American Society for Engineering Education Annual Conference, 25–29 June 1995, Anaheim, CA.

Nonaka, I., & Takeuchi, H. (1995) The Knowledge-Creating Company. New York, Oxford University Press.

Palmer, B., Terenzini, P. T., McKenna, A. F., Harper, B. J. & Merson, D. (2011) Design in Context: Where do the Engineers of 2020 Learn this Skill? Proceedings of the American Society for Engineering Education Annual Conference and Exposition, 26–29 June 2011, Vancouver, Canada.

Paulus, P. B. & Yang, H. C. (2000) Idea Generation in Groups: A Basis for Creativity in Organizations. Organizational Behavior and Human Decision processes, 82 (1), 76–87.

Reid, F. J. M. & Reed, S. (2000) Cognitive Entrainment in Engineering Design Teams. Small Group Research, 31 (3), 354–382.

Schön, D. A. (1983) The Reflective Practitioner: How Professionals Think in Action. New York, Basic Books.

Schön, D. A., & Wiggins, G. P. (1992) Kinds of Seeing and Their Functions in Designing. Design Studies, 13 (2), 135–156.

Sheppard, S., Macatangay, K., Colby, A. & Sullivan, W. (2009) Educating Engineers: Designing for the Future of the Field. San Francisco, CA, Jossey-Bass.

Valkenburg, R. & Dorst, K. (1998) The Reflective Practice of Design Teams. Design Studies, 19 (3), 249–271.

Waite, W. M., Jackson, M. H., Diwan, A. & Leonardi, P. M. (2004) Student Culture vs Group Work in Computer Science. Proceedings of the SIGCSE 2004, 3–7 March 2004, Norfolk, VA.

Yasuhara, K., Morozov, A., Kilgore, D., Atman, C. & Loucks-Jaret, C. (2009) Considering Life Cycle During Design: A Longitudinal Study of Engineering Undergraduates. Proceedings of the American Society for Engineering Education Annual Conference and Exposition, 14–17 June 2009, Austin, TX.

Chapter 5

Working together across disciplines

Robin S. Adams & Tiago Forin
School of Engineering Education, Purdue University, West Lafayette, IN, US

I INTRODUCTION

There is growing awareness that single discipline approaches to engineering problems will not be able to meet the demands of today or the future. Engineering problems have always been complex, but our awareness of this complexity has dramatically evolved. Through developments in technology our world has become smaller, improving our ability to quickly and efficiently connect across the globe and with different disciplines to experience diverse views of the world. Due to these connections, engineers are learning about the need for developing both depth in engineering knowledge and breadth in different kinds of knowledge that shape approaches to complex engineering problems. This is changing ideas about what it means to be an engineer and what it means to prepare engineers for the profession. For example, statements on the future of engineering education in the United States stress goals for preparing engineers to become 'emerging professionals' who can deal with complexity, flexibly adapt to new situations, and bridge disciplinary boundaries to integrate technical and non-technical considerations, manage trade-offs between interdisciplinary criteria, and innovate (Bransford, 2007; Jamieson & Lohmann, 2009; National Academy of Engineering, 2004).

The engineering problems facing society today and tomorrow are global grand challenges that require integrating diverse perspectives into a collective whole. These include providing access to clean water, restoring and improving urban infrastructure, advancing health informatics, engineering better medicine, preventing nuclear terror, securing the cyberspace, enhancing virtual reality, advancing personalised learning, engineering the tools of scientific discovery, reducing the impacts of burning fossil fuels, creating a sustainable food supply that will not degrade the environment, and finding or creating renewable, economical, and environmentally conscious solutions for producing energy (Grand Challenges for Engineering Committee, 2008). Using the global challenge of providing access to clean water as an example quickly reveals the scale and complexity of this challenge and why it requires working together across different perspectives. Focusing on the idea of 'access', technical issues such as the location of a water treatment system, the size of the system, and the design of what goes in and what comes out are all affected by social issues. Engineers who focus on water treatment and hydrology may understand how water flows from a source to its user and how clean water is important, but may not be aware of the many ways water scarcity is viewed and treated from culture to culture, location to location, and discipline to discipline. Perspectives such as geology and climatology can help illustrate how water scarcity issues differ for an area that receives a solid season of rain versus an area that is mainly arid, but perhaps may not provide depth of understanding about water scarcity issues for urban and rural settings. Perspectives such as sociology, anthropology, political science and resource management may help unpack how different cultures experience water security, in particular the ways water scarcity and equitable water distribution is addressed for different conditions such as a readily available source of water as compared to scare resources. Perspectives that live within a community are also important. For example, determining an appropriate site may involve working with local leaders to find a site that is far enough away from an informal settlement to limit or enable local government involvement and support, which can have consequences for determining the size of a system so that it adequately provides service. Focusing on the idea of 'clean water', technical issues for effectively treating water are also shaped by social issues. Disinfectants such as chlorine, that are common in some parts of the world, may not be acceptable to stakeholders unfamiliar with the ways chlorine alters the taste or look of water. Similarly, adding disinfectants may be perceived as adding toxins to water rather than working to clean the water for safe consumption. Working with experienced local nongovernmental organisations may make the difference between failure and success by providing critical insights into ways to treat water that work within a community culture. Finally, while the criticality of providing access to clean water may be obvious, it may overshadow needs that emerge from the community itself such as how access to clean water can create time and resources for parents to send their children to school (Forin, 2011). The global challenge of access to clean water is just one example of how global engineering work involves working together across different kinds of knowledge, learning through these experiences, and being open to stepping outside the comfort zone of disciplinary boundaries and transforming beliefs, assumptions, and values about the nature of engineering work (Mezirow, 2000). Deeply engaging those who have a stake in the issue at hand provides opportunities to link the technical and social elements of global engineering work and develop as a cross-disciplinary professional.

In this way, becoming a cross-disciplinary professional involves working together across disciplines, cultures, organisational functions, and lived experiences. Depending on what people do in these situations, the results can range from overwhelming success to devastating failure. While there has been substantial investment in promoting cross-disciplinary work, the level of empirical attention has been considerably less (Bromme, 2000). Existing research is situated in the humanities and social sciences (Klein, 1990; Lattuca, 2001) and in some cases research laboratories (Galison, 1997; Nersessian, 2006). This chapter presents findings from a study on the ways people experience working with others from different disciplines, and the ways they make sense of these experiences as distinct ways of cross-disciplinary thinking, acting, and being. The study was conducted in the US, and used interview techniques to elicit authentic experiences from individuals who were working in cross-disciplinary academic, industrial, and global engineering settings. Study findings are presented as four qualitatively different ways individuals experience cross-disciplinary practice in terms of what they've come to understand about cross-disciplinary work, how and why they approach this kind of work, and how they see themselves as cross-disciplinary professionals. For those who engage in cross-disciplinary work, this chapter may provide an opportunity to critically reflect on and make sense of your own experiences. For those who seek to enable cross-disciplinary work, this chapter may provide a language and set of ideas for designing cross-disciplinary projects or learning environments whether they are formal (such as curriculum) or informal (such as mentoring and advising).

2 BACKGROUND

Whereas the term *disciplinary* signifies a particular set of tools, methods, exemplars, concepts and theories, in this chapter the term *cross-disciplinary* characterises a collection of practices associated with thinking and working across disciplinary perspectives. Movements between and across disciplinary boundaries are marked by significant challenges such as language differences, clashing paradigms or ways of seeing the world, and borrowing and translating ideas into new contexts. As such, it is easy to understand how and why cross-disciplinary work may be incredibly difficult and fraught with failure, as well as how productive conflict can open up new insights and transform engineering work.

There are two sets of ideas that are useful for investigating critical differences in how people experience a phenomenon such as cross-disciplinary practice. One set of ideas involves understanding important differences in how cross-disciplinary situations are structured such as multidisciplinary, interdisciplinary, and transdisciplinary situations (Table 1) (Lattuca, 2001; Klein, 1990; 2004). Comparing across these suggests critical differences for the ways people experience cross-disciplinary work including: how the problem is framed and approached (from a thematic orientation to a problem setting orientation), the mode or structure of knowledge production (from juxtaposition of perspectives to an overarching and transformative synthesis), the outcome (from no real change in knowledge to knowledge fusion through critical reflection), interaction structures (from collaborating as disciplinarians to moving beyond disciplinary structures), communication and discourse practices (from a perception of common ground to the creation of new language and paradigms), and impacts on participants (from retaining a disciplinary identity to critical reflection on pluralistic identities) (Adams *et al.*, 2009).

The second set of ideas focuses on understanding individuals in cross-disciplinary situations through a lens of learning and development. For example, those who develop cross-disciplinary abilities are often referred to as 'T people', signifying competency in a vertical axis of depth (the stem of the 'T') and a horizontal axis of breadth, comprehensiveness, and synthesis (the top of the 'T') (Klein, 1990). What does it take to spread out from a core discipline to understand other disciplines? Is it made up of embedded personality traits (something you are born with) or is it a particular way of becoming that involves learning through experience?

For this project, we used a framework of becoming a professional (Dall'Alba, 2009) as a way of describing the unfolding and open-ended process of becoming cross-disciplinary. In this framework, an "embodied understanding of professional practice integrates not only our knowing and how we act (giving meaning to the knowledge and skills being developed), but also who we are as professionals (embodying an understanding of the practice itself)" (Adams *et al.*, 2011, p. 603). It is embodied because knowing, acting, and being are enacted within the dynamic and intersubjective flow of activity that is professional practice. As such, development is not a stepwise process of

Table 1 A synthesis and comparison of cross-disciplinary practices.

	Multidisciplinary	Interdisciplinary	Transdisciplinary
Definition	Joining together of disciplines to work on common problems; split apart when work is completed	Joining together of disciplines to work on or identify common problems; interaction generates new knowledge	Beyond interdisciplinary combinations to new understanding of relationships between science and society
Problem orientation	Thematically oriented projects where several disciplines contribute to a theme	Instrumental (pragmatic problem solving) or conceptual (philosophical enterprise) problem solving orientation	Problem setting orientation of collectively formulating problems in highly heterogeneous environments and including the experiences of affected persons
Mode of knowledge production	Juxtaposition of perspectives as separate voices – emphasis on concatenating, adding breadth without impacting status quo	Integrative synthesis, holistic mixing of perspectives, methods, tools, etc. Renegotiation of frameworks	Integrative and action-oriented transformation that transcends disciplinary views through an overarching synthesis
Outcome of knowledge production	No new cross-disciplinary knowledge	New interdisciplinary knowledge	Knowledge fusion characterised by critical reflection
Interaction and communication structures	Collaborate as disciplinarians with different perspectives; no shared home; divide and conquer approach. Common ground assumed to exist	Beyond academic disciplinary structures; Close collaboration and development of common ground – overcoming problems created by disciplinary and paradigmatic differences	Participatory – science and society. Close and continuous collaboration; elaboration of new language, logic, and concepts that permit genuine dialogue
Impact on participants	Retain disciplinary identity	Involves some learning of other discipline and reflecting on epistemological assumptions	Beyond either/or thinking; a move beyond problem solution to problem choice

moving though a fixed sequence of stages, but includes both continuity and change as an understanding of practice develops (Dall'Alba & Sandberg, 2006). As professionals learn to deal with new situations, their embodied understanding of practice evolves in qualitatively different ways – revealing multiple development trajectories that lead to more comprehensive understandings of practice or refinements of an existing understanding of practice. These different ways of being professionals are limited in number and logically related to each other, because they are based on the same practice. When multiple development trajectories open up as possibilities, we shape and form our own development – taking "up those opportunities that are consistent with or advance our sense of self, while resisting those that undermine our sense of who we are" (Dall'Alba, 2009, p. 55).

3 STUDY DESIGN

This study uses phenomenography to investigate critical differences in the ways people experience and make sense of cross-disciplinary practice in engineering contexts. This technique is well suited for investigating how people experience professional practice and is consistent with a framework of 'an embodied understanding of practice'. Phenomenography is an empirically derived research approach that is used to capture variations in understanding an aspect of the world, while revealing the critical components that comprise those variations (Bowden, 2000; Marton & Booth, 1997). The outcome of a phenomenographic study is a representation that characterises the (1) 'outcome space' - a collective and comprehensive record of the different ways a phenomenon is experienced or understood within a particular context, and (2) 'categories of variation' – the distinctly and qualitatively different categories that capture the essence of variation within that outcome space (Case & Light, 2011; Marton & Booth, 1997). Because these representations capture differences in learning and awareness, they may be used to inform the design of learning environments (both formal and informal), activities, goals, and assessments (Daly et al., 2008).

Phenomenographic research emerged in the 1970s and has been used broadly in education contexts to investigate such issues as variations in the ways teachers understand teaching (Prosser et al., 1994), students understand science, math, and problem solving concepts (Case & Light, 2011; Marton & Booth, 1997), and students understand their own learning (Entwistle & Peterson, 2004). While relatively new to engineering, there are a number of studies that focus on different aspects of engineering work such as design (Daly et al., 2012), sustainable design (Mann et al., 2007), human-centred design (Zoltowski et al., 2012), and social justice (Kabo & Baillie, 2009), as well as competencies in professional work environments (Sandberg, 2000). Case and Light (2011) provide an overview of phenomenography and how it compares to other forms of inquiry. As a form of inquiry, phenomenography is similar to phenomenology and grounded theory. Each are forms of qualitative inquiry where findings emerge from the data as compared to fitting the data to a set of predetermined categories. However, there are important differences that impact sampling procedures and data analysis. For example, in phenomenology the unit of analysis is the phenomenon and the focus of analysis is the researcher's point of view that reveals the 'essence' of the phenomenon. In phenomenography the unit of analysis is the experience and the focus of analysis is the individual's point of view that reveals critical variations across a diverse set of experiences. Phenomenography is also similar to grounded theory, where both are non-dualist approaches that emphasise reality as made up of relations between the individual and an aspect of the world; but they are different because phenomenography seeks to identify the range of experiences in a particular setting and how these are related as critical variations (Trigwell, 2000).

A unique characteristic of phenomenographic research is its goal of identifying critical variations in experiences and the meanings associated with those experiences. This requires strategically maximising the variation of experience in a study sample in a way that will enable an inclusive and representative view within the aims of the study (Åkerlind, 2005). For this study, twenty-two (22) engineers and non-engineers who work in engineering contexts were strategically recruited to maximise diversity. Rather than focus on individuals who shared similar training, such as engineering, we focused on individuals who shared similar work contexts – complex engineering projects involving people with formal training in a variety of disciplines. This sample size and strategy is consistent with typical phenomenographic studies. Target variations in the sample were identified through a literature review and include (see Table 2): context of work (i.e., academia, private industry, and community service), years of cross-disciplinary experience, gender, and nature of experience in terms of project scale (i.e., size of teams), complexity (i.e., number of disciplines involved), and epistemological distance (i.e., scale of disciplinary similarity or difference). Epistemological distance is the extent to which an individual has experience working with others who have different ways of knowing, such as views on what counts as good evidence and the value of subjective or objective knowledge. For example, a person trained in engineering and a person trained in social science would be expected to use different techniques and ideas that draw on different frameworks of good practice and good evidence.

Data was collected using a semi-structured interview protocol to provide deep, reflective, and contextualised data about participants' experiences and the meanings associated with those experiences. Interviews lasted at least 30 minutes and were audio recorded, then transcribed. In the course of the interview, participants were asked to describe a particular experience they felt was cross-disciplinary in nature. The term 'cross-disciplinary' was used in the interview to provide opportunities for participants to talk about one of many possible forms of cross-disciplinary practice (such as multidisciplinary, interdisciplinary or transdisciplinary) and to compare or critique these different practices. As participants shared their experiences, they were asked a common set of prompts to help elicit aspects of these experiences, such as the setting in which they occurred and the number and kinds of people involved, and what they did within these experiences such as how they approached cross-disciplinary work or their roles and responsibilities within the team. They were also asked follow up questions to get details of their experiences and their associated meanings. These follow up questions were based on ideas participants expressed during the interview and were not based on predetermined ideas or leading questions from the interviewer's perspective. For instance if a participant mentioned that she got involved in cross-disciplinary practice because "it is the future of engineering", then the interviewer would ask follow up questions to probe what she meant by "the future of engineering" and how this idea is important for her. At the end of each interview, after the participant had opportunities to reflect on the experiences they described, they were asked, "what

Table 2 Characteristics of study participants in terms of background and prior experiences. Participants are listed by categories of variation (the first rows represent Category 1 and the last rows, Category 4).

Study Name	Gender	Experience Setting Edu=academic Ind=industry Res=research center	Engineering Training	Years of Experience	Project Scale and Complexity (Very low to Very high)	Epistemological Distance (Very low to Very high)
Emily	F	Edu	No	<5	Low	Medium
Grace	F	Res/Edu	Yes	6–>15	Medium	Low
Isabella	F	Res/Edu	No	6–>15	Medium	Low
Brianna	F	Edu	Yes	<5	High	Medium
Nadia	F	Edu	Yes	<5	Low	Low
Olivia	F	Res/Edu	Yes	>15	Medium	Medium
Pablo	M	Res/Edu	Yes	>15	High	High
Ryan	M	Ind	Yes	>15	High	Low
Uri	M	Edu	Yes	<5	Low	Low
Daniel	M	Ind	Yes	>15	Very high	Low
Fergus	M	Ind	Yes	>15	Very high	Medium
Jacob	M	Ind/Edu	Yes	>15	High	Medium
Michael	M	Ind/Edu	Yes	>15	High	High
Tyler	M	Res/Edu	No	>15	Medium	High
Yvonne	F	Res/Edu	No	>15	Medium	Very high
Anthony	M	Ind	Yes	>15	Medium	Medium
Hannah	F	Ind	Yes	>15	Very high	Very high
Kelvin	M	Edu	Yes	6–>15	Low	Very high
Logan	M	Ind	Yes	>15	High	Low
Samantha	F	Edu	Yes	6–>15	Low	Very high
Wendy	F	Res/Edu	No	>15	High	High
Xavier	M	Res/Edu	No	6–>15	High	Very high

does cross-disciplinary practice mean to you?" This allowed an opportunity for each participant to articulate their own definition as it related to their own experiences.

In phenomenographic research, analysis focuses on characterising the outcome space – the fewest (logically related) categories that describe the totality of variation in the pooled data – and the critical dimensions that differentiate these categories (Case & Light, 2011). Because the unit of analysis in phenomenography is the experience, this involves iteratively reading whole transcripts and sorting them into distinct ways of experiencing or understanding an aspect of the world. Overall, this analysis was a rigorously iterative co-evolutionary process of discovery of categories and construction of structural relationships between categories. The process began with members of the research team individually reading each transcript and then discussing each transcript as a team. This involved sharing observations about how participants experienced cross-disciplinary work and the meanings they associated with these experiences, and coming to a tentative consensus on interpretations based on evidence within the transcripts. Transcripts were then sorted into piles of similarities and differences and assessed by identifying substantive excerpts from the transcripts that supported the patterns in each pile. Multiple iterations occurred over a six-month period as the team revisited the piles and renegotiated interpretations of what each pile represented and how it related to other piles. This often resulted in moving a transcript to another pile, and in some cases removing or adding a pile. Over time, this process stabilised until there were no more changes. The final piles became the four categories discussed later in this chapter; they collectively represent the outcome space of this study and the qualitatively different ways of experiencing cross-disciplinary practice in engineering contexts. Subsequent passes through the data focused on articulating and substantiating the relationships among categories, drawing from theoretical frameworks to explore the logic underlying observed relationships.

At the end of this process, the research team assessed the quality of interpretations using three criteria identified by Marton and Booth (1997). First, the categories met a criterion of being clearly related to the phenomenon being researched. In addition, the categories emerged from the data and not a predetermined framework. We found that the four categories and their relationships often both supported and challenged anticipated distinctions such as those summarised previously in Table 1. Second, the categories were logically related to each other and in this case the relationships were hierarchical. For example, transcripts associated with the highest categories illustrated substantive and unique features of that category while also showing some evidence of ideas in the lower categories. In this way, the relationships among categories mapped out increasingly more comprehensive understandings of cross-disciplinary practice (moving from the first to the fourth, and highest, category). Finally, the categories were parsimonious so that they became a reasonable explanation of the variations experienced in the phenomenon. In our case, four categories were observed and all transcripts mapped uniquely to one of those four categories.

4 FOUR WAYS OF EXPERIENCING CROSS-DISCIPLINARY PRACTICE

The outcomes of this study revealed four hierarchically related categories of variation (Figure 1). Collectively, the four categories provide an outcome space of the qualitatively different ways the participants in this study experience cross-disciplinary practice in engineering contexts. Relationally, the progression from Category 1 to Category 4 represents increasingly complex ways of experiencing cross-disciplinary practice and a growing awareness and comprehension of cross-disciplinary practice.

The following sections describe each category using supporting evidence. The data is signified using quotation marks, and substantive excerpts (shown in italics) include the assigned study name of the participant (in brackets) as referenced in Table 2. To maintain authenticity, data is provided in its original form and therefore may include grammatical errors. For those who engage in cross-disciplinary work, these findings may provide language and ideas for you to critically reflect on and make sense of your own experiences as well as self-assess or plan your own progress towards becoming a cross-disciplinary professional. For those who seek to enable cross-disciplinary work, these findings may provide insights into the unique challenges and opportunities of this kind of work as well as language and ideas for how to design effective cross-disciplinary collaborations and learning environments whether they are formal (such as curriculum) or informal (such as mentoring).

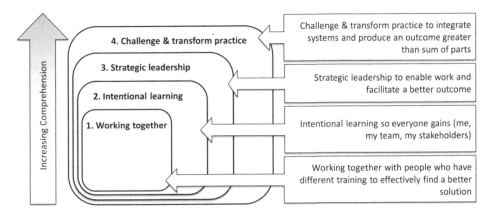

Figure 1 Four hierarchically related categories of variation in ways of experiencing cross-disciplinary practice as increasing awareness and comprehension of cross-disciplinary practice from surface (1) to deep comprehension (4).

4.1 Ways of experiencing cross-disciplinary practice as Working Together

The first and lowest level category illustrates *cross-disciplinary practice as working together with people who have different training to effectively find a better solution.* Critical attributes of the experiences associated with this category include: (1) knowing what you and others contribute and points of synergy, (2) recognising disciplinary differences in what people do and how they communicate; an iterative process of asking questions, challenging assumptions, and listening for understanding, (3) being comfortable with asking for information that might seem obvious to an expert in that domain, and (4) taking personal responsibility to be an effective collaborator. Participants associated with these experiences include Emily, Grace, and Isabella (see Table 2).

Experiences associated with this category focus on communication and collaboration, illustrating how interactions with people are a crucial aspect of cross-disciplinary practice. For Emily, this involved "working with someone of another discipline in a rather intricate way":

> "In a way that's as brief as possible so that each party can continue their own work, but as thorough as possible when you get together so that you have very good information for each other. From the engineering side what they are doing, and from the other side, what needs to be improved, what details need to be in there.... It's kind of a question, try something, go back and talk to them about it, question process." [Emily]

Emily notes that the importance of this iterative communication process is "to be sure that we're going the right way. So that you don't end up going too far down the wrong road". For her, "good feedback goes both ways", where the engineers need to be "willing to ask the questions" and listen as well as "sharing what their idea might be". Her concern was that by not having everyone share what they know, expertise would be missing and the project might fail.

"But if they hadn't shared that with me, I never would have known. And I never could have pointed it out right off, and they would have started and done the whole project and that would have been a waste of time and money." [Emily]

Similarly, Grace notes "probably the biggest challenge is getting the two to communicate". She emphasised a need "to listen and be very clear as to what you're trying to say and to make sure that the people listening to you are understanding that so, listening, asking questions to follow up is the biggest thing". Grace puts this responsibility on herself by developing an awareness of "hey this person doesn't understand me, I need to make it more clear . . . seeing on people's faces that they weren't understanding me". She also expects this of her collaborators:

"Just say what capacitance means to a biologist is totally different than what it means to an electrical engineer. So say in a meeting where both groups were represented, the idea of capacitance came up and the electrical engineer just continues with what they were trying to say, and right away the two go along different paths because the electrical engineer knows what they are trying to say, but the biologist hears capacitance and thinks in a totally different direction. And without that engineer knowing that the two don't mesh, they just start separating right there. So it's up to both parties to make sure that they're talking about the same thing." [Grace]

For Isabella, the process involved "baby step(s)" where group members would "go away and work on a little bit of something and bring it back and ask 'how about this' and then we'd talk about it some more". She describes an intricate questioning-listening process of asking seemingly simple questions to elicit implicit knowledge and taking the risk of looking "ignorant" while maintaining a culture of respect within the group. The outcome is new knowledge about what is possible or impossible.

"So it became trying to maintain the respect for the other discipline but at the same time trying to communicate what you need or want was really tricky . . . . Asked a lot of questions. Not be afraid to say, 'I don't know what that means, can you explain it' rather than you know acting like I already understood just to save face which I think a lot of people who are working on multi-disciplinary stuff do just because it's very intimidating to work with somebody that talks something you don't understand. So you have to be really ready to sort of play the ignorant person and have the other person explain it to you." [Isabella]

". . . and we hit this impasse because I didn't know what was feasible in terms of their discipline and they didn't know what I wanted in terms of this web utility. And so just getting to the point, because I didn't want to suggest something that was going to be very difficult or impossible, but I didn't know about the constraints and they couldn't communicate them to me." [Isabella]

All the experiences for this category were directed towards a shared goal of producing a better outcome that leveraged a cross-disciplinary approach. Participants did vary in terms of what they considered a 'better outcome'. For Emily it was defined in terms of meeting the client's needs: "If you get something that truly fits the client, you have just a stellar, stellar thing for them". For Grace, it was creating new applications that extended the current practice: "If we can now take that information or that knowledge and apply it to medical and biological applications, we can extend our medical facilities, our medical practices so much more ... that's the great potential". For Isabella, a better outcome was finding elegant solutions to problems she couldn't have conceived of on her own: "It would have been a lot less elegant if I wasn't working with them because ... they have ways of dealing with problems that I couldn't conceive of ... and I knew that going in and that's what I wanted from them".

Recognising differences in disciplinary training (goals, methods, language and communication styles, etc.) and appreciating those who can move beyond disciplinary mindsets, are qualities which link an iterative communication approach to a goal of creating a better outcome.

> "Everybody's got a different twinge in their discipline. But let's just use engineering and biologists. They're brought up in two different ways. When you look at say the biology curriculum from my stand point, it's a lot of memorisation, a lot of here's the facts, this is how you use them. For engineering it's more like here's a problem, here's how you figure it out. And so you're kind of brought up or taught or trained in different ways. And so we really get into being the biologist, your mindset is in that way as opposed to an engineer who might look at something and say, well, I can kind of figure this out, I'll figure out how to make it better. So they both have their pluses and minuses, but I just think they're trained in different ways and they get into those mindsets too much. ... That has a huge impact as far as you know, number one how they communicate but how they approach the problem or the challenge I should say. How each individual will approach a challenge makes the impact as to how it will get done and if it can get done." [Grace]

Underlying this understanding was a belief that everyone has something to contribute or as Emily notes that everyone is "smart but in a whole different way"; a respect for how everyone has "their own special way of knowledge and I've come to appreciate that a lot". Emily notes that when you encounter others what they know may look like "common knowledge like take chicken soup for a cold or something" but more "intricate". Leveraging these different perspectives to address truly complex problems was the main driver for spending considerable time learning how to work together.

> "I think it's vital ... that's the only way to do it. If you don't have value for the other discipline, what are you even doing working with them? If you don't value what they are doing, you are not going to care how they do it, why they do it. You're not going to ask enough questions, and the questions are going to be important for the way you design whatever you're designing." [Emily]

Overall, ways of experiencing cross-disciplinary practice as Working Together emphasise particular ways of thinking, acting, and being. This category emphasises ways of thinking as (1) an awareness of how disciplinary differences can make cross-disciplinary collaboration complicated, (2) having respect for different kinds of disciplinary knowledge, (3) understanding complexity as driven by client and technology needs, and (4) having a goal of producing a better outcome that is feasible, meets the client's needs, and expands the current function. Ways of acting emphasise an iterative communication process of listening and asking questions, where partners are a source of information and conversations focus on exploring feasibility limits and needs across different disciplinary perspectives. The experiences associated with this category suggest ways of being that highlight taking individual responsibility for being an effective collaborator and providing disciplinary expertise on solution feasibility and desirability. These distinct ways of thinking, acting, and being are interrelated. For example, an awareness of differences in disciplinary training, and how these differences complicate the process of working together towards an effective outcome, supports an iterative communication process of asking questions and listening for understanding to determine what is desired or feasible within a bounded application space.

In addition, the experiences representing the Working Together category are distinct because of their focus on the experience of collaborating and communicating with people who have different perspectives, language, interaction styles, and ways of thinking. In terms of the relationships among categories, Working Together appeared to act as a foundational category. Aspects of this category, particularly the language difficulties associated with cross-disciplinary work, were evident in transcripts associated with higher-level categories, suggesting that each higher category speaks to increasingly complex ways of experiencing and understanding the collaborative aspects of cross-disciplinary practice.

4.2 Ways of experiencing cross-disciplinary practice as Intentional Learning

Where the experiences in the first category, Working Together, focus on the dynamics between people, the experiences that represent the next higher category focus on individuals' intentional learning. Here, *cross-disciplinary practice is intentional learning so that everyone gains (me, my team, and my stakeholders)*. Intentional learning involves: (1) creating opportunities to learn new perspectives or ways of knowing, (2) purposefully educating each other to collectively enable a systems perspective, (3) learning *through* experience and failure, (4) learning *how* to negotiate meanings across perspectives and formulate or investigate problems through multiple lenses, and (5) having a passion and appreciation for continual learning. Participants associated with these experiences include Brianna, Nadia, Olivia, Pablo, Ryan, and Uri (see Table 2).

An appreciation for learning was a consistent and prevalent theme in all the transcripts for this category:

> "I love it. I'd rather do that than anything else. . . . I mean there's just so much to learn out there and so many cool things, so many talented people at a place like UNIVERSITY. . ." [Pablo]

> "I guess I'm pretty comfortable with knowing that I don't know everything and having areas of expertise. . . . So I think it's just a willingness and an openness to just

not know everything and being willing to learn about some of the other things."
[Brianna]

"It's part of the me that I like, learning about what you do and if there's some kind
of synergy." [Olivia]

Intentionality also plays out in being proactive in creating opportunities to learn.
For Ryan, being "naturally open" and having a "curious mind" is an important personal motivator. Nadia did her own needs-assessment on her university experience and
realised that she wasn't "getting enough science and math ... getting a comprehensive
view on what's going on", and decided to explore different kinds of courses that were
not required for her degree. Similarly, Olivia started with reading a sophomore level
textbook to train herself in a new discipline and through this process developed respect
for a different way of thinking. Intentional learning also involves understanding how
complex problems require not just new knowledge but also new approaches to learning. Pablo, a self-professed learner, notes "societal problems that need – you can't solve
them in a single discipline anymore. I mean you could make incremental steps and that
then can come in and help some area, but these problems are way too complicated ...
and I think the most successful people in the future at universities are going to be those
that really aspire, want to, and work really hard at being interdisciplinary." Olivia
uses metaphors of the "tail wagging the dog", "immunisations", and "injections" to
explain why she continually expands her skill set.

"You know, here comes the tail wagging the dog because now you've got this
problem that is now kind of motivating your research but for which you develop
this skill-set. And this really was where I would say there was this major broadening
of my career because I had the tools for doing the analysis and I developed tools
for doing analysis but I had this major interest in terms of really understanding
an application as well ... you need your immunisations, you need to have these
points in life of injection of something and I think that it's important to plant those
seeds along the way and I haven't really thought about that before. I still think
and I'm pretty convinced and I don't know if this has changed or not I don't think
maybe is hadn't; but you need to have this disciplinary strength; but I would say
that definitely I've gotten more respect for you know how difficult it is and how
much effort it is to learn the other person's field or something about it." [Olivia]

Learning and teaching others are interconnected, such as working together to co-
develop a cross-disciplinary awareness and skill set. Ryan describes this process as
learning "enough of someone else's job" to know how two jobs can "mesh". Similarly,
Pablo notes that cross-disciplinary practice involves educating others about what you
do, how you think, and broadening points of view in order to imagine new possibilities. Olivia purposely puts her students into situations where they must learn how to
work cross-disciplinarily in order to be successful. She places students with different
disciplinary perspectives in a project team:

"The first difficult problem for the class project was, 'well what problem do we
do?' 'I'm not telling you what problem, you go out and find a problem of interest
and it can't be either one of your dissertations and you know ... there's got to be

something from both of you. And so then they have to propose that, so that was a major issue. Just what could they do to communicate and then of course you always have a strong student dominating weak students in sense of telling them 'okay, we're going to do this and you go do that', but you know it's all part of the learning experience too for them to be able to work with people who are different and ask the right questions." [Olivia]

The experiences for this category emphasise learning through immersion or "on the job" and from reflecting on failures. Ryan describes a situation in which the company he worked for got sued and how he needed to quickly learn about patents: "If that other company hadn't sued us, then I wouldn't know what I know. I wouldn't be doing patents for this group". Nadia talked about the importance of "hands-on learning by mistake". Similarly, Brianna observed that her students had difficulty "understanding outside their discipline" which reflected a "reluctance of students to want to venture out of their comfort zone". For Olivia, learning was an immersion process that involved iterative cycles of monitoring the knowledge needed to fully collaborate:

"Those experiences involved field work; significant field work which I didn't as an engineer have any experience in so I had to learn about that. In Australia, I worked with a collaborator actually. I had collaborators at New South Wales and we went out to sites and I learned to do vegetation surveys and I learned what they were doing . . . so we had kind of a you help me/I help you kind of thing but and I would come back and I would never know enough so you'd have to read more and the best way of doing it was in an immersion kind of mode I think . . ." [Olivia]

Learning outcomes of these intentional learning activities span new insights regarding how to communicate and interact with people who have different backgrounds (particularly outside of engineering), seek out and find connections among what may seem at first glance very different viewpoints, improve the conditions needed to work in cross-disciplinary situations, and improve an understanding of the problem by integrating social and environmental issues. As Ryan notes "I inherently know things that I wouldn't know otherwise".

Participants associated with this category perceived the benefits of intentional learning as "everyone gains": the individuals, the team, and stakeholders. As Brianna notes, "I think it's essential. It allows to get the benefit, if we're working in parallel and not learning from each other, the product is not going to be better if we do that. It's essentially to really capture those different ideas." Olivia notes that even society gains by "capacity building in a way with bright people very fast that didn't have the same background". This capacity building includes a new appreciation for different points of view, a respect for how "hard the other person's field is", and a new open-mindedness towards other disciplines that "strengthens everybody".

"I want to say I just became more open minded towards other disciplines. I used to think, we're engineering, it's much more difficult than everything else and although we can't communicate as well as the business people, but we're more technical, but when you start working with them, everyone sort of has their own set of abilities and weaknesses that they have to work together. I think by working cross-disciplinary, it sort of strengthens everybody." [Uri]

"... you underestimate how hard the other person's field is. You've spent all this time and effort developing your capability; and oh, by the way you'll pick that up too and you learn pretty quickly if/that if you're not just going to do this superficially; that you know they've traveled their long road too. And so you have to meet as equals but you have to, and respect each other, but you have to learn about their problem if you're really going to be successful in at least some amount and commit to that. But that's fun ..." [Olivia]

Overall, ways of experiencing cross-disciplinary practice as Intentional Learning emphasise distinct ways of thinking, acting, and being. They are distinct from the Working Together category because of their focus on the experiences of self-directed learning processes fuelled by a passion for learning and an appreciation of the multi-level benefits of learning. As such, the Intentional Learning category emphasises the process and outcomes of collaborative and situational learning. In particular, ways of thinking such as (1) an awareness of how difference creates opportunities to learn, (2) seeing the value of disciplines beyond science and engineering and having respect for disciplinary training in which no field is more difficult than another, (3) understanding complexity as having societal and global dimensions, (4) recognising the limits of your own knowledge and the need to engage with others, and (5) valuing learning as its own goal with outcomes that yield benefits for everyone (me, my team, my stakeholders). Ways of acting emphasise a self-directed intentional learning process of creating opportunities to learn through interaction, immersion, and failure as well as purposefully educating others about your own discipline. This involves being open to complexity and emergence, and developing abilities to think and synthesise across disciplines as well as to collaborate and form learning partnerships. The experiences associated with this category suggest ways of being that emphasise an identity as a learner and educator deeply grounded in a passion and openness to learning through differences.

The Intentional Learning category builds on the Working Together category because it represents a process of improving the conditions needed to work together with people with different training to address complex problems of social and global significance. Here, a focus on cooperation and collaboration expands to include social learning, a focus on complex problems expands to include social and global elements, and a role evolves from being a collaborator to being a self-directed learner. More specifically, an awareness of differences changes from recognising disciplinary differences to respecting the difficulty of different forms of training and learning at the intersection of different perspectives. Creating opportunities to address complex challenges emerges through intentional learning experiences that involve immersion in other disciplinary ways, seeing failure through an opportunistic mindset and being open to discovering alternative ways of seeing the world. As seen in the next section, the relationship between this category and the next higher category speaks to how intentional learning can lead to ways of understanding and addressing the challenges of enabling cross-disciplinary work.

4.3   Ways of experiencing cross-disciplinary practice as Strategic Leadership

Where the experiences of the second category, Intentional Learning, focus on learning, the experiences associated with the next higher category focus on applying prior learning to enable the conditions for successful cross-disciplinary work. In other words, *cross-disciplinary practice is strategic leadership to enable cross-disciplinary work and synergy for the best outcome.* Leadership is central in that it involves being the 'interface', 'connector', or 'communication specialist' to cross disciplines, organisational functions, and cultures to proactively create an environment for innovation. Unique ways of experiencing cross-disciplinary practice for this category include: (1) making or enabling conceptual connections, (2) building allegiances, shared ownership, and trust, and (3) managing differences to create new paradigms or frameworks that leverage diverse perspectives. Participants associated with these experiences include Daniel, Fergus, Jacob, Michael, Tyler, and Yvonne (see Table 2).

All of the experiences in this category involve a common approach or identity: being an interface or boundary worker to help a cross-disciplinary team work together towards a 'worthwhile goal'. For Jacob, his experiences involved being the communication specialist, someone who helps "people understand each other" and has a "sense of what they mean when they're telling me something. And so I would use that capability to get minds met. I would use a lot of other techniques, but I was a bridge." Fergus' experience involved interfacing with lawyers and marketing people to "translate their needs to a technical engineering perspective". Similarly, Daniel described one of his first experiences as a developmental "guinea pig experiment" where the company he worked for "took one level, pushed it apart and stuck me in between layer A and layer B". One aspect of being the interface is the capability to lead and do what is necessary to get a job done well. For Michael this involved being the person "willing to step forward and, to a certain extent, take the risk" in "orchestrating the answers". Daniel notes that sometimes being a leader is letting "up on ego and individual things like that". Fergus describes leadership as a broad thinking approach where "at the end of the day, if all else fails you have those leadership capabilities, communications skills, where you can ask the right questions, kind of like you're doing right now, to come up with the answers. That's really the key and for me I think that's really what cross-disciplinarity is".

Building networks of expertise and enabling system connections was a crucial aspect of these experiences, and much of this emphasised the importance of problem formulation over pushing "too far into implementation". When discussing why he brought on certain people for a specific project, Fergus describes, "So very early on I realised, man we have to have all of those individual pieces connected if we're going to be successful with getting something to launch". Similarly, Tyler notes that one of his strategies involves knowing the "set of skills and maybe even opinions that you think need to be brought to bear to resolve a problem efficiently and effectively". Yvonne's experiences exemplify this idea of building systems-oriented networks of people. In her situation she was working "in an area of study that has no disciplinary ownership" where she wanted to get as many different kinds of experts involved "because the complexity of the area demands it". For Yvonne, it is this complexity that drives a need for building networks of expertise: "it's complex when domain specificity can't answer all the questions for you. It's complex when even knowing who the community is, is not clear. When you have more questions than answers, it's complex".

"I built a network of people, constructed a network of people who are interested in serious games … from computer graphics, from electrical and computer engineering, from engineering education, from communications studies, from computer science … from visual and performing arts, people from, who are interested in nanotechnology, people from the veterinary school … you can't do interdisciplinary research if you can't see those intersections; if you can't see relevance of one domain of knowledge to another. There are not connections everywhere, but I think you have to be able to know if there is." [Yvonne]

Yvonne notes that not everyone has this capability to make connections and turn "your lens in a way you couldn't have" before. As she notes "extremely bright people sometimes cannot see intersections".

"… I've had these one-on-one conversations and I'll tell them what I'm interested in … I'd ask them about their work, they'd tell me about their work and then they'd say, 'I don't think I have anything to contribute'. And I would say, 'how? But look, you do this and we're looking at this and so what you're doing can be applied to this'. Some people can't do that … you can see it almost right away. … He just kind of sat there and didn't know. I said 'well I would love it, would you come and join us'. Anyway, he said, 'I don't really think my work has anything to do with what you're talking about'. I said 'have you ever done any applied work?' And he said, 'yes', he had worked for a helicopter design firm in designing displays for helicopter dashboards. And I said 'and you don't see that relevant?' " [Yvonne]

Another aspect of the experiences associated with this category involves proactively building allegiance and trust, a strategy that was often described as a lesson learned from failed projects. Fergus notes that the biggest issues were people issues not technical issues – getting "the right people involved in the decision-making at the appropriate time" and "project management skills, driving things forward and managing risk, communicating risk". Others also described strategies for enabling communication and collaboration across differences. Jacob's strategy is to "seek first to understand, then be understood". He also works to persuade people to learn from each other, expand "their minds to understand that there were things outside of their discipline that they needed to understand" and change "their parameters of how they approached the problem". The goal of this process was to "find common ground to build before they actually met in person, because there were some very confrontational meetings". Daniel's strategy to "get a bunch of people to play together well" and "have fun doing it" involves creating a trust-based system:

"I get a set of people excited and positive about what the job is. So I'm not a detail person actually. I'm bad at that. But I'm a good judge of other people. My management style is a trust-based system. I mean if I have someone working for me, I have a high trust level. Basically I just convey the big picture and stay out of their way, if they have a problem they come and ask me but I'm really not involved very much. It scares a lot; I mean I've had bosses that think this is going to be a train wreck.... For most people, they enjoy working in an environment where they have a trust level but it also means you have to earn and hold that trust. You can't break it." [Daniel]

Jacob also believes that establishing trust is crucial: "you have to be trustworthy yourself. So it really starts with you and then you model the behaviour you want". Similar to these trust issues was a perception of a need to combat a "human tendency to blame the new thing when everything else worked right before we changed to put the new thing in". To deal with this, Jacob worked with individuals to develop "the value of synergy" and appreciate "treading in a new space". This involved developing shared ownership so that "everybody's in on the ground floor" and everybody feels "like a father to what they're doing and then of course they own it". Fergus described similar experiences and his strategies for "involving people, partnering with people, getting alignment around, having people feel like they're a part of it, it's theirs".

Processes for managing and leveraging differences were also described as strategies to help generate cross-disciplinary frameworks. For Tyler, an "interdisciplinary framework" is the "only appropriate way of dealing" with the significant problems facing society. Fergus described one experience where he was in his "own little shell" designing a product where he didn't talk to others who represented the client's needs. He explained, "the product manager wasn't overly thrilled because it didn't really meet their need so there were a lot of things that were sort of missing. So that was probably the biggest wake up moment for me from a standpoint that you just can't assume that what you're putting together is going to meet everyone's needs". He learned that "you've got to ask questions, you've got to engage folks, you've got to really start developing some systems engineering type skills and understand requirements, before you just jump and put a solution together". Similarly, Jacob described how a balanced approach to problem solving in which all voices are heard and respected, and all share ownership in a process can enable the best outcome:

> "... a balanced approach to the problem-solving which I happen to believe, statistically you're more likely to have a better solution if you have more voices that are reasonable and respected sitting in the room.... So balance, to me, says that you, you and I and three or four other people, whatever the size of the team is, are sitting around and we respect each other, ok, and we're willing to listen to each other and consider that these things, these issues that are raised or solutions that are raised are real and should be considered." [Jacob]

As with the earlier categories, ways of experiencing cross-disciplinary practice as Strategic Leadership emphasise distinct ways of thinking, acting, and being. They are distinct because of their focus on proactive strategies for cross-disciplinary discovery and innovation. This emphasis on "orchestrating" is explicit in a self-identified role of being a facilitator at the cross-disciplinary interface and taking the risk of leading projects towards shared and valued outcomes. This is a leadership role that is directed toward facilitating synergy and enabling the team rather than promoting individual egos. Distinct ways of thinking associated with this category include (1) an awareness of how differences need to be managed and productively leveraged, (2) a broader view of difference that encompasses global and organisational cultures, (3) a respect for the interface between different perspectives and the challenges of working at this interface, (4) a view of complexity that integrates an awareness of the complexity of social interactions, (5) being aware of the limits of existing single disciplinary paradigms in supporting discovery and innovation, and (6) the need for building common ground, trust, shared ownership and allegiance towards a worthwhile goal. Ways of acting emphasise a proactive approach to managing and leveraging difference to formulate problems through multiple perspectives versus jumping too quickly to a final solution. This involves creating expert networks to change the parameters and orchestrate discovery and innovation. It also involves building common ground by attending to the human and social dynamics of cross-disciplinary work. The experiences associated with this category suggest ways of being that involve taking the risk and leading, letting up on the ego to enable the team, and facilitating synergy at the interface by being an interface, bridge, or boundary worker.

The experiences associated with the Strategic Leadership category imply a logical relationship with the next lower category, Intentional Learning. In particular, learning through experience about the ways cross-disciplinary work can break down can provide opportunities to develop strategies or mindsets for working together across differences. This category also expands on the Working Together category by including issues of trust, shared ownership, and inclusivity such that disciplinary, organisational, and cultural perspectives can be managed and leveraged to open up new ways of thinking.

4.4 Ways of experiencing cross-disciplinary practice as Challenging and Transforming Practice

Where the experiences of Category 3 (Strategic Leadership) focus on enabling cross-disciplinary work, the experiences of Category 4, the highest category, represent transformative reflective practice that challenges prior training and ways of thinking. For this category, *cross-disciplinary practice is challenging and transforming practice to integrate systems and produce an outcome greater than the sum of its parts.* This transformative reflective practice involves challenging prior training and beliefs about what counts as 'good practice', attributes of good solutions with respect to stakeholder risks, and how organisational cultures support or inhibit professional growth and social justice. This involves (1) critically challenging disciplinary practice and investing in the ways conflict can be transformative, (2) integrating stakeholders as collaborators, (3) attuning to the human aspect of complex systems, (4) advocating less visible perspectives by taking into account the broader context, and (5) embracing cross-disciplinarity as an everyday practice. Participants associated with these experiences include Anthony, Hannah, Kelvin, Logan, Samantha, Wendy, and Xavier (see Table 2).

Category 4 experiences are situated in an awareness of human-sociological-technological interactions within cross-disciplinary problems and the ways these interactions can enable transformative critical reflection. For Wendy, "the great societal problems that we do ... all take interdisciplinary approaches". In developing a cancer prevention expertise network she prioritised "collaborating with the department of English so in that; but still that can be a scientific approach in the sense that many physicians will tell you and I think many social scientists as well, and there's some evidence for that how you react emotionally to cancer may have a bearing on how well you do". For Hannah, cross-disciplinary problems involve not "just technical things but attorneys, financial people, controllers, all of that" because it brings "so many different facets together". Xavier notes that when you "insert human behaviour in the whole thing" you mix "quantitative and qualitative approaches (which) is sometimes straightforward but often times very difficult".

Productive conflict, either within a team or within the self, creates opportunities for transformative critical reflection that challenge ideas about the nature of practice, including what constitutes 'good' practice, and the influence of structures and cultures on cross-disciplinary work. As an example, Logan notes, "If everyone thinks like I do then the world is not going to be too good in the sense that you need some diversity, you want someone with other backgrounds to bring to the party. You need some conflict or some diversity.... If we all think the same way, you're going to miss a lot ... Now would things run a little smoother? Maybe because everyone is agreeing, but that's not a good thing either". For Logan, cross-disciplinary practice involves productive conflict, taking risks, challenging boundaries and values, and most importantly be willing to "give up some turf" to enable new ways of thinking.

Participants who worked primarily in academic or research centre settings tended to challenge disciplinary boundaries and the hegemony of science. Xavier reflects, "I see a lot of people claiming to be doing, what they would say, multidisciplinary or disciplinary science ... I kind of look at it and say, well, not really". He described cross-disciplinary practice as not "taking stuff from me and stuff from someone else, and putting them together" but rather something that helps you understand as well as challenge your own way of thinking. For Xavier, a sign of substantive cross-disciplinary collaboration is "when you sit down across the table from someone else and you listen to the way in which they do their science and you become very suspicious at first" because you don't "share common ideas, share common language", and then you come to understand how the "gap between those are huge, they're, I mean, to the point where if either of the two disciplinary scientists aren't willing to give up some turf, it's never going to happen, you know, it's never going to happen productively ... we need everyone together as a team. No one thing is better than the other". Taking a cross-disciplinary approach requires deep investigation, "looking at the science between the disciplines" to move from an exclusive or reductionist view of "good science" to a more inclusive and pluralistic perspective.

> "And when I began to do more reflection I thought that, okay, why is that we had these problems? You know, and that's where I started thinking more about what I called practice of science. What they valued most – well it's the reality versus. what a model can do. So what they value is good science, I value it much differently. My terms are very different, sometimes extremely technical. I can't be using those to communicate what it is I want to do.... Practice of science is – what is good science? How do you actually collect information? How do you test your ideas? Where do your ideas come from? How are they developed? What is the nature of theories?" [Xavier]

Kelvin experienced a critical transformation that began with enrolling in a life-changing course on the nature of science that changed his "perception of what I thought science was, how I thought science moved" and he "realised that this was something I needed to deal with". He became curious about "why African American students were so unable to be successful in ... learning science ... liking science and eventually getting into professions with a scientific base". He "took a step back and said, well, all of the research that's gone into the nature of science that's produced this body of knowledge" is "only based on one group of individuals" and that this has "hurt people of colour". Motivated by "the depths of inequity that children of colour, particularly African Americans, have always had in the US education system", Xavier explored how a more inclusive picture of the nature of science may transform the experiences of African Americans in education systems.

> "One of the things we don't teach is that, first of all, the scientific method is not the only way to do science, but one way. We don't teach about the cultural embeddedness of science. We do teach that science is strictly without value, human value. We teach that it's objective and that it's pure and what I know from my own research is that the same perspective has been used to hurt people of colour because if the science says it, it's objective, I didn't have anything to do with it ... science was developed by you, so you had a lot to do with it. It had your personal biases..." [Kelvin]

> "... how you can start introducing these kids and saying 'yea, you are valued, you are part of it'. Science is a human endeavour, it's a human construct, it comes from you. Whatever stories and whatever you may believe, you are science, you produce science, you do science." [Kelvin]

Like Kelvin, Samantha explored a nagging question about why engineering faculty "are never taught how to teach". An undergraduate feminist theory course challenged her views on the nature of engineering and had her mind "totally blown open" with "new insights" because people "talked about things I had never considered about the framework of what science was" and how that extended into engineering. Rather than turn away as if this new insight had "nothing to say to me" she went "whole hog" into exploring "other traditions or other areas or other methodologies that may not be represented or may not flow from peoples' experiences, but which have something to say". Her journey to integrate engineering and social justice involved using tools "from all kinds of stuff" which on their own couldn't provide an inclusive view "without the help of other disciplines".

Participants who were working in industry settings tended to challenge ideas about good engineering practice. For example, Hannah characterised good practice as a system level view that attunes to the human issues and places stakeholders at the centre of the system. She challenges a narrow view of good practice and argues for understanding "what some of the gaps were from the very high level perspective, so initially you were working at, what in industry they call the 50,000 ft. level" because "one solution may disadvantage another part of the project". This high level view involves "working with a national government" and attuning to social and political issues. To ensure stakeholders are present in decision-making Hannah acts "as an advocate for the other company" because "at some point you have to be able to bring everything together, so you have to be able to come from those different perspectives". Hannah describes this as a participatory relationship with key stakeholders because they "ultimately own the result of the problem", have "to pay to solve the problem", and "need to decide" if the results are "worth doing ... is it a risk that they, as stakeholders, can accept?" Similarly, Xavier notes, "working with stakeholders is not just, you know, something we take lightly. It's actually part of the science".

> "I've always thought that it's always a good interaction from the standpoint of a multi-disciplinary scientist, that when you go in you work with the stakeholders you're at, you're on equal ground, you're on equal footing. There is an exchange of information. They're kind of framing the problem in a much different way than what you see it. You tend to be more isolated as a scientist, as someone who is at a university. They're dealing with this on a day-to-day basis, they have the muddy-boots kind of experiences. So the way in which they articulate the problem ... the constraints that they have, is very, very valuable to listen to that." [Xavier]

For Anthony, a system level view involved challenging ideas about engineering work practices such as build-test-fail approaches to product development in order to create a "new model for work ... what I would call analysis-led design or analysis-led discovery". This new practice begins with getting the "business questions clear". He notes, "nine times out of ten, when we get involved in the wrong modelling work, it's because we're not modelling the right thing". The second step involves translating the business question into an "abstract science question". Next, the "hard-core modelling begins" but the last step is to "simply and clearly, shape decisions", "changing the outcomes in innovation and at the end of the day, making money by delighting consumers and getting it in the market faster".

A common theme of experiencing cross-disciplinary practice for this category is a sense of significant personal investment – of time, hard work, sacrifice, putting the greater good first, and a lifelong endeavour.

> "I'm making the personal investment to sit down and learn what I can about their practice of science and how it integrates with my traditional practice of science ... if people like me don't make this investment, then we're basically screwed." [Xavier]

> "My point is, you do need a bit of the altruism; you do need to be able to do this for the greater good and not for something that's going to glorify you as a leader..." [Wendy]

> "I decided that this wasn't going to be a one-time thing." [Samantha]

Hannah reflects that this is an investment that impacts not only "a team perspective" but also an "individual's perspective" that integrates disciplinary training with perspective that "comes from the experiences that one has had". For her, cross-disciplinary practice involves "bringing all your aspects of your life together". Logan notes he's been cross-disciplinary "since kindergarten", that "this is not my 'day' personality" but rather his everyday practice. Samantha notes that this new identity has advantages and disadvantages:

"I'm wrestling with this idea of homes now. And how I really feel that you have to have people that you can talk with, and so a cognitive home. But does the home need to have institutionalisation or institutional markers or some kind of outward existence or can it be a group of people who identify in certain ways are, that don't find their cognitive home somewhere else? I don't know. But it's a more dynamic place. It is a more revolutionary place. It is a more explosive place. It's like, to use a boundary metaphor, the reactive place on the edge, where you have to make tools and ideas come from all kinds of stuff around, but none of which will be able to describe without the help of other disciplines." [Samantha]

Ways of experiencing cross-disciplinary practice as Challenging and Transforming Practice also define distinct ways of thinking, acting, and being. They are distinct because of their focus on the process of questioning practices and boundaries, and emphasising an explicit identity of being cross-disciplinary. Where the Strategic Leadership category involves leading teams to enable common ground and discovery and innovation, this category is about critical reflection to enable transformative learning and potentially transdisciplinary outcomes. In particular, this category expands an awareness of difference to include lived experiences and recognising how boundaries between differences are socially constructed and negotiable. This awareness supports critical analysis of the idea of difference and a critical exploration into similarities across different perspectives, which may lead to new and inclusive practices that honour differences in perspective and value productive conflict as a pathway towards transcending prior values, beliefs, and assumptions. This involves attuning to the human and contextual aspects of complex problems through participatory strategies that engage diverse stakeholders as partners, not just information resources. When human and contextual factors are integrated into the system, the limits of prior assumptions about good practice or good science are revealed to enable new ways of thinking about system performance.

In summary, distinct ways of thinking associated with this category include (1) an awareness of how diverse perspectives can create productive conflict and enable new paradigms of thought, (2) a broader view of difference to include lived experiences that transcend disciplinary boundaries, (3) honouring other ways of knowing, particularly situated ways of knowing, and challenging the idea that one way of knowing is better than another, (4) an awareness of system boundaries and the need to attune to human systems and stakeholder risks, (5) a belief that giving up some disciplinary turf is necessary for shaping new practices and enabling participatory system level performance. Ways of acting emphasise a process of challenging and transforming views on practice towards an integrative and inclusive synthesis. This involves challenging the nature of differences, boundaries, and good practice. It also involves both working at the interface and working at the 'mile high' system level to link all aspects of a system, and engaging stakeholders as partners. The experiences associated with this category suggest ways of being in which cross-disciplinarity is an everyday practice and an attribute of individuals, not just teams, that integrates work, life, and activism. While ways of being emphasised living at the interface as a transformational learning site, there may also be feelings of homelessness or loss of respect within disciplinary structures or cultures.

4.5   Relationships among different ways of experiencing cross-disciplinary practice

The categories of description that emerged from the analysis are *distinct*. Each captures a qualitatively different way of experiencing cross-disciplinary practice in engineering contexts. These can be mapped to Dall'Alba's embodied understanding of professional practice framework in terms of interrelationships among ways of thinking, acting, and being. Moving from the first to the fourth category, qualitatively different ways of experiencing and comprehending cross-disciplinary practice may be described in terms of:

– Ways of thinking: a growing awareness of 'difference', situation complexity, and goal intentions
– Ways of acting: increasingly comprehensive approaches for engaging with 'difference' and situation complexity
– Ways of being: an embodied understanding of cross-disciplinarity practice as moving from a self-perceived role in cross-disciplinary situations to an identity as a cross-disciplinary professional

Relationships among the four categories are also *hierarchical* (see Figure 1). Attributes of each higher category encompass attributes of prior categories; critical variations of the more comprehensive categories are not evident in less comprehensive categories. This hierarchical relationship does not mean that the four categories are stages of development that all people go through in the same order, but rather qualitative changes in understanding cross-disciplinary practice that people may traverse in any order, depending on the nature of their experiences and the ways they learn through these experiences. In this way, the four categories map out multiple trajectories of becoming cross-disciplinary.

This hierarchical progression is data-driven because the relationships emerged from the analysis and not an existing framework. It is also logical because attributes of the higher categories illustrate ways of experiencing cross-disciplinary practice as increasingly complex and comprehensive. For example, a review of Table 2 illustrates that the experiences associated with the highest categories were associated with the highest degree of epistemological distance. This suggests that the experience of working with others with very different worldviews about the nature of knowledge can have high potential for transformative critical reflection, a view supported by theories of transformative learning (Mezirow, 2000).

5   CROSS-DISCIPLINARY PRACTICE AS WORKING WITH OTHERS

While the goals of this study were not specifically about global engineering practice as 'working together', the findings provide useful insights and perhaps a language for talking about increasingly complex ways of experiencing and understanding what it means to work together across disciplines in engineering contexts. Each of the categories described earlier provide a way for talking about what it means to work together across differences and the transformative potential of learning *about* and *through* differences – differences in disciplinary training, cultures, organisational functions, and lived experiences.

Aspects of working together were evident in all four categories and collectively illustrate an arc of increasingly complex ways of experiencing and comprehending collaborative cross-disciplinary work in engineering contexts. For example, those who recognise differences in perspective are likely to engage in an iterative learning process where they learn how to communicate and make sense of each other's perspectives. Those who take the initiative to deeply learn another perspective (including deep immersion in the actual work) are likely to develop a high level of respect for how difficult another's practice is and a deeper respect for what it means to bring together different points of view. Those who understand the collaborative aspects of doing this will likely attend to building trust and shared ownership, bringing diverse voices to the table, managing and leveraging differences, and focusing on the core of the problem before jumping too soon to solutions. This will likely be characterised by at least one person actively orchestrating these processes – someone who takes on the identity of a bridge or translator – someone who is a networker bringing in expertise. In other words, *translation is more than the act of translating – it involves building collaborations to facilitate changes in how people work or think*. Finally for those who can articulate their perspective and reflect on it in relation to others, there should be an opportunity for transformative learning where the outcome is a more integrative and inclusive system level perspective that opens up the space of possibility for new paradigms that are more representative of how the world really works.

6 IMPLICATIONS FOR PREPARING CROSS-DISCIPLINARY PROFESSIONALS

Just as cross-disciplinary professionals need to develop expertise in a domain, they also need to develop expertise in cross-disciplinary knowledge practices so that they can create the conditions for successful cross-disciplinary work as well as recognise and self-manage ways to overcome obstacles to this kind of work. They need to become cross-disciplinary professionals who can attune to and leverage differences, see the broader system, and imagine new frameworks for problem setting and solving. This maps out an arc of learning that connects ideas about cross-disciplinary practice with an evolving cross-disciplinary identity. This learning trajectory is influenced by the extent to which a learner is able to critically reflect on experiences, is committed and open to learning through experiences (successes and failures), has opportunities to participate in as well as construct cross-disciplinary projects, and has opportunities to work with others who differ considerably in their views, values, beliefs, and assumptions. Collectively, these represent critical conditions for learning how to become a cross-disciplinary professional. In addition, the four qualitatively different ways of experiencing and comprehending cross-disciplinary practice presented in this chapter provide a framework for individuals to self-assess their personal growth as cross-disciplinary professionals able to work together across disciplines and create the conditions for successful cross-disciplinary work.

ACKNOWLEDGEMENTS

This work is supported by a grant from the National Science Foundation (EEP-0748005).

REFERENCES

Adams, R.S., Daly, S., Mann, L.L. & Dall'Alba, G. (2011) Being a professional: Three lenses on design thinking, acting, and being. *Design Studies*, 32, 598–607.

Adams, R.S., Mann, L., Jordan, S. & Daly, S. (2009) Exploring the boundaries: language, roles, and structures in cross-disciplinary design teams. In: McDonnell, J. & Lloyd, P. (eds.) *About Designing: Analysing Design Meetings*. Chapter 16, London, Taylor and Francis Group.

Åkerlind, G. (2005) Learning about phenomenography: Interviewing, data analysis and the qualitative research paradigm. In: Bowden, J.A. & Green, P. (eds.) *Doing developmental phenomenography*. Melbourne, RMIT University Press.

Bowden, J. (2000) The Nature of Phenomenographic Research. In: Bowden, J. & Walsh, E. (eds.) *Phenomenography*. Melbourne, RMIT University Press.

Bransford, J. (2007) Guest Editorial: Preparing People for Rapidly Changing Environments. *Journal of Engineering Education*, Jan, 1–3.

Bromme, R. (2000) Beyond One's Own Perspective: The Psychology of Cognitive Interdisciplinarity. In: Weingart, P. & Stehr, N. (eds.) *Practising Interdisciplinary*. Toronto, University of Toronto Press.

Case, J.M. & Light, G. (2011) Emerging Methodologies in Engineering Education Research. *Journal of Engineering Education*, 100 (1), 186–210.

Dall'Alba, G. (2009) *Learning to be Professionals*. Dordrecht, Springer.

Dall'Alba, G. & J. Sandberg (2006) Unveiling Professional Development: A Critical Review of Stage Models. *Review of Educational Research*, 76 (3), 383–412.

Daly, S., Adams, R.S. & Bodner, G. (2012) "What does it mean to design? A qualitative investigation guided by design professionals' experiences." *Journal of Engineering Education*, 101 (2), 187–219.

Daly, S., Mann, L. & Adams, R. (2008) A new direction for engineering education research: Unique phenomenographic results that impact big picture understandings. *Proceedings of the 2008 AaaE Conference, 7–10 December 2008, Yeppoon, Australia.*

Entwistle, N.J. & Peterson, E.R. (2004) Conceptions of learning and knowledge in higher education: Relationships with study behavior and influences of learning environments. *International Journal of Education Research*, 41, 407–428.

Forin, T. (2011) It's gonna be a long trip. *Proceedings of the annual American Society for Engineering Education Conference, 26–29 June 2011 Vancouver, June.*

Galison, P. (1997) *Image and Logic*. The University of Chicago Press, Chicago.

Grand Challenges for Engineering Committee (W. Perry, Chair) (2008) Grand Challenges for Engineering. Washington DC, National Academy of Sciences on behalf of the National Academy of Engineering. Available from: http://www.engineeringchallenges.org [Accessed 10th February 2013].

Jamieson, L. H. & Lohmann, J. R. (2009) Creating a Culture for Scholarly and Systematic Innovation in Engineering Education. American Society for Engineering Education,

Washington, D.C. Available from: http://www.asee.org/about-us/the-organization/advisory-committees/CCSSIE/CCSSIEE_Phase1Report_June2009.pdf [Accessed 16th May 2011].

Kabo, J. & Baillie, C. (2009) Seeing through the lens of social justice: A threshold for engineering. *European Journal of Engineering Education*, 34 (4), 317–325.

Klein, J. T. (1990) *Interdisciplinarity: History, theory, and practice.* Detroit, Wayne State University.

Klein, J. T. (2004) Prospects for transdisciplinarity. *Futures*, 36, 515–526.

Lattuca, L. (2001) *Creating interdisciplinarity: Interdisciplinary research and teaching among college and university faculty.* Nashville, Vanderbilt University Press.

Mann, L. M. W., Radcliffe, D. & Dall'Alba, G. (2007) Using Phenomenography to Investigate Different Ways of Experiencing Sustainable Design. *Proceedings of the ASEE Annual Conference & Exposition, 23–28 June 2007, Honolulu, Hawaii.*

Marton, F. & Booth, S. (1997) *Learning and Awareness.* Mahwah, NJ, Lawrence Erlbaum Associates.

Mezirow, J. (ed.) (2000) *Learning as transformation: Critical perspectives on a theory in progress.* San Francisco, Jossey-Bass.

National Academy of Engineering (2004) *The Engineer of 2020: Visions of Engineering in the New Century.* Washington DC, National Academy Press.

Nersessian, N. J. (2006) The Cognitive-Cultural Systems of the Research Laboratory. *Organizational Studies*, 27 (1), 125–145.

Prosser, M., Trigwell, K. & Taylor, P. (1994) A phenomenographic study of academics' conceptions of science learning and teaching. *Learning and Instruction*, 4 (3), 217–231.

Sandberg, J. (2000) "Understanding human competence at work: An interpretive approach". Academy of Management Journal, 43 (1), 9–25.

Trigwell, K. (2000) A phenomenographic interview on phenomenography. In: Bowden, J. & Walsh, E. (eds.) *Phenomenography.* Melbourne, RMIT University Press, pp. 62–82.

Zoltowski, C.B., Oakes, W.C. & Cardella, M.E. (2012) Students' Ways of Experiencing Human-Centered Design. Journal of Engineering Education, 101 (1), 28–59.

Chapter 6

Engineering problem-solving in social contexts: 'Collective wisdom' and 'ba'

Rachel Itabashi-Campbell & Julia Gluesing
Industrial & Systems Engineering, Wayne State University, Detroit, Michigan, US

I INTRODUCTION

This chapter is based on an empirical study that explores one aspect of engineering practice: knowledge creation and sustained learning resulting from solving engineering problems. The chapter focuses on a particular finding of the study, 'cognitive convergence'. We explore this concept in depth and argue that *deliberate* context-building promotes unification of technical interpretations leading to higher engineering performance.

2 BACKGROUND

Our qualitative study of US-based product engineers tasked to solve product-related problems (Itabashi-Campbell *et al.*, 2012; Itabashi-Campbell *et al.*, 2011) has highlighted the social aspects of engineering problem-solving. The rich descriptions of the engineers" lived experiences were collected through semi-structured interviews and were subsequently analyzed using thematic development techniques (Eisenhardt, 1989b). The findings illuminated a set of conditions and process phenomena that contrast 'successful' problem-solving stories with the "less successful".

Using these findings as a basis, we develop an argument for a meaning of shared cognition in engineering problem-solving contexts. We introduce the concept of *ba* as a mechanism for achieving a superior collective interpretation, i.e., a "single unified *phronesis*" (Baba *et al.*, 2004). Viewing engineering problem-solving as epistemic collaboration, we first characterise knowledge that flows through the problem-solving process using the Aristotelian distinction of knowledge types (Aristotle, 1934; Baba *et al.*, 2004; Grint, 2007). We then introduce the concept of "*ba*" (Itami, 2010a, 2010b; Nonaka *et al.*, 2000; Nonaka *et al.*, 2006; Yamaguchi, 2006) and extend it to engineering problem-solving contexts. Our goal is to build a contextual framework that is conducive to high engineering problem problem-solving performance.

After a brief presentation of study findings, we begin our discussion by first describing the nature of engineering problem-solving. Here, we describe the engineering problem-solving process from three perspectives: cognitive, epistemological, and behavioural. Subsequently, we turn our attention to the contextual factors that influence the process. Finally, we integrate the process and contexts to form a unified view of engineering problem-solving as a system.

We then focus our attention to on our 'cognitive convergence' phenomenon: Working with others to correct problems, challenging and negotiating each other's perceived 'boundaries' until all stakeholders' views converge. Combining our study results and theories, we launch a discussion to develop a conceptual framework of 'collective wisdom,' which we argue is the ultimate epistemological state to be achieved for successful engineering problem-solving. We borrow Barrett's (1998) seven principles that create a highly effective organisational environment, one that promotes well-coordinated improvisation in a manner similar to that of a jazz band. We interpret what would be a highly effective *ba* for engineering problem-solving using these principles.

3 OUR STUDY

Our study was conducted from August 2009 through August 2010. The primary method of data collection was semi-structured interviews, which were all digitally recorded and were later transcribed. Our sample consisted of 31 product engineers – 27 men and four women – primarily based in US Mid-West (approximately half of them were employed in the automotive industry), having from three to over thirty years of experience with varying industry backgrounds. We adapted Schein's (1996) definition of "engineers" to define "product engineers" as "designers of products and systems that have utility" (p. 14) and recruited engineers who were in a position to take ownership of design, application, integration, or manufacture of physical goods, software, or information technology (IT) architecture. A majority of them were in their mid- to late-career stages and had multiple industry, as well as product, experience. Twenty-seven respondents were currently employed full time. The remaining four had either recently retired or were temporarily out of the workforce.

Each interviewee was asked to recount a case of product-related problem-solving that, from his or her perspective, went particularly well and then to narrate a second, less effective, case. At our encouragement most respondents narrated their experiences from a relatively recent past – say, within the last two to three years. A few felt compelled to tell stories from further back in time, as many as ten years ago. These were accepted. The interviewee was then asked to compare and contrast the two cases. All 31 product engineers were able to narrate a successful case each. Two of them could not come up with unsuccessful examples, making a total of 29 less successful stories. Finally, each interviewee was asked to discuss, based on his or her experience, how engineering knowledge can effectively be captured and disseminated in an organisational context. Probes were used to elicit rich elaboration. The data analysis followed the grounded theory protocol of Corbin and Strauss (1990) and systematically progressed from 'open coding' to 'axial coding' to, finally, 'selective coding' (Corbin & Strauss, 1990; Douglas, 2003; Walker & Myrick, 2006). By inductively traveling to and fro among the data and literature, themes emerged, eventually resulting in a set of findings.

Our data demonstrated stark differences between 'successful' and 'less successful' product-related problem-solving efforts. These differences were found in the environmental conditions, processes, and outcomes of the problem-solving – and are to be expected given the problem-solving literature. Successful problem-solving efforts, at least in our sample, received clear leadership guidance, were unconstrained, were associated with controlled urgency, had adequate resources, and utilised an available framework for sharing. They moved with a clear cadence, leveraging stakeholders' knowledge and skill sets and ultimately unifying everybody's point of view. They culminated in positive system changes and conclusions that "made sense". Unsuccessful problem-solving efforts, on the other hand, had limited leadership guidance, were more constrained, were more chaotic or inertial, had inadequate resources, and showed limited sharing. They were less systematic and were not sufficiently able to unify stakeholders' cognitive perspectives. They did not effect positive system changes or deliver closures satisfactory to the product engineers.

One of the most interesting findings is that the interviewees' narratives portray product engineers as being not only technical investigators but also managers of relationships among stakeholders. Their ability to effectively manage stakeholders' viewpoints – which ultimately converged to reach concurrence on the problem definition and analysis direction in successful problem-solving stories – appears to be correlated with system changes and knowledge distribution beyond the immediate workgroups. We equate this phenomenon to "cognitive convergence", the process by which the cognitive structures of individual group members become more similar through knowledge sharing (Baba *et al.*, 2004). In the sections that follow, we discuss this concept in depth, in conjunction with other findings noted in our study.

4 NATURE OF ENGINEERING PROBLEM-SOLVING

4.1 Engineering problem-solving process: Cognitive, epistemological, and behavioural views

The literature defines "problem" as the gap between the existing state and the desired state (Corti & Storto, 2000; Tucker *et al.*, 2002) and "problem-solving" as a set of rational activities to reduce or eliminate this gap (Corti & Storto, 2000). If a car operates with a higher than expected noise level, then there is clearly a gap between the expected and observed performances. The implication, from the perspective of problem-solving, is that how the "desired" state is defined will set the course for the problem investigation. Since "problem-solving framing naturally fosters identification of new interpretations of the situation" (Corti & Storto, 2000, p. 251), this framing is bound to influence problem-solving outcomes.

Defining successful problem-solving outcomes as those that effect positive system changes – such as improvements in product, work designs, or routines – we can equate engineering problem-solving to organisational learning. Organisational learning is about "a process of improving organizational action through better knowledge and understanding" (Tucker *et al.*, 2002, p. 124) and has taken place when changes occur in response to newly gained knowledge or insights that can enhance the organisation's performance (Dodgson, 1993; Tucker & Edmondson, 2003). Product features, blueprints, specifications, and design approaches are all engineering artefacts that reflect past system changes, which were likely to have been prompted by previous engineering problem-solving.

Engineering problem-solving as a knowledge-acquisition process may be framed in different ways. Cognitively speaking, problem-solving can be an attempt to deal with uncertainty or ambiguity by trying to resolve or reduce this state (Corti & Storto, 2000). For example, an engineering problem, when first reported, may be as vague as "the car makes a lot of noise". The problem can be caused by any number of factors in the drivetrain, chassis, or anywhere else. Through iterative diagnostics, more insight is gained that infuses clarity into the picture. Thus, problem-solving proceeds as actors' cognitive perspectives move progressively from fuzzy to less fuzzy – and eventually to "completely understood" if the root cause analysis is successful.

Epistemically, knowledge morphs from being "more tacit or less articulated" to being an explicitly expressed form during the course of engineering problem-solving (Corti & Storto, 2000). Cognitive fuzziness in the beginning of a problem-solving journey lends itself to problem investigation activities that are highly exploratory. Confronting the unknown, the problem-solving team is less inclined to formalise its knowledge until further clarity is gained. Once the root cause is found and a solution identified, that knowledge is implemented through hardware, software, or both. In other words, the knowledge becomes explicit.

Finally, from a "behavioural" perspective, engineering problem-solving can be conceptualised as an attempt to transit from an individual and local focus to collective and organisation-wide aims. Specifically, in order for a system change to occur, the relevant knowledge generated during the problem investigation must eventually be translated into organisation-wide routines. Routinisation is achieved via codification of the learning and knowledge gained, i.e., making the individual and group-level knowledge explicit, so the organisational members outside that group are able to adopt it. For example, new design guidelines and specifications enable the deployment of codified engineering knowledge for the benefit of future product development. Likewise, standardisation of improved manufacturing processes helps prevent future defects for the rest of the operation.

Thus, a successful engineering problem-solving process should take engineers "through a subtle game going from the creation of highly uncertain and ambiguous situations to the reduction of these" (Corti & Storto, 2000, p. 253), during which the problem-solving focus becomes more collective in orientation, with the transfer of new knowledge from individuals to the group and to the organisation. At the end of a successful problem resolution, a new set of standards is created, through which "contradictions are resolved and concepts become transferable" (Nonaka, 1994, p. 21) – hence, a successful routinisation.

4.2   Engineering problem-solving environment: Contextual factors

An engineering problem-solving environment does not just make the transition, on its own, from fuzzy to clear, tacit to explicit, or individually-focused to collectively-aimed. If the problem-solving effort is to be successful, that is to culminate in learning, stewardship must be provided to make these transitions. Organisational routines can create contexts and mechanisms that facilitate the problem-solving process.

Organisational routines are "forms, rules, procedures, conventions, strategies, and technologies" but also include more intrinsic factors such as "the structure of beliefs, frameworks, paradigms, codes, [and] cultures" (Levitt & March, 1988, p. 320). Some routines are oriented more towards exploration of new ideas than exploitation of existing premises (Bartel & Garud, 2009; March, 1991) and vice versa. For engineering problem-solving, how the routines are applied at various phases of the root cause analysis may be critical. Specifically, it makes more sense to enact an environment that fosters a forward-looking, explorative outlook in early phases of the investigation when engineers must confront ambiguity. As the root cause investigation moves toward completion, the routines should exercise regimented control to standardise and deploy across the organisation those new solutions that the investigation has discovered.

In fact, routines can be structurally induced. A problem-solving practice at a Japanese transplant facility in the US Midwest chronicled by MacDuffie (1997), for example, includes an "accounting system [that] is deliberately designed to minimize the time spent figuring out who's to blame" (p. 488). Openness in the information exchange resulting from this measure to endorse the "Five Whys" problem-solving routine – rather than "Five Whos" – is dramatic (especially in comparison with another plant studied in MacDuffie's case that required booking charges to a specific party for initial causal assignment before problem investigation could begin). Alongside such deliberate measures to enhance an explorative mindset, structured problem-solving routines such as 8D, Kepner-Tregoe, and Five Whys (Handley, 2000; Smith, 1998) can provide operational and behavioural control to ensure consistency in approach. At the end of a root cause analysis, transitional routines such as experimentation followed by a pre-/post-data comparison to ensure efficacy of the chosen solution can prepare the system for complete standardisation (MacDuffie, 1997).

Viewed from the vantage point of knowledge acquisition, we can see that these routines function as a conversion mechanism that churn the bits and pieces of data generated from the root cause investigation into a cohesive set of facts. Knowledge acquisition, or learning, is about human action (Senge, 2006). This aspect of learning contributes to the subjective and context-specific nature of knowledge (Corti & Storto, 2000), which points to the need for validation. In other words, for knowledge to become an institutional-level logic set, it needs to be gauged and polished against a set of "standards" or "justification mechanisms" as Nonaka (1994) and Nonaka and Takeuchi (1995) conceptualise. Justification deeply affects engineering problem-solving since the quality of a root cause analysis depends on how causality is ascribed in light of what is deemed abnormal versus normal (Smith, 1998). If the problem-solvers operate with the established premise (usually developed through iterations of past problem-solving experience) that, for instance, design-related changes are nearly impossible to implement due to organisational constraints, such a preconceived notion invariably limits the scope of the investigation. As reported in MacDuffie's (1997) case study, at one US Midwest automotive plant, the "norm" of problem-solving was to stay within the boundary of product features that could be controlled without involving design engineering. Having learned how difficult it was to seek help from the company's design headquarters, the problem-solving team was seen assigning causality exclusively to the aspects that could be worked out directly between the plant and its suppliers – eliminating opportunities for potential design improvements. Thus, causality can largely be a mental construction and is influenced greatly by the organisational justification logic provided.

4.3 Engineering problem-solving as a system

A "system" analogy is one way to portray the interaction between an engineering problem-solving process and its environment. Root cause investigation, in this framework, is an exercise that plays out in a socio-technical system. The implication is that the effectiveness of the investigation is dependent on the extent to which this socio-technical system supports (or inhibits) it. This view is reasonable because every non-conformance is ultimately traced to system design (Crosby, 1979; Deming, 1982). Process improvement initiatives, such as Lean, fail to take root unless implemented with a concerted effort to appropriately alter the organisation's culture, which a socio-technical system engenders (Carroll *et al.*, 2002; Senge, 2006). In fact, Leveson (2011) argues that, from a product safety management perspective, safety is an emergent property of a socio-technical system. Likewise, the Toyota Production System (a.k.a. Lean) emerges from a purposefully constructed system architecture (Liker & Morgan, 2006; Womack *et al.*, 1990). Following this logic, effective engineering problem-solving, i.e. resulting in positive system changes, is also an emergent property of the socio-technical system that engenders the root cause investigation.

A system is an aggregate of routines which provide justification logic for organisational action. We can then argue that organisational contexts that prompt different routines should be deliberately created, especially for facilitating engineering problem-solving efforts because a swift resolution of the problem is usually desired. Routines that seek to evoke innovative ideas, as well as those that target consistency, should be purposefully applied. Because effective and speedy completion of engineering problem-solving requires opportunity framing to assure freedom to define the problem, routines should be applied carefully so as not to unduly constrain the problem-solving efforts by costs, politics, or other organisational factors. The freedom thus garnered actively steers information processing and keeps the momentum going toward problem resolution (Cannon & Edmondson, 2001, 2005; Carroll *et al.*, 2002). Such an environment is a deliberately engineered "system".

5   VOICES OF OUR ENGINEERS

5.1   Problem-solving environment

In our sample, product engineers portray environmental conditions differently in successful versus less successful problem-solving narratives. Five conditions from our textual analysis stood out.

5.1.1   Clarity of external leadership vision

'External leadership' in this study, refers to a source of direction or to the guiding force that is external to the work group engaged in solving the problem. Depending on the nature of the problem-solving, the firm's management or other constituents such as customers and suppliers played external leadership roles. Our analysis results show that external leaders are more likely to provide clarity of vision to team members in successful problem-solving efforts than in unsuccessful efforts. All 31 of the successful problem-solving stories included descriptions of the positive impact of their leadership in effective resolution of the problem. Instances of leadership influence included a "you shall solve this" decree by senior management, as well as descriptions of close status monitoring by members of management. Many product engineers spoke of "a very strong management team" and "getting things resolved quickly by having our management's support". In one story, a vice president cut his personal vacation short and showed up at the plant to attend to the problem-solving effort. One product engineer whose problem-solving efforts encompassed several plant locations described his experience as knowing that "whatever we came up with was going to be supported by the management team" not just at his facility but elsewhere in the company. In the instances of problem-solving that were initiated by the customer, the "we are in this together" message conveyed by the customers greatly facilitated the customer-supplier joint teamwork.

In contrast, not only was any mention of clear leadership oversight or vision absent in the 29 less successful problem-solving narratives, 27 of them explicitly cited instances that were contrary. For example, "lack of firm leadership" caused "wheel spinning" or allowed "multiple competing interests" to prevail. In a few cases, managers scrapped or wrapped up the problem-solving effort without discovering root causes. Instances in which members of management were not entirely truthful – or in some cases ignorant – about the nature of the problem were also mentioned. One product engineer described the firm's management as being one that "minimised the value of its workforce". Instances of customer-instigated problem-solving degraded into a "bring-me-a-rock exercise" because the customers' expectations and requirements were unclear or confusing.

5.1.2 Autonomy

Twenty-nine of 31 successful problem-solving examples were described as "open" and free from "outside influences that might have created road blocks". They involved "very open brainstorming", "open and honest communication", and "open-mindedness to new ideas" of stakeholders. Product engineers had "authority to ask for help" and "direct control over design or process changes if needed".

In contrast, none of the 29 less successful problem-solving stories chronicled autonomous characteristics of the problem investigations. In fact, 25 of them cited instances of having limited choices or constraints imposed on the team's decision making. For example, the existing contracts with their suppliers limited their options, culminating in "a deteriorating relationship" and eventually "resourcing the business to another company at a higher cost". Customers "forced" materials or designs that were not of the engineers' first choices. The management created one-sided directives that "were forced down on engineering's throat". A politically-motivated agenda, having "a lot of hidden agendas, cross purposes, contrary goals, contrary objectives", constrained the problem-solving environment.

5.1.3 Controlled urgency

Successful problem-solving efforts tend to be characterised by a positive tension or a sense of urgency that is controlled, as opposed to less successful problem-solving that carry a sense of either inertia or chaos. All 31 successful problem-solving stories were "rather hot" and "high-paced", prompted by high pressure from the customer or management, the safety-critical nature of the problem, or imminent deadlines for resolution imposed by the organisational process, such as product development time-lines. The problem-solving efforts were driven by "a controlled sense of urgency" and this urgency was perceived as "exciting" by many engineers.

In contrast, none of the 29 less successful narratives conveyed such a positive sense of urgency. All of them were associated with helplessness arising from either very high tension or inertia due to organisational disinterest. High-tension situations degraded into chaos – "getting a lot of help and advice from people you don't need", "going into all kinds of escalation modes" or "lots of emotions and stress and then ... panicking". At the other end of the tension scale was a total lack of urgency. Problem-solving failed to gain momentum because "this is the way that we have always done it", the problem had "a low occurrence rate", or the problem was simply "not visible to the customer". Lack of "financial pressure" also failed to push for solutions.

5.1.4 Resources

All 31 successful problem-solving stories had instances in which the right mix of resources – such as technical experts, technicians, laboratory equipment, and manufacturing lines – were provided internally by the firm or by the external stakeholders such as customers and suppliers. Engineers reported success in "getting the right people involved at the right level quickly" to "put some brainpower into coming to a solution". They attributed success of their problem-solving efforts to "a combination of all these people in addition to people who design tooling [or] who specify the metallurgy of the metal" because "it is always a team effort". The major success factor, to them, was that they "had most of the needed expertise [or] were able to, at least, have access to the needed skills".

Contrarily, of the 29 less successful stories only one described resources as being satisfactory. Thirteen of them explicitly cited difficulty in accessing personnel and expertise due to assorted organisational reasons. Corporate restructuring caused "degradation of service" because "when you've cut back, [by] maybe 15 per cent [,] that makes it difficult ... to give you the luxury to look at things with a clean mind". Prompted by workforce reductions, "so many people had left ... had taken a package and left [so the right] people were not available". Getting "more supplier involvement" was difficult because suppliers "don't have the time, are not supportive". Across narratives, people who could have helped "were spread too thin in terms of this problem" or were "now on three continents and [so] didn't have enough resources to go around and ... really get to the heart of understanding why it was happening". Sometimes, corporate politics put in place a wrong resource, such as a project lead who "wasn't an engineering person [and] didn't really know" the problem.

5.1.5 Framework for sharing

In 26 of the 31 successful problem-solving stories, interviewees reported that information or knowledge sharing was accomplished in many forms at many organisational levels. They may be informal or formal, ad hoc or structured, and intra- or inter-group level. Engineers shared knowledge via existing organisational routines or forums, such as "weekly technology meetings". They also communicated information informally using a "person list" that pointed users to the right person for the information being sought. Engineers reported "communication among all levels", frequently and actively "to just go over the current issues". Special trips to other company locations to present the problem-solving outcomes were also cited, such as going "to Mexico to present to the [customer's] management ... and distributed and shared with the people there [the entire presentation]" and "a few times to Europe to share the methodology and the process".

Contrarily, in less successful problem-solving stories, only one story reported sharing lessons by inter-company e-mail. Twelve examples explicitly cited instances in which sharing information was difficult. Collaboration was difficult because "their system for change control wasn't set up to be rapid enough to allow all engineers to understand what is going on". Structurally or politically induced mental walls also inhibited communication and created "layers" between stakeholders. Engineers' effort to "break down those walls and to drive communication" was "sometimes successful [and] sometimes it wasn't" so "there were no daily meetings ... little communication between the different groups ... little pockets of people working but it was hard for them to interface with each other" In one instance, management was cited as having no interest in enhancing communication so "people didn't even know how to use computers [and] things weren't shared very well".

5.2   How problem-solving ends

Problem-solving started because there was a deviation from the norm in some aspects of the product, which needed to be fixed. Successful problem-solving is not only more likely to correct the problem but to take it to a higher level of understanding, which manifests itself in system changes and outcomes that truly 'make sense'.

5.2.1   Positive system changes

In our sample, virtually all 31 successful problem-solving efforts were associated with positive system changes. The efforts not only corrected the immediate problems but profoundly affected the fundamental premises surrounding the product, reflected in such engineering artefacts as product design, manufacturing processes, and testing protocol. Engineers said, "Not only did we get [to] our root cause but we also built up a bigger picture of information about the product and the components and the process so that we understood the sensitivity to variation [leading to] application to other programs". The problem-solving teams "were able to come up with a solution that even better matched or optimised the customer's experience than the initial part". They effected "some design changes to subsequent products that made the products more, simpler ... easier to verify, and more cost efficient" and improved product performance to be "even better than the conventional system [which] surprised all the management team". Problem-solving resulted in "a good long-term learning that changed everything that we did", such as changes in process parameters and supplier management protocols. Findings from the team's root cause analysis "triggered a change to standard procedure" and often yielded "benchmark information [that got deployed] across platforms". Teams "got to put two different technologies together for the first time ever and ... were able to sort out and standardise all of the investigative tools" winning "a technology award over that". One of the reported outcomes of the investigation was the creation of a new engineering function to address "all the interfaces and interactions [which] became a new business opportunity".

Contrarily, none of the 29 unsuccessful problem-solving narratives contained instances of such learning or system changes. Product engineers explicitly reported that there wasn't "anything [that was] really learned or changed" from their problem-solving, nor were the stakeholders "any happier than before the project started". The problem "was never studied properly [and] never understood properly", resulting in no "benchmark system or anything changed because of this one analysis". Engineers do not believe that "there was anything really learned from this" seeing that "the original design group ... seems to be still perpetuating some of these designs that are going to be too complex and require extraordinary control plans and are not really going to be that cost efficient".

5.2.2   Sense of closure

Very much intertwined with the positive system outcomes is a sense of complete and satisfactory closure conveyed from the interviewees' narratives. All 31 successful problem-solving stories ended with everything "falling into place" and "making sense". The entire experience was looked back upon as being "fulfilling", "fascinating", or "something that you felt you had really achieved something". For example,

> "[The lessons learned are] still used as 'Gosh, this is a nice one, way to go!'" (Male, 15–20 years' experience)

> "You clearly achieved the goal and you solved the problem and you fixed it and you moved forward, and everyone is high-fiving and you move on." (Male, 20+ years' experience)

> "Finally ... we fixed that problem, and since then, this part has run reasonably well.... There was a series of mistakes [that] we had to go through in order to learn the right way[,] and to me, it exemplified everything about product development." (Male, 10–15 years' experience)

None of the 29 less successful stories, in contrast, conveyed a sense of closure that was satisfactory and that made sense. Instead, their endings were associated with unresolved questions, feelings of regret, and even downright frustration.

> "[Management] was not interested in solving problems so much ... I never really understood why they wouldn't give [funding], it wasn't a whole lot ... it was in the $100,000 range." (Female, 20+ years' experience)

> "Just that we didn't really close the [investigation] the way we wanted. Right now we consider it closed but in my mind, it's not ... we don't really have a clear story [here]." (Female, 15–20 years' experience)

> "I was never satisfied with this, because they never fixed the manufacturing process.... I wanted to go down there, I wanted to beat up the supplier that was shipping these ... and I wanted to beat up the plant that wasn't using this packaging. The stuff that had to get fixed didn't." (Male, 20+ years and just retired)

Hence, at least in our sample, successful and less successful problem-solving sagas are distinguished by the different ways in which they come to an end.

5.3 How problem-solving proceeds

The final area of notable discovery is the problem-solving process. Successful problem-solving is likely, more so than when unsuccessful, to be facilitated by a systematic and disciplined five-step process that culminates in knowledge distribution. Further, this process appears to correlate with the degree to which multiple stakeholders' views and beliefs eventually converged.

5.3.1 Rigorous five-step routines

Problem-solving routines as described by engineers differ in analytic rigour between successful and less successful problem-solving examples. The greater richness of detail presented in successful problem-solving dynamics brings out clearer images of the problem-solving process transition. Patterns in the data clustered around visibility of five phases in engineers' narratives: (1) problem discovery/communication; (2) problem investigation; (3) root cause discovery and identification of potential solutions; (4) selection and execution of the team solution – as opposed to measures dictated without the team's buy-in; and (5) active attempts to communicate the lesson at least within, and often across, the immediate workgroups. In all of the 31 successful examples, five elements of the problem-solving routines are prominent in their narratives. Although iterative, problem-solving steps pressed on to eventually clear these five stages one by one. A narrative typically starts out with a detailed description of how a failure event set the problem-solving in motion, e.g., "This part failed. Which allowed this to go in, and this to go in, and all of a sudden that component, they were saying, caused the problem". Subsequently, the problem-solving team was "meeting at 6:30 or 7:00 a.m. in the morning for several weeks to resolve the issue" and "would run through the list of issues . . . create a time and the team would go out and investigate those and . . . would be responsible to come back and report the status" The team "looked at the theory . . . went back to theory and then listed all the potential areas, which could provide this unexpected bad performance". Following that, the team "started attacking one by one and then basically identified the root cause". Once the team "had design solutions that [the team] thought might be good", it "would then send those off for the analysis [and] would evaluate them". The team discussed in terms of "what we can do in the short and long term and not getting . . . all sprung up [by] some feedback from executives", doing "a good job of hearing the feedback and not trying to rush in a solution . . . to really optimise this experience [by] taking an iterative approach". In the end, the team "did get to the root cause . . . defined it, came up with a way of making it better [and] documented it and carried the lesson learned by correcting the problem permanently by updating the standard". Often, the lesson "was communicated through meetings and . . . in our cross functional meeting within our engineering group or weekly or monthly meetings", as well as "[in] a corporate-wide steering meeting where we were exchanging knowledge that is part of the engineering centers around the world".

In contrast, all 29 less successful examples were missing one or more of these five elements. In other words, none of the 29 stories were "complete" with all five. Three of the 29 stories did not get past the first phase "because the data was hard to get . . . it was kind of like a phantom", correlating with general lack of interest in starting the investigation. Eight of the 29 cases stopped after the second stage, i.e., failing to find root causes, because the problem investigation was "closed" or "scrapped" before it had a chance to identify root causes. Eleven of the 29 stories did not move past the third stage or had an "incomplete" fourth phase. They reached the root cause but resulted in either no solution or "suboptimal" countermeasures. They are sometimes a consequence of a deliberate attempt to hide a "failure" of investigation that disclosed flaws in the product design that could not be corrected before a "deadline" or until the current iteration of the program was over, spanning two or so years that the problem had gone uncorrected. Six of the 29 examples were not able to go past the fourth phase. The problem-solving teams ultimately implemented optimal solutions, but the product engineers felt that the process was extremely "painful" and "resulted in no learning". No learning occurred because of "the politics of troubleshooting" which made the publicised solution "half true and … half lie" or the team "didn't want to tell anyone that they had messed up … or to let the other part of the business know what happened because they didn't want to admit a mistake … didn't want to dwell on this as a learning point". Sometimes the learning was only "partial", e.g., "beyond some manufacturing learning, any learning on the design side in terms of how we are going to avoid this in subsequent products" was missing. In one particular problem-solving story, the engineer reported that the investigation found the root cause but could not proceed to identification of solution; nevertheless, he transmitted his learning by company e-mail to another group within the firm engaged with development of a similar product.

5.3.2 Cognitive convergence

Finally, the last phenomenon that draws a distinct line between successful and less successful problem-solving stories is the manner in which the process dynamics of multiple viewpoints and perspectives play out. The extent to which these views converged, or stayed diverged, characterise successful versus less successful problem-solving efforts. Such cognitive convergence of stakeholder beliefs, actively managed by product engineers, is likely to facilitate successful resolution of problems in our sample. In all 31 examples of successful problem-solving, product engineers reported the ability to harness and leverage the cumulative input and knowledge of a wide variety of stakeholders (e.g., management, customers, suppliers, fellow engineers, etc). Engineers made purposeful and proactive efforts to manage relationships among relevant stakeholders by "getting the right people involved at the right level quickly to really put some brainpower into coming to a solution", "having everybody on the same page", and "learning to develop confidence in each other". Several stated that "the value of getting everybody together" was in itself a worthwhile learning that came out of the problem-solving because "if we had tried to do it in a small group or solve the issues with the two or three people that were directly involved, I don't think we ever would have resolved it". The learning can be taken "forward as part of being more effective team efforts in the future".

In 12 of the successful problem-solving examples, convergence was reported to be swiftly accomplished as all stakeholders were "willing to work together … and established a mission quickly". In the cases that involved the customers, "supplier-customer lines kind of disappeared there for a period" making it possible to have "a cross-functional in between the customer engineering team and their development, product development validation, safety team, and their plant team and ourselves". In the remaining 19 narratives, convergence was more protracted and difficult – even "heated" or "tumultuous at the beginning" – but was ultimately achieved. Engineers "went to manufacturing management and explained the importance of this DOE to resolve this issue in a timely manner and ... got their acceptance". They "also went to the manufacturing floor" by themselves and "worked with the operators [and] explained the importance" so that they "got priority". They pulled together "a quality department, a vibration analysis department, a CAE and ... PhDs and Fellows [who] were knocking heads" to "bridge the difference between their theories". They had "to be very patient" at first so as not to "have people build walls against each other, and especially against me". Their suppliers "were very reluctant on providing anything", but by forcing collaboration they "started having a team where they [= suppliers] were involved and where our customer was involved ... felt like we are on the same team". Engineers tried to help their suppliers by going "to their facility at night or on the weekend". At first, "it was very difficult, it required a few sacrifices, but, then, they started feeling like I was part of their team and then, they gave [me] a lot of help". Engineers and their immediate team members would clearly broadcast the "We are not here to lay blame, we are here to fix the problem" message "every time anyone tried to lay blame or point fingers". Once everyone agreed to "Let's get past that", personal interactions became more constructive to form "a very unique bonding experience", "partnership", and even a "symbiotic relationship".

When reporting less successful problem resolution, however, the same engineers often referenced the disinterest or confusion of stakeholders about the problem and emphasised the difficulty of aligning their divergent viewpoints. In 28 of the 29 less successful cases, convergence failed to occur. The problem-solving efforts were described as "definitely not a team atmosphere", "all around a bad situation", "one region telling another region how things should be", and "lots of blaming sessions". Inter-personal relationships tended to be "adversarial", "strained", or "almost hostile". "[A] little mistrust" between team members precluded "a complete agreement on when and where the problem happened, whether it was our problem or their problem" and "contributed to the friction" so the problem "wasn't really clearly owned". Stakeholders "didn't want to be doing what they were asked to do. Engineering didn't want to open up the tolerances, manufacturing didn't want to live with the way the situation was, the parts supplier didn't want to change his process." Some problems had a high enough visibility that involved "a lot higher management from multiple groups" ending up with "like three different vice presidents basically telling me what to do" yet their views "don't line up". Often times "emotions got in the way and halted addressing the problem", and the problem-solving "became an emotional issue". Because "people gave up" trying to synchronise their viewpoints, the problem "became a political and emotional problem ... a political and emotional and personal battle between people involved". Unable to cope with the dynamics of having "lots of different people involved", the team "ultimately scrapped [the problem] instead of solving the problem ... just eliminated, changed the design, changed the whole process to a different process".

6 ENGINEERING PROBLEM-SOLVING AND SHARED COGNITION

We now come back to cognitive convergence, one surprising characteristic of engineering problem-solving that our study has highlighted. As touched upon earlier, in our sample successful problem-solving efforts are distinguished from less successful cases by the higher frequency of cognitive synchronisation. This is not too surprising as shared beliefs have been reported to positively influence group performance in previous empirical studies (e.g., Baba *et al.*, 2004; Cannon & Edmondson, 2001; MacDuffie, 1997). Further, Baba *et al.*'s (2004) longitudinal case study performed on global work groups reports that shared cognition is likely to be achieved through two parallel paths: a straight convergence of multiple view points and overcoming of cognitive divergence. This dual-path convergence was also observed in our study. What is surprising, however, is that in their interview narratives the product engineers are found operating almost as "agents" of diverse interest groups (Eisenhardt, 1989a; Hill & Jones, 1992), actively working to establish common ground among various stakeholders and promoting congruence of their views through technical discourse (Suchman *et al.*, 1999). To better comprehend this somewhat unconventional image of "engineers", we first delve into the process by which knowledge is transformed through the problem-solving process. We then take a step back and put our findings in the perspective of system controls.

6.1 Ways of knowing and shared cognition

Continuing with the premise that engineering problem-solving is essentially a knowledge creation process, we take a constructionist view of knowledge, which associates knowledge development with social interactions within the organisation and thereby with the cognitive characters of individuals (Corti & Storto, 2000). This view implies that the collective learning starts with individuals' tacit knowledge, and that the manner in which these knowledge holders socialise greatly influences collective knowledge output. Because tacit knowledge is embodied in each individual and does not easily transfer to others (Edmondson *et al.*, 2003; Nonaka, 1994; Nonaka & Takeuchi, 1995), it needs to be transformed into a more portable form so it can be elevated to an organisational level. Tacit knowledge is closely tied to procedural knowledge or know-how; explicit knowledge, on the other hand, is often referred to as know-what or declarative knowledge (Edmondson *et al.*, 2003; Nonaka, 1994). Engineers use both theories and practical know-how to design, test, and manufacture products. Thus, both types of knowledge – explicit and tacit – have close affinity with engineering.

Knowledge resides at various ontological levels, from individuals to groups to organisations (Nonaka, 1994; Nonaka & Takeuchi, 1995). Engineering artefacts demonstrate that knowledge is not only transformable but is also transferrable. Drawings, specifications, physical prototypes, and finished products all have their genesis in someone's tacit knowledge that was eventually codified and broadcast (Edmondson *et al.*, 2003). Conversely, individual cognitive development is facilitated by processing declarative knowledge into one's procedural knowledge (Anderson, 1993), as virtually all practicing engineers have learned to operationalise textbook learning into skills that accomplish day-to-day tasks. Indeed, Nonaka (1994) argues that knowledge conversion is multi-directional, morphing from tacit to explicit and vice versa, as well as from tacit/explicit to amplified tacit/explicit as knowledge traverses to a wider audience.

Knowledge can also be viewed from the standpoint of "ways of knowing". Aristotle's (1934) view of knowledge has several dimensions, three of which continue to serve as useful points of reference for modern scholars of organisational studies (e.g., Baba *et al.*, 2004; Grint, 2007). They are *episteme*, *techné*, and *phronesis*. In engineering problem-solving, *episteme* and *techné* are analogous to declarative and procedural knowledge, respectively, with the former guiding "what" needs to happen in order to find the root cause of a part failure and the latter directing "how" to do it. The root cause investigation for a failed microelectronic component may start with a decision to pursue a cross sectioning and metallurgical analysis based on the lead engineer's knowledge (*episteme*). Subsequently, a technician trained to perform scanning electron microscopy and energy dispersive spectroscopy can implement the analysis (*techné*). However, it is the third dimension of knowledge, which scholars of organisational studies point out to be the most important yet often lacking in discourse (e.g., team dynamics by Baba *et al.*, 2004 and leadership study by Grint, 2007), that completes the knowledge triad. That is *phronesis*, practical wisdom or prudence (Aristotle, 1934), that helps the engineering problem-solving team know "whether" to acquire more information or try a different technique when the initial diagnostics to test for the hypothesised root cause turn out inconclusive.

While Aristotle makes it clear that all three are critical elements of one's intellectual virtues, it is *phronesis* that ultimately sets the direction for the next raft of tasks in problem-solving (Grint, 2007) and eventually brings the process to a closure that "makes sense" to all who are involved. *Phronesis* stands above *episteme* and *techné*, encompassing the characteristics of both, but different from them. *Phronesis* evokes the intellectual power of *episteme* but not in abstraction; *phronesis* is situation-specific like *techné* but goes beyond being able to perform the task on hand (Grint, 2007). *Phronesis* is the reservoir of interpretive power that delivers the most appropriate and reasonable judgment in the context in which the problem is situated. *Phronesis* is essentially about value judgment, which cannot be reduced or abstracted into a set of problem-solving procedures (Grint, 2007), and this concept may be vital in understanding cognitive convergence in engineering problem-solving.

On an individual basis, *phronesis* may be equated to Senge's (2006) "personal mastery" that "goes beyond competence and skills, though it is grounded in competence and skills" (p. 131), akin to the artistry of champion ice skaters and prima ballerinas. Their "skills have been developed through years of diligent training, yet the ability to execute their artistry with such ease and seeming effortlessness" (Senge, 2006, p. 151) transcends. Whereas amateurs may equally possess such potential, it is the "deliberate practice" for acquiring expert performance (Ericsson, 2003; Ericsson *et al.*, 1993) that cultivates one's potential with discipline.

In organised activities such as engineering problem-solving, however, *phronesis* becomes more of a collective discipline. It is about amassing various interpretive schemes and unifying them into the best-fitting solution to the given situation. "In today's rapidly changing environments, the complexity of problems requires solutions that combine the knowledge, efforts, and abilities of people with diverse perspective" (Hargadon & Bechky, 2006, p. 484); and in modern engineering, complexity is the norm rather than the exception. Presented below is a sample quote from our field study. It is the voice of a senior engineer tasked to resolve a noise-vibration-harshness (NVH) problem in an automotive powertrain system:

"I forced a collaboration because, originally, I would have had to accept one [view] and reject the other. How could two people get a PhD and one be very wrong and one be very right? So, I think there was more of a misunderstanding.... Through collaboration I was able to use brainpower from a wide range of intelligence.... Everybody brought a lot to the table, so I learned a lot.... That would have been missing if I handled it alone or only chose one side. And once they got beyond the conflict stage, they were actually more than willing, I got a lot of firepower from departments that weren't even on my budget that might not have been available to me [had I chosen sides]." (Male, 20+ years' experience)

As the quote succinctly demonstrates, this example is a clear case of cognitive convergence after a protracted period of divergence. The author of this quote had to exert a considerable effort to get his team members, all of whom had come with years of NVH experience yet with divergent opinions, to view the problem in what he labelled as the "same boundary conditions". His team was an assemblage of high-power *episteme* and *techné* (PhDs and highly experienced engineers), but these knowledge holders were initially viewing the problem from disparate angles based on their respective past experiences with NVH. Finally, through his persistent effort, every member of the team came to agree that the culprit of their NVH problem was not one particular component, as originally presumed to be, but the interactions of several components within the assembly. The end of this story is that the team not only fixed the immediate problem but also developed a new technology for designing and testing future-generation assemblies.

Phronesis – the wisdom to prudently combine "insights from theories and research that draw upon diverse premises" (Nonaka *et al.*, 2006, p. 1196) – does not develop overnight, let alone collective wisdom that seamlessly combines multiple perspectives. Shared cognition, especially in the dynamically shifting and contingency-ridden world of engineering, "means more than simply exchanging declarative and procedural knowledge. It means suspending our own judgment as we learn the cultural logic and rationality of others' divergent beliefs and values, while also allowing those others to call our own beliefs and values into question as they learn about us" (Baba *et al.*, 2004, p. 583).

Huber (1991) proposes that "more learning has occurred when more of the organisation's units understand the nature of the various interpretations held by other units" (p. 102), and that is essentially what we observed in our study. Engineering is a highly context-sensitive discipline (Goldman, 2004) and so, by their nature, interpretations are everything. Getting together a group of experts to solve a problem, however, is only the first step. Their views must be aligned with the context in which the product is applied, used, and interpreted. Goldman (2004) argues, "Engineering is contingent, constrained by dictated value judgments and highly particular. Its problem solutions are context sensitive, pluralistic, subject to uncertainty, subject to change over time and action directed" (p. 163). Engineering practice, from the epistemological perspective, is essentially about developing useful interpretive schemes, a form of collective wisdom.

6.2 Shared cognition in contexts

We now turn our attention to organisational factors that enhance shared cognition in engineering problem-solving. The engineering problem-solving environment is a socio-technical system, so the quality of engineering knowledge creation depends heavily on the social interactions that take place within a cultural milieu (Corti & Storto, 2000; Dodgson, 1993; Nonaka, 1994; Nonaka & Takeuchi, 1995; Simon, 1991). Just as learning "does not mean acquiring more information, but expanding the ability to produce the results we truly want in life" (Senge, 2006, p. 132), a single unified *phronesis* (Baba *et al.*, 2004) is not likely to be achieved solely with logistic (e.g., providing meeting rooms and technology) or infrastructural (e.g., information system database) considerations.

Following Leveson's (2011) application of system controls theory to a human socio-technical structure, the role of actors engaged in engineering problem-solving may be conceptualised as a controller that dynamically controls the problem-solving process using a set of algorithms. In an embedded control system, the microprocessor is where these algorithms would reside. We make an analogy that the role of microprocessor equates to that of mental models, shared by people on the problem-solving team. As is the case of a microprocessor, mental models need to be kept healthy for an optimal system performance. In this regard, it is crucial to understand a unique property of human mental models in order to safeguard the power of mental logic. Humans are boundedly rational, i.e., they can be rational only to the extent that their surrounding environment allows them to be (Leveson, 2011; Senge, 2006; Simon, 1991).

Bounded rationality places "limits upon the ability of human beings to adapt optimally, or even satisfactorily, to complex environments" (Simon, 1991, p. 132). The implication of this theory is that people rarely attempt to go beyond the perceived system boundaries so "the learning process is generally conservative and sustains existing structures of belief" (Dodgson, 1993, p. 385). In engineering problem-solving, therefore, the root cause analysis must be framed in such a manner as to help the members overcome their own mental boundaries. If contextual factors are overly restrictive (for example, forcing the problem into pre-set categories and/or restricting the repertoire of corrective actions from the beginning) this will further inhibit the problem-solvers' mental frontiers and in turn cause their investigation scope to shrink (MacDuffie, 1997). Bounded rationality is closely tied to sensemaking (Weick, 1979, 1993; Weick *et al.*, 2005). The existence of a threshold beyond which is off limits to human rationality essentially implies that people will not do what does not make sense to them (Leveson, 2011; Senge, 2006). Further, when the environment changes, as a socio-technical system does because it tends to be dynamic, what made sense at one time may no longer do so (Leveson, 2011). Consequently, as Leveson (2011) argues from a safety management point of view, better learning is achieved by framing accidents not as events stemming from operator errors but as 'sensemaking' events, that is to say we change "our emphasis ... from what they did wrong to why it made sense for them to act the way they did" (Leveson, 2011, p. 39).

In stark contrast to many successful problem-solving stories such as the NVH case presented earlier, there were also a host of examples presented by the interviewed engineers as being "less effective" or even "how not to do it". One such story involved a sealant, the odour of which was found acceptable by the European team members but objectionable to their American counterparts within the same firm. Unable to unify the stakeholders' divergent viewpoints on the disposition of the sealant, the team gave up its autonomy and relegated the decision to the senior management.

The situation-dependency of engineering practice also implies that the original meaning associated with an artefact can change when transferred to another context

(Brannen, 2004). Such re-contextualisation is a dynamic process as it is propelled by sensemaking that is context-bound. If the context is not shared by all stakeholders, meanings can easily shift in transfer (Brannen, 2004). Whereas in the NVH case the actors were eventually able to put their respective knowledge and experience into the present context and so jointly made sense of the problem, this sealant case somehow fell short of such unification of interpretations. Akin to a host of environmental measures that designers of an embedded control system implement to protect the health of its controller, especially its microprocessor, an appropriate sensemaking forum where the new context was better understood and opposing viewpoints worked out should have existed but did not.

We now turn to the discussion of *ba*, a deliberately constructed forum where such sensemaking activities can take place.

7   BUILDING BA OF ENGINEERING PROBLEM-SOLVING

7.1   Nature of work and sensemaking

Our world is connected with a set of forces, forming relational structures such as the electromagnetic field. Society and cultures, in this sense, are networks of clustered relationships (Goldman, 2004). Following this logic, organisations are the reality that we created, and how they function goes beyond the structures documented in organisation charts (Orr, 1998; Senge, 2006; Weick, 1979). An enacted world is a result of daily sensemaking, which is essentially a form of information processing to deal with equivocality in the observed data, thus reaching or maintaining a desired system state (Weick *et al.*, 2005). From this enactment perspective, engineering problem-solving efforts are a form of organised sensemaking, in which product engineers try to "make sense of equivocal inputs and enact this sense back into the world to make that world more orderly" (Weick *et al.*, 2005, p. 410).

Technical work, such as engineering, also involves sensemaking. Theories, techniques, and procedures that can be abstracted into a context-free schema are a very small part of engineering practice (Goldman, 2004; Jonassen *et al.*, 2006; Trevelyan, 2007). Much of what goes on in day-to-day engineering is grounded in particular contexts and involves the development of situated logic. Sensemaking occurs by default in engineering because there is often a great distance between what is prescribed and the nature of technical work in practice (Orr, 1998). There is always a cognitive distance between the designer's mental model and the way the technical crew interfaces with the designed equipment (Leveson, 2011). One of us who is a practicing engineer still remembers a lesson taught by a senior quality engineer from her early automotive career: that you always "play" with the measuring gauge yourself before shipping it to your plant because there always are "idiosyncrasies" in it even though it is your design and the paperwork shows it "passed the gauge R & R and meets your specifications". Further, time constantly advances forward, so there always is a temporal distance between implementations of the previous and the current artefacts. In a nutshell, there are rarely completely duplicative contexts in engineering, so engineers must "solve problems by remembering similar cases and applying the lessons learned from those cases to the new one" (Jonassen *et al.*, 2006, p. 140). That is sensemaking.

The cognitive and temporal distances necessitate improvisation before a set of technical routines can be applied, i.e., engineering sensemaking cannot be avoided. Although in our narratives of engineering problem-solving the engineers mentioned the use of such structured problem-solving methodologies as 8D, Kepner-Tregoe, and Six Sigma, it was clear that moments of direction-turning revelation came for them while actively interacting with co-team members, irrespective of the methodologies being used – especially in problem-solving that was presented as "successful" examples. Because 8D and other methods show up in both "successful" and "less successful" narratives, they are likely to be no more than a guiding framework that helps to establish a structure for the group activities.

What, then, gives rise to superior team performance that not only resolves the problem but takes the existing premises to a higher level as seen in many of the 'successful' problem-solving stories?

7.2 Shared experiential space

While sensemaking is improvisational, it can also have elements of 'coordination'. In the same way that "[t]he championship sports team and great jazz ensembles provide metaphors for acting in spontaneous yet coordinated ways" (Senge, 2006, p. 219), a problem-solving framework can be engineered to promote healthy and constructive interactions among stakeholders. Such a context is one in which all players trust each other to work together in ways that complement each other's contribution, or what Senge (2006) calls a "collective discipline". To build such an environment, we now introduce the concept of "*ba*".

Closely related to the concept of enacted environment, *ba* is a Japanese word roughly translated as "place, space, or forum". A Japanese philosopher, Kitaro Nishida, originally proposed the concept, which was further developed by Hiroshi Shimizu (Nonaka & Konno, 1998). Scholars have since adapted *ba* to various organisational contexts. For example, Itami's (2010a, 2010b) *ba* is a spatial structure that facilitates interconnectedness which, he argues, is indispensable for understanding a broad spectrum of organisational phenomena to effectively manage an enterprise. Nonaka and colleagues (2000) conceptualise *ba* as an integral component of their model of knowledge creation. They define *ba* as a "shared context in motion for knowledge" (Nonaka *et al.*, 2000, p. 13) and systematically classify it into four different categories based on the nature of personal encounter that takes place in the shared space (face-to-face or virtual) and the mode of knowledge exchange (tacit or explicit). Yamaguchi's (2006) framework for "paradigm-disruptive" innovation entails a "field of resonance", which he portrays as a special type of *ba* in which engineers and decision makers intimately share technical tacit knowledge. Put together, the core ideas of *ba* are that (1) the shared space can be, but need not be, physical; (2) it must be purposeful; and (3) its structure should enhance meaningful interactions among members. Further, because it is a forum of sensemaking, a *ba* must engender and encourage spontaneous exchange. This last point implies that merely providing a meeting room or an intranet forum for potential *ba* participants is not sufficient for effective *ba* operation; neither is prescriptively following a set of abstracted methodologies.

A *ba* of engineering problem-solving may be envisioned as an improvisational theater in which engineers deftly operate as "agents" of diverse interest groups (Eisenhardt,

1989a; Hill & Jones, 1992) to unify different personal visions. "One of the most reliable indicators of a team that is continually learning is the visible conflict of ideas" (Senge, 2006, p. 232) and so a fully functioning *ba* is likely to first witness the exchange of divergent viewpoints followed by their gradual convergence – as was the case for the NVH example and many other successful stories provided by the interviewees.

Our problem-solving stories chronicled formation of, as well as absence of, *ba* where collaboration took place. Unsuccessful problem-solving was lacking *ba*. In several cases, despite existence of expertise and resources, they could not be pulled together to form a shared space for constructive dialog. In one such example, the lead engineer narrates:

> "We should get expert[s] involved. In the first [case] expert support did make a difference when we faced difficulties ... but this [less successful] one, we drove it pretty hard but we didn't really have a chance to really talk to the bearing supplier and also our internal expert. Well we had a review with the expert [but] didn't really get him on board every time [we needed him] in this investigation process." (Female, 15–20 years' experience)

A successful problem resolution is a collective *phronesis*, an amalgamation of everyone's evaluative knowledge. A set of beliefs that found this evaluative knowledge may be key to understanding divergence and convergence of views (Baba *et al.*, 2004). Beliefs are formed through experience, and so *phronesis* – unlike theories that can be taught in lectures or techniques acquired through practice – can only be secured through the experience of applying them in specific contexts (Grint, 2007). Baba *et al.*'s (2004) empirical study suggests the role of parallel or similar experiences in a common context as potentially a driver for gaining a common understanding from "distributed" team members spread across geographical and cultural divides. The role of experience in engineering problem-solving, especially shared experience, may similarly facilitate recalibration of members' past understanding to fit the new context in which the problem is being investigated. In the problem-solving stories reported as "successful", engineers recounted "going to manufacturing plants to make face-to-face contact", visiting customer sites, or being engaged in experiments.

We, hence, posit that *ba* of engineering problem-solving is a "shared experiential space".

7.3   Forming *ba* of engineering problem-solving

How do we, then, form a *ba* that encourages every stakeholder to get "on board"?

Barrett's (1998) jazz band metaphor offers seven principles that help create "contexts that require an improvisatory mindset [for] interdisciplinary project teams formed to address a specific problem" (Barrett, 1998, p. 617). We project these principles onto the dynamics of engineering practice and conjecture how they may help to construct a highly effective *ba* for successful engineering problem-solving.

1   *Provocative competence for pattern breaking:* Perhaps the most significant of the seven principles is "provocative competence", which is about the deliberate

interruption of habit patterns to reach something new. This attribute is what gives jazz music a dialectic tension, in a sense, between discipline and free-flow. It is the "tension between constancy and change" (Fiol & Lyles, 1985, p. 805) or Nonaka's (1988, 1991, 1994) "creative chaos" that serves as a trigger for self-reflection and gives impetus for challenging the status quo. Such a coordinated improvisation occurs in an environment that harbours the "creative tension" that is a key element of Senge's (2006) systems approach, as well as a characteristic of Toyota's lean production system as described by Womack *et al.* (1990). In our study, as described earlier, one of our findings was a "controlled urgency", a situational characteristic akin to a "high-velocity" environment that is "in the continuously unstable state" of the "edge of chaos" (Eisenhardt & Martin, 2000, p. 1113). Successful problem-solving stories recounted by engineers pointed to an environmental characteristic that was neither inertial nor chaotic but was well controlled by a positive sense of urgency projected by higher authority (e.g., management or customers).

A creative tension is essentially a juxtaposition of vision and the current reality, by holding steady to the vision and pulling reality to it (Senge, 2006). The pulling is what sets the direction and is the job of leadership. Leadership vision in problem-solving provides goal clarification, guiding principles, and commitment to seeing that the problem is solved. Nonaka and Takeuchi (1995) underscore this point in their knowledge-creation construct, "intention", which is tied to corporate vision and ultimately provides the yardstick with which to measure the knowledge output. Literature on learning organisation also links leadership to the notion of "wisdom" (Bierly *et al.*, 2000; Boal & Hooijberg, 2001; Carroll & Edmondson, 2002; Grint, 2007), essential for guiding learning with clear purpose. The role of leadership vision to provide goal clarification, guiding principles, and support was demonstrated to be crucial in our engineering problem-solving narratives. In many of our successful problem-solving stories, management or customers explicitly demonstrated their commitment to seeing that the problem was solved. Cultivation of "provocative competence" ultimately relies on the effectiveness of *ba* leadership (Barrett, 1998).

2   *Error as a source of learning:* An accomplished jazz band makes playing "wrong notes" look intentional because, rather than treat it as an error, "often jazz musicians ... repeat it, amplify it, develop it further until it becomes a new pattern" (Barrett, 1998, p. 610). Engineering problem-solving is normally instigated by a performance failure. While mistakes and errors are opportune triggers for problem-solving and thereby set a launch pad for learning (Scott & Vessey, 2000; Tjosvold *et al.*, 2004) – more so than the experience of success (Baumard & Starbuck, 2005) – impediments exist that can inhibit learning from failure. Potential barriers to problem-solving and learning are embedded in both the technical and social systems and will need to be addressed through measures such as information systems, systematic reviews, training, availability of needed expertise, and deliberate postulation of "failure as opportunity" to ensure psychological safety (Cannon & Edmondson, 2005). Empirical evidence demonstrates a link between an effective root cause analysis and open, fact-based management (Carroll *et al.*, 2002; Handley, 2000; MacDuffie, 1997; Uhlfelder, 2000). Specifically, resources and psychological safety ensured by management are likely to lead to trust and

confidence among team members, which in turn enable input of quality data and superior analysis.

3   *Minimal structures that allow maximum flexibility:* A minimum unconstrained structure for a coordinate project enhances temporal updating from start to finish, "like the chord changes of a song, and increases the likelihood that people can achieve a successful joint awareness throughout the life of the project" (Barrett, 1998, p. 613). The unconstrained problem-solving environment noted in our interview narratives is in sync with the concept of underspecified organisational structure, which is known to promote flexibility of action (Barrett, 1998; Dodgson, 1993; Levinthal & Rerup, 2006), to facilitate rapid information flow (Nonaka, 1994; Nonaka & Takeuchi, 1995), and to more ably accommodate "shifts of beliefs and actions" (Fiol & Lyles, 1985, p. 805). Virtually all successful problem-solving narratives in the qualitative study involved self-organizing or cross-functional work groups of varying sizes – exemplary of Nonaka's autonomy concept (Nonaka & Takeuchi, 1995).

4   *Distributed task:* Jazz, in the end, is not an individual but a social accomplishment achieved through "continual negotiation toward dynamic synchronisation [through a] … remarkable degree of empathic competence, a mutual orientation to one another's unfolding" (Barrett, 1998, p. 613). Grant (1996) argues that "cross-functional product development teams, TQM (Total Quality Management), and organisational change programs … can be viewed as attempts to change organisational structure and processes to achieve better integration across broad spectra of specialised knowledge" (p. 384). In our study, successful problem stories capitalised on requisite variety in the available resources, an appropriate mix of technical expertise for a given context that is needed to cope with the complexity of problems (Nonaka & Takeuchi, 1995; Weick, 1979). Further, skill sets from the available mix of resources were leveraged through effective "interactions across individuals who each possess diverse and different knowledge structures that will augment the organisation's capacity for making novel linkages and associations … beyond what any one individual can achieve" (Cohen & Levinthal, 1990, p. 133).

5   *Retrospective sensemaking:* Action is always a tad faster than reflection (Weick, 1988), and so we are always having to make sense of what has just taken place. Not relying on a pre-developed plan, jazz musicians keep their music from going totally out of order by reflecting while acting (Barrett, 1998). As these are people of high artistic mastery, however, they approach this sensemaking with discipline rather than haphazardly exploiting the "level of rapport between their normal awareness and … the subconscious" (Senge, 2006, p. 151). Successful problem-solving narratives depict a systematic and disciplined process that moves with clear cadence from problem communication to problem resolution leading to subsequent knowledge distribution. The investigation process asked many "what else?" questions and traveled to and fro between theories and experimental results. Above all, successful problem-solving efforts resulted in "team solution" (rather than management- or customer-dictated measures without the team's buy-in) that made sense to all.

6   *Communities of practice:* Jazz players "hang out" in jazz communities – that is essentially how they learn to play, i.e., by having a membership in a community

of practice (Barrett, 1998). Organisations are a form of community of practice (Orr, 1998, 2006), engineering organisations in particular because of their high concentration of specialised knowledge and skills. Jazz music creation is a form of team learning, which is "the process of aligning and developing the capacity of a team to create the results its members truly desire" (Senge, 2006, p. 218). All of our problem-solving narratives that were presented as 'successful' have endings that entailed at least some level of intentional knowledge distribution beyond individual learning. From the viewpoint of team learning as a quest for a single unified *phronesis*, true learning will not be achieved by transmitting de-contextualised information back and forth. In order for useful knowledge to materialise from engineering problem-solving, there needs to be a shared context in which collective value judgments occur. The interviewees' problem-solving stories illuminated many pathways through which the knowledge possessed by the stakeholders was leveraged to solve problems. The sharing frameworks (such as discussion forums, special meetings, and information technology networks) chronicled by the engineers are precisely this platform of value synchronisation.

7   *Leading and supporting:* Jazz band members take turns soloing and supporting, thereby rotating leadership (Barrett, 1998). This "deceptively simple practice of taking turns creates a mutuality structure that guarantees participation ... shared ownership without insisting on consensus" (Barrett, 1998, p. 617). In a sense, this deliberate role swapping arrangement is similar to the "parallel or similar experience" concept discussed earlier. As we already touched upon, actors in our successful problem-solving stories were often found doing things together, experimenting and having dialog often at the site where the problem was believed to have generated. Such sharing frameworks could also be a platform to develop "redundancy" that "connects individuals and the organisation through information, which converges rather than diffuses" (Nonaka, 1994, p. 29). From a sensemaking perspective, redundancy or information overlap creates a structure of "interconnections" among actions and "relations and networks [that] determine outcomes" (Weick & Putnam, 2006, p. 285). In other words, deliberate attempts to create overlapping knowledge help to increase inter-subjectivity among team members.

Engineering problem-solving is better facilitated by organisational conditions that promote cross-disciplinary information sharing and teamwork to drive investigation. Such a shared experiential space, or *ba*, should be deliberately created rather than being left to haphazard chances or accidental opportunities. An effective *ba* is enabled through an interplay of dialectic forces, such as those that prepare jazz musicians to be spontaneous (Barrett, 1998). A *ba* of engineering problem-solving is where rich epistemic exchange occurs, out of which new insights about the problem emerge. If the *ba* functions as a virtuoso jazz band, the experience of deriving such insights will be transcendental (Barrett, 1998). *Ba* is a place for joint sensemaking so by its nature is dynamic – and fragile. Cognitive convergence is achieved by a series of interactions, each of which introduces players to new perspectives to ponder. Such a moment of "reflective reframing" (Hargadon & Bechky, 2006) is "a rare and fleeting phenomenon even in the most creative of organisations ... and is not easy to predict when [it] might happen" (p. 494) and so it is valuable.

8   RECOMMENDATIONS FOR THEORY AND PRACTICE

The state of our research is still exploratory in nature. As we contemplate future studies to continue our course, we pause and offer tentative recommendations for theory and future research, as well as for practice for managing engineering problem-solving efforts from the perspective of *ba*.

8.1   For theory and research

Our study has, thus, highlighted the sociality of the engineering knowledge construction process stemming from problem investigation. From this point, there are a number of possibilities that future research can take. Within the framework of *ba* – an empirically under-explored construct (Nonaka *et al.*, 2006) – we present a few of them.

At a micro level, the knowledge creation routines can be explored in further depth to really understand the "how" of collective learning. Adams and Forin's chapter in this book illuminates what "working together across disciplines" really means and entails. Their four-tier categorisation of cross-disciplinary interactions provides a useful clue for judging the progression of cognitive synchronisation among engineering actors. Further, as these actors working together learn from others and eventually attain higher learning through difference (Adams & Forin from this book), do they "traverse" or "transcend" their knowledge differences? Majchrzak *et al.*'s (2012) study deftly addresses this point, so further exploration of *ba* dynamics using their findings should benefit theory.

At a meso or macro level, the actor network theory (Latour, 2005) is a possible application for better understanding the effects of bounded rationality on the socio-technical network configuration. Engineers trying to solve technical problems constantly build and test hypotheses as forms of argument. It is through complex technical discourse that a lead engineer tries to convince other actors and solicit their "enrolment" (Harty, 2008). How their aligned interests become an actor-network, and, most importantly, leveraging its heterogeneity (Law, 1992) to form a competitive advantage will be of great interest to engineering firms. Along this line, a deliberate construction of contextual settings that harbour power for effective problem-solving can benefit from an analysis of engineering practice at large. An environment that facilitates collective sensemaking in problem-solving efforts may be put in Bourdieusian perspective of organisation as a field (Emirbayer & Johnson, 2008; Swartz, 2008). That is to say, engineering is a cultural milieu that is situated in a larger field, configured in power clusters. Given that problem-solving stakeholders (management, customers and suppliers, for example, as well as engineers) reside in various parts of the field, each entitled to a varying intensity of "capital", what routines are at play when they come together? What is the underlying logic of the field that is closely connected to its "habitus"? Such characterisation may shed further insight into a strategy to create a "field of resonance" (Yamaguchi, 2006) capable of yielding breakthrough solutions.

8.2 For practice

Our recommendations for practicing engineers are as follows:

- Actively engage stakeholders – fellow engineers, customers, suppliers, or whatever the case may be – in the task of finding a solution.
- Work to create bridging skills to effectively collaborate with diverse talent, such as engineers from other disciplines or cultures. Doing so will require suspending judgment from time to time and being open to a different set of logics that may be at play in the exchange.
- Learn to develop the "same language" by openly asking questions.

Only after understanding differences in what people know and how they communicate, can one start to constructively challenge others' assumptions. These skills are particularly important for engineers who are in a position to facilitate or lead solution-finding for a problem.

By extension, it is the manager's duty to enable cultivation of such skills. First, he or she needs to create "conditions that encourage members to bring mindfulness to their task that allows them to imagine alternative possibilities heretofore unthinkable" (Barrett, 1998, p. 610). Similar to managing a jazz ensemble, such an engineering *ba* must be carefully nurtured. New perspectives tend to form through unanticipated connections, often facilitated by personal interactions (Hargadon & Bechky, 2006). Such interactions, however, are difficult to predict beforehand and can only be monitored and steered as they progress. Engineering practice is dynamic, both contextually and temporally, thus it needs care and attention as it develops – which is always a challenge to project.

Further, with specific reference to a *ba* to facilitate engineering problem-solving, the process needs to be managed *in situ*. Merely handing out a set of prescriptive steps to follow – be they 8D, Five Whys, or Six Sigma – is not adequate. An effective root cause analysis is more than just following a set of routines. While problem-solving methodologies can facilitate a basic procedural structure for those engaged in the problem investigation, they alone will not facilitate constructive interactions and knowledge exchange. Cognitive synchronisation is a result of on-going reframing to evaluate the problem in new contexts as they surface. Encouragement of healthy technical exchange by on-going management participation in the dialog is one of the ways to keep *ba* productive.

By the same token, the management of engineering problem-solving *ba* also needs to look beyond prescriptive documentation and computer infrastructure for knowledge sharing and collaboration. Computers are not a substitute for this socialisation. While helpful as a repository of past problems and solutions, the IT database does not perform sensemaking for the user since it can neither reflect on new questions nor reframe a past problem in new contexts (Brannen, 2004). How people use organisational metrics or intranet collaboration sites, and what happens when they do, is what matters (Senge, 2006) – and needs watching.

Last, but not least, while we emphasised the importance of evaluative knowledge in problem-solving, we must not forget that this knowledge is grounded on both the declarative and procedural knowledge. At the end of each interview, prompted by the question of how a company could enhance organisational learning, the interviewee often mentioned in-house training classes and mentoring programs to enhance the technical knowledge and skill levels of engineers. A few younger engineers complained about lack of mentoring and coaching in their workplaces, which, according to them, was once a norm but is rapidly disappearing in the companies' attempts to rationalise workforce structure and protocol. This criticality of *episteme* and *techné* as the two legs upon which *phronesis* stands is consistent with the findings from Trevelyan's (2007) empirical study that while non-technical, inter-personal coordination is a major aspect of engineering practice, its effectiveness relies critically on the engineer's technical expertise. Therefore, it is in the best interest of engineering organisations to make available venues for continuous skill improvement.

REFERENCES

Anderson, J. R. (1993) Problem-solving and learning. *American Psychologist*, 48 (1), 35–44.

Aristotle. (1934) *The Nicomachean Ethics* (H. Rackham, Trans.). Cambridge, MA, Harvard University Press.

Baba, M. L., Gluesing, J., Ratner, H. & Wagner, K. H. (2004) The context of knowing: Natural history of a globally distributed team. *Journal of Organizational Behaviour*, 25 (5), 547–587.

Barrett, F. J. (1998) Coda: Creativity and improvisation in jazz and organizations: Implications for organizational learning. *Organization Science*, 9 (5), 605–622.

Bartel, C. A. & Garud, R. (2009) The role of narratives in sustaining organizational innovation. *Organization Science*, 20 (1), 107–117.

Baumard, P. & Starbuck, W. H. (2005) Learning from failures: Why it may not happen. *Long Range Planning*, 38(3), 281–298.

Bierly, P. E. I., Kessler, E. H. & Christensen, E. W. (2000) Organizational learning, knowledge and wisdom. *Journal of Organizational Change Management*, 13 (6), 595–618.

Boal, K. B. & Hooijberg, R. (2001) Strategic leadership research: Moving on. *Leadership Quarterly*, 11 (4), 515–549.

Brannen, M. Y. (2004) When Mickey loses face: Recontextualization, semantic fit, and the semiotics of foreignness. *Academy of Management Review*, 29 (4), 593–616.

Cannon, M. D. & Edmondson, A. C. (2001) Confronting failure: antecedents and consequences of shared beliefs about failure in organizational work groups. *Journal of Organizational Behaviour*, 22 (2), 161–177.

Cannon, M. D. & Edmondson, A. C. (2005) Failure to learn and learning to fail (intelligently): How great organizations put failure to work to innovate and improve. *Long Range Planning*, 38 (3), 299–319.

Carroll, J. S. & Edmondson, A. C. (2002) Leading organisational learning in health care. *Quality and Safety in Health Care*, 11 (1), 51–56.

Carroll, J. S., Rudolph, J. W. & Hatakenaka, S. (2002) Lessons learned from non-medical industries: root cause analysis as culture change at a chemical plant. *Quality and Safety in Health Care*, 11 (3), 266–269.

Cohen, W. M. & Levinthal, D. A. (1990) Absorptive capacity: A new perspective on learning and innovation. *Administrative Science Quarterly*, 35 (1), 128–152.

Corbin, J. & Strauss, A. (1990) Grounded theory research: Procedures, canons, and evaluative criteria. *Qualitative Sociology*, 13 (1), 3–21.

Corti, E. & Storto, C. L. (2000) Knowledge creation in small manufacturing firms during product innovation: An empirical analysis of cause-effect relationships among its determinants. *Enterprise and Innovation Management Studies*, 1 (3), 245–263.

Crosby, P. B. (1979) *Quality is Free: The Art of Making Quality Certain*. New York, NY, McGraw-Hill.

Deming, W. E. (1982) *Out of the Crisis*. Cambridge, MA, MIT.

Dodgson, M. (1993) Organizational Learning: A Review of Some Literatures. *Organization Studies*, 14 (3), 375–394.

Douglas, D. (2003) Grounded theories of management: a methodological review. *Management Research News*, 26 (5), 44–52.

Edmondson, A. C., Winslow, A. B., Bohmer, R. M. J. & Pisano, G. P. (2003) Learning how and learning what: Effects of tacit and codified knowledge on performance improvement following technology adoption. *Decision Sciences*, 34 (2), 197–223.

Eisenhardt, K. M. (1989a) Agency theory: An assessment and review. *Academy of Management Review*, 14 (1), 57–74.

Eisenhardt, K. M. (1989b) Building theories from case study research. *Academy of Management Review*, 14 (4), 532–550.

Eisenhardt, K. M. & Martin, J. A. (2000) Dynamic capabilities: What are they? *Strategic Management Journal*, 21 (10/11), 1105–1121.

Emirbayer, M. & Johnson, V. (2008) Bourdieu and organizational analysis. *Theory and Society*, 37, 1–44.

Ericsson, K. A. (2003) The acquisition of expert performance as problem-solving: Construction and modification of mediating mechanisms through deliberate practice. In: Davidson, J. E. & Sternberg, R. J. (eds.) *The psychology of problem-solving*. New York, NY, Cambridge University Press. pp. 31–83.

Ericsson, K. A., Krampe, R. T. & Tesch-Römer, C. (1993) The role of deliberate practice in the acquisition of expert performance. *Psychological Review*, 100 (3), 363–406.

Fiol, C. M. & Lyles, M. A. (1985) Organizational learning. *Academy of Management Review*, 10 (4), 803–813.

Goldman, S. L. (2004) Why we need a philosophy of engineering: A work in progress. *Interdisciplinary Science Reviews*, 29 (2), 163–176.

Grant, R. M. (1996) Prospering in dynamically-competitive environments: Oranizational capability as knowledge integration. *Organization Science*, 7 (4), 375–387.

Grint, K. (2007) Learning to lead: Can Aristotle help us find the road to wisdom? *Leadership*, 3 (2), 231–246.

Handley, C. C. (2000) Quality improvement through root cause analysis. *Hospital Materiel Management Quarterly*, 21 (4), 74–78.

Hargadon, A. B. & Bechky, B. A. (2006) When collections of creatives become creative collectives: A field study of problem-solving at work. *Organization Science*, 17 (4), 484–500.

Harty, C. (2008) Implementing innovation in construction: Contexts, relative boundedness and actor-network theory. *Construction Management and Economics*, 26, 1029–1041.

Hill, C. W. L. & Jones, T. M. (1992) Stakheolder-agency theory. *Journal of Management Studies*, 29 (2), 131–154.

Huber, G. P. (1991) Organizational learning: The contributing processes and the literatures. *Organization Science*, 2 (1), 88–115.

Itabashi-Campbell, R., Gluesing, J. & Perelli, S. (2012) Mindfulness and product failure management: An engineering epistemology. *International Journal of Quality & Reliability Management*, 29 (6), 642–665.

Itabashi-Campbell, R., Perelli, S. & Gluesing, J. (2011) Engineering problem-solving and knowledge creation: An epistemological perspective. In: IEEE *International Technology*

Management Conference (ITMC) 2011, 27–30 June 2011, San Jose, CA, USA. IEEE International. pp. 777–789.

Itami, H. (2010a) 場のマネージメント実践技術 [*Application techniques for management of ba*]. Tokyo, Japan, Toyo Keizai.

Itami, H. (2010b) 場の論理とマネージメント [*Theory and Management of Ba*] (5th ed.) Tokyo, Japan, Toyo Keizai.

Jonassen, D., Strobel, J. & Lee, C. B. (2006) Everyday problem-solving in engineering: Lessons for engineering educators. *Journal of Engineering Education*, April 2006.

Latour, B. (2005) *Reassembling the social: An introduction to actor-network-theory*. Oxford, NY, Oxford University Press.

Law, J. (1992) Notes on the theory of the actor-network: Ordering, strategy and heterogeneity. *Systems Practice*, 5, 375–393.

Leveson, N. G. (2011) *Engineering a safer world: System thinking Applied to Safety*. Cambridge, MA: MIT.

Levinthal, D. & Rerup, C. (2006) Crossing an apparent chasm: Bridging mindful and less-mindful perspectives on organizational learning. *Organization Science*, 17 (4), 502–513.

Levitt, B. & March, J. G. (1988) Organizational learning. *Annual Review of Sociology*, 14 (1), 319–338.

Liker, J. K. & Morgan, J. M. (2006) The Toyota way in services: The case of lean product development. *Academy of Management Perspectives*, 20 (2), 5–20.

MacDuffie, J. P. (1997) The road to "root cause": Shop-floor problem-solving at three auto assembly plants. *Management Science*, 43 (4), 479–502.

Majchrzak, A., More, P. H. B. & Faraj, S. (2012) Transcending knowledge differences in cross-functional teams. *Organization Science*, 23 (4), 951–970.

March, J. G. (1991) Exploration and exploitation in organizational learning. *Organization Science*, 2 (1), 71–87.

Nonaka, I. (1988) Creating organizational order out of chaos: Self-renewal in Japanese firms. *California Management Review*, 30 (3), 57–73.

Nonaka, I. (1991) The knowledge-creating company. *Harvard Business Review*, November–December 1991, 2–9.

Nonaka, I. (1994) A dynamic theory of organizational knowledge creation. *Organization Science*, 5 (1), 14–37.

Nonaka, I. & Konno, N. (1998) The concept of "ba": Building a foundation for knowledge creation. *California Management Review*, 40 (3), 40–54.

Nonaka, I. & Takeuchi, H. (1995) *The Knowledge-Creating Company*. New York, NY, Oxford University Press.

Nonaka, I., Toyama, R. & Konno, N. (2000) SECI, *ba* and leadership: A unified model of dynamic knowledge creation. *Long Range Planning*, 33 (1), 5–34.

Nonaka, I., von Krogh, G. & Voelpel, S. (2006) Organizational knowledge creation theory: Evolutionary paths and future advances. *Organization Studies*, 27 (8), 1179–1208.

Orr, J. E. (1998) Images of work. *Science, Technology, & Human Values*, 23 (4), 439–455.

Orr, J. E. (2006) Ten years of talking about machines. *Organization Studies*, 27 (12), 1805–1820.

Parkman, M. (2000) Thematic foreword: Reflective practices: The Legacy of Donald Schön. *Cybernetics & Human Knowing*, 7 (2–3), 5–8.

Schein, E. H. (1996) Three cultures of management: The key to organizational learning. *Sloan Management Review*, Fall 1996, 9–20.

Schön, D. A. (1987) *The Reflective Practitioner: How Professionals Think in Action*. New York, NY, Basic Books.

Scott, J. E. & Vessey, I. (2000) Implementing enterprise resource planning systems: The role of learning from failure. *Information Systems Frontiers*, 2 (2), 213–232.

Senge, P. M. (2006) *The Fifth Discipline: The Art and Practice of the Learning Organization.* New York, NY, Doubleday/Currency.

Simon, H. A. (1991) Bounded rationality and organizational learning. *Organization Science*, 2 (1), 125–134.

Smith, G. F. (1998) Determining the cause of quality problems: Lessons from diagnostic disciplines. *Quality Management Journal*, 5 (2), 24–41.

Suchman, L., Blomberg, J., Orr, J. E. & Trigg, R. (1999) Reconstructing technologies as social practice. *American Behavioural Scientist*, 43 (3), 392–408.

Swartz, D. L. (2008) Bringing Bourdieu's master concepts into organizational analysis. *Theory and Society*, 37, 45–52.

Tjosvold, D., Yu, Z.-Y. & Hui, C. (2004) Team learning from mistakes: The contribution of cooperative goals and problem-solving. *Journal of Management Studies*, 41 (7), 1223–1245.

Trevelyan, J. (2007) Technical coordination in engineering practice. *Journal of Engineering Education*, 96 (3), 191–204.

Tucker, A. L. & Edmondson, A. C. (2003) Why hospitals don't learn from failures: Organizational and psycholoical dynamics that inhibit system change. *California Management Review*, 45 (2), 55–72.

Tucker, A. L., Edmondson, A. C. & Spear, S. (2002) When problem-solving prevents organizational learning. *Journal of Organizational Change Management*, 15 (2), 122–137.

Uhlfelder, H. F. (2000) It's All About Improving Performance. *Quality Progress*, February 2000, 47–52.

Walker, D. & Myrick, F. (2006) Grounded theory: An exploration of process and procedure. *Qual Health Res*, 16 (4), 547–559.

Weick, K. E. (1979) *The Social Psychology of Organizing* (2nd ed.) New York, NY, McGraw-Hill, Inc.

Weick, K. E. (1988) Enacted sensemaking in crisis situations. *Journal of Management Studies*, 25 (4), 305–317.

Weick, K. E. (1993) The collapse of sensemaking in organizations: The Mann Gulch disaster. *Administrative Science Quarterly*, 38 (4), 628–652.

Weick, K. E. & Putnam, T. (2006) Organizing for mindfulness: Eastern wisdom and western knowledge. *Journal of Management Inquiry*, 15 (3), 275–287.

Weick, K. E., Sutcliffe, K. M. & Obstfeld, D. (2005) Organizing and the process of sensemaking. *Organization Science*, 16 (4), 409–421.

Womack, J. P., Jones, D. T. & Roos, D. (1990) *The Machine that Changed the World*. New York, NY, Rawson Associates.

Yamaguchi, E. (2006) イノベーション破壊と共鳴 [*Innovation: Paradigm disruptions and fields of resonance*]. Tokyo, Japan, NTT Publishing.

Chapter 7

Finding workable solutions: Portuguese engineering experience

Bill Williams[1,2] & José Figueiredo[1]
[1]*ESTBarreiro, Polytechnic Institute of Setubal, Portugal*
[2]*CEG-IST, Technical University of Lisbon*

1 INTRODUCTION

Up to fairly recently, the answer to the question "what do engineers do?" was seen as being fairly self-evident: "engineering, that is technical design and problem solving". However as more empirical studies have begun to appear, it would seem that a more nuanced answer is called for and indeed more empirical research needed. Accordingly, we decided to collect data in our national context to complement the research being carried out in other countries.

This chapter presents an empirical study of the work of engineers in Portugal. The data comes from an online survey of 177 engineers about the time they spend on various activities and from in-depth interviews with a sample of 18 engineers about their work.

The interview data were gathered over the course of the year 2012 and as we listened to the interview participants and subsequently transcribed and analysed the recordings we were faced with the task of finding a way to capture these engineers' working lives – tables of data and exemplary quotes did not seem sufficient to do justice to the descriptions they shared. Seeking to encapsulate their perceptions of workplace practice into a visual representation we eventually turned to a study of maritime technology in the fifteenth century by sociologist John Law. Law proposed the concept of heterogeneous engineering to understand how technical innovation allowed Portuguese navigators to find the solutions that allowed the country to achieve technological and commercial dominance in the 15th and 16th centuries. He describes the early Portuguese technology users as heterogeneous engineers operating in a network of actors and we have adapted this approach to outline the portrait of contemporary engineering practice emerging from our data.

In structuring the chapter, we have opted to follow a chronological logic: we begin with Law's depiction of 15th century heterogeneous engineering and subsequently present our study of contemporary engineers.

2 HETEROGENEOUS ENGINEERING

The term heterogeneous engineering was originally coined by John Law in his analysis of the role of technology in the Portuguese Expansion of the 14th and 15th centuries (Law 1987, 2003). He points out that prior to 1433 the farthest limit for all European

Se queres aprender a orar, entra no mar[1]

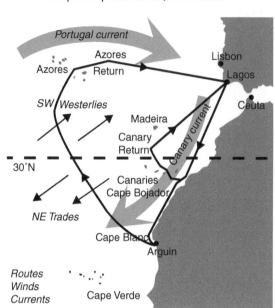

Figure 1 A volta do mar largo[2].

south-bound ships was the much-feared Cape Bojador, on the Northwest coast of Africa[3]. Law describes the development of technological innovation in the form of the galley and caravel vessels and of the navigational instruments which resulted in Portuguese mariners solving the problem of Cape Bojador by proposing a return path called a *volta do mar largo* or "turn of the open sea" (Figure 1). This depended on information regarding the prevailing winds and on a sophisticated knowledge of how instruments such as the astrolabe, quadrant and magnetic compass could be used to translate astronomical data into an effective navigation course. Such knowledge was coded in documents called *regimentos* and ephemerides (see example in Figure 2). The return from Cape Bojador was achieved by a counterintuitive indirect course which involved the Portuguese vessels striking out into the Atlantic, away from the African coast, to catch the winds and currents that would eventually bring them back safely to Portugal, as shown in Figure 1.

---

[1] "He who would go to sea, should learn to pray" Proverb cited by Diffie and Winius (1977) and found in Portuguese and seven other European languages (Strauss, 1994, p. 970).

[2] General depiction of the winds, currents and the approximate sailing routes of Portuguese navigators during the era of Henry the Navigator (c. 1430–1460). Map created by Walrasiad based on the description in Gago Coutinho, 1951.

[3] Located in what is now the Western Sahara region, it was common knowledge among mariners of the time that ships which rounded this cape never returned, one explanation being that there were sea monsters which attacked and destroyed them.

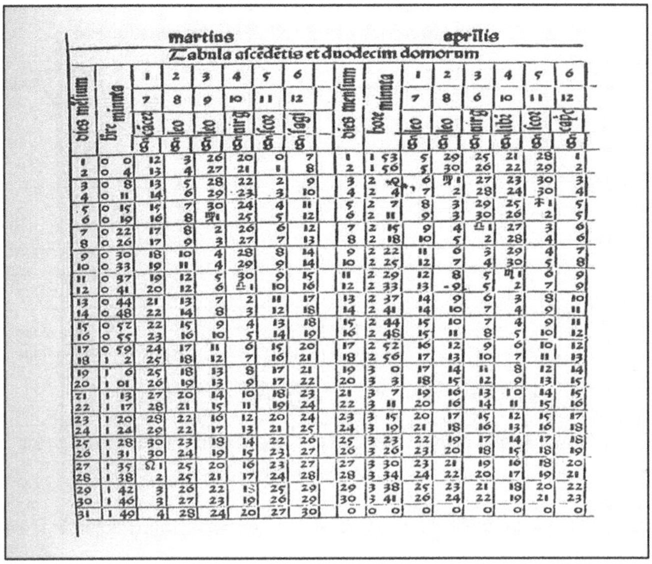

Figure 2 Ephemeride (astronomical table) from Abraham Zacuto's Perpetual Almanac (Zacuto, 1496).

The fact that Portuguese mariners were the only ones who had the knowledge and training to successfully carry out this navigation gave Portugal a crucial advantage over its European competitors and allowed it to dominate the newly opened trade routes to India, China and Japan, as can be seen in Figure 3. This dominance was to prove tremendously lucrative over the centuries that followed, as it opened up exclusive access to the spice trade and meant that Portugal controlled not only the trade between Asia and Europe, but also much of the trade between different regions of Asia (Fernandez-Armesto, 2007, pp. 480–481).

Law claims that Portugal's competitive advantage can be traced to three main factors: documents, devices and drilled people[4]. He argues that earlier systems-building metaphors used to characterise the extraordinary technological and engineering advance of this period were inadequate in that system-builders "associate elements in what they hope will be a durable array" and thus only described a limited part of the reality faced by these sea-farers (Law, 1987, p. 120). He points out that the Portuguese explorers of the period were faced with a complex and fluid mesh of factors in their constant struggle to master nature, technology and commercial opponents and proposes a description based on an actor network view of the technological process, where we find the navigator "standing at the heart of his or her network". The success of Portuguese

---

[4]We have already mentioned the documents and technological instruments involved. The training of mariners to apply the knowledge took place in the Casa da India later destroyed by the Great Lisbon Earthquake of 1755. Modern-day visitors in downtown Lisbon sipping a coffee as they look out over the Tagus river from the Terreiro de Pacos square may not realise that they are just around the corner from what was once the maritime technology training school which enabled Lisbon to be the center of a trading empire extending to almost all corners of the planet.

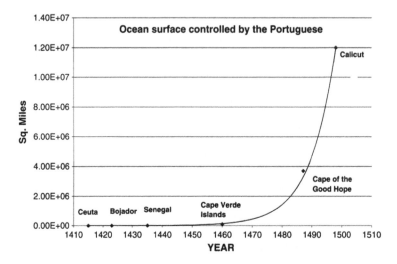

Figure 3 Exponential growth of Portuguese maritime expansion (Magee & Devezas, 2011).

mariners, he claims, was due to the application of "a combination of social and technical engineering in an environment filled with indifferent or overtly hostile physical and
social actors". This combination he classifies as heterogeneous engineering, an activity
taking place in a network where the process of developing solutions and overcoming
scientific and technological challenges was decisively influenced by a heterogeneous
mix of components ranging from the harsh natural forces of the maritime world to
the economic and political factors confronting maritime explorers. Law stresses that
to comprehend this complex interplay "it is not good enough to add the social as an
explanatory afterthought" but rather it "has to be placed alongside everything else if
the collisions and closures between forces and entities are to be understood".

Law argues that the Portuguese had generated a system that allowed them to control half the world for 150 years by creating a stable network made up of actors which
included: ships, sails, mariners, navigators, ores, spices, winds, currents, astrolabes,
stars, guns, ephemerides, gifts, and merchants' drafts; all being linked together in a
precarious but durable web. Taking an Actor Network Theory perspective (Latour,
1987), he considers the network as being made up of both human and non-human
actors and posits that decisions about which elements are included as actors should be
based on identifying those which make their presence felt by influencing the network
in a noticeable and individual way[5].

In this analysis, the description of the individual heterogeneous engineer probably
best fits the *Piloto*, the crew member with the most advanced technological training
and the one to whom the ship captain and armada captain-major would defer on
technical and navigational matters (Eça, 1894, pp. 36–37; Diffie & Winius, 1977,
pp. 137–142). At the same time, Law stresses that the network of actors around the

---

[5]See chapter 9 and the Glossary for a more detailed theoretical description of Actor Network
Theory.

Actors:

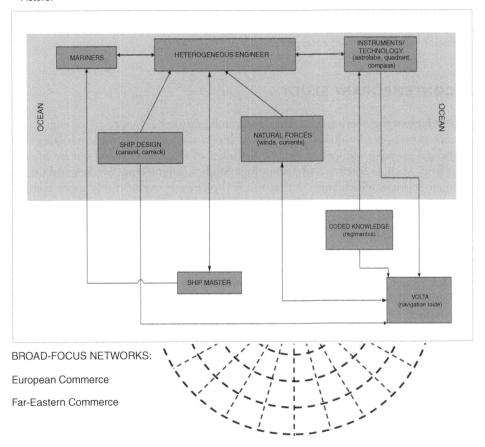

Figure 4 Our view of the actors in 15th century Portuguese Maritime Expansion (after Law).

Piloto as heterogeneous engineer is but one network within many. If we shift the focus back from those on board the ships to the masters of these vessels, or step back further again to consider those who commissioned the fleets and their policies, we are required to consider other network structures with their respective actors.

In Figure 4 we have represented the primary network of actors Law describes. The upper part depicts the different actors of the network around the heterogeneous engineer while the lower section shows how such a network is in turn located within other broader networks such as, for example, those of European and Far Eastern commerce. The connecting arrows show the principal lines of influence the actors exert on each other.

To sum up, the network of actors in Figure 4 shows how early Portuguese heterogeneous engineers were unique at that time in their use of technology to find the solutions they needed to meet the challenges of successful maritime exploration. Later in this chapter we will argue that contemporary engineers are confronted with less violent, but nonetheless very challenging, demands of the complex network of actors within which engineering professionals operate. Moreover, based on the descriptions of workplace practice gathered from a range of engineers in Portugal, we will propose a similar network to visually represent the interactions commonly found in the contemporary engineering enterprise.

3   CONTEMPORARY STUDY

Although most engineering education research literature has tended to see engineering practice as being predominantly made up of design and technical problem solving there has been a growing number of empirical studies of engineers carried out by scholars from fields such as Science and Technology Studies, Engineering Studies and Organisational Sciences which suggest that the real picture is considerably more complex and that the social and the technical are inextricably interlinked in the workplace (Bucciarelli, 1994; Vinck, 2003; Faulkner, 2007; Downey, 1998). These have tended to be fairly small-scale studies but they were complemented in 2007 by a larger scale study of engineers in Australia, Pakistan and India carried out by James Trevelyan and colleagues (Tilli & Trevelyan, 2008) in the University of Western Australia (UWA). These researchers identified a significant social dimension of engineering practice which they term technical coordination – informally securing willing and conscientious cooperation of other people in technical contexts. They also proposed a visual model to capture engineering practice as revealed in their study and identified a need for more research into engineering as a sociotechnical process requiring constant distributed learning by practitioners in the workplace (Trevelyan, 2009, 2010).

The initial objectives of the research underlying the study we report here were threefold:

1   To gather and analyse empirical data on the time Portuguese engineers spend on activities in the workplace, both those done individually and those involving interaction with others;
2   To gather and analyse empirical data on Portuguese engineering workplace settings as seen from the perspective of practicing engineers;
3   Given that the Australian researchers had proposed a model of engineering practice where the principal focus was on the engineering enterprise as a provider of reliable service, we wished to use qualitative data analysis to develop a complementary representation where the focus would be on the role of individual engineers within the enterprise.

3.1   Methodology

A mixed methods approach with a sequential explanatory design was chosen for the study. This design is considered appropriate for a study of this kind, one where initial analysis of quantitative data is followed by qualitative data collection and analysis to explain the findings from the quantitative phase (Creswell & Plano Clark, 2011, chapter 3). The first objective was addressed through a survey of 177 engineers and findings from the survey were then studied in more depth by carrying out interviews of

18 practitioners to provide qualitative data to meet objectives two and three. Whereas the first two objectives are addressing what Shavelson and Towne describe as "What is happening?" research questions (Shavelson & Towne, 2002, p. 99), the third aims to contribute to answering more complex questions in the related category they categorise as "Why or how is it happening?"

For the survey stage, the authors used an online alumnus survey of engineers to enquire about the time they devoted to various activities in the workplace. To meet the second objective we gathered data on individual engineers' perceptions of their workplace practice and used qualitative data analysis to construct a broadly applicable characterisation of their workplace practice. Qualitative research has been preferred for many systematic investigations of engineering practice (Bailey & Barley, 2011; Faulkner, 2007; Downey, 1998) and is appropriate for constructing a qualitative description based on interviews with engineers, such as we sought. To allow comparison with the Australian studies, the interview structure used by the UWA researchers was adopted and the sample was chosen, as in their case, so that the participating engineers represented a range of engineering fields and included professionals at various career stages.

Finally, after analysing and coding the data from the 18 practitioner interviews with the aid of qualitative analysis software, we identified a number of factors which consistently emerged in the engineers' descriptions of their workplace practice and sought to visually represent them.

3.2 Perception of working time survey

3.2.1 Procedure

A quantitative study of junior engineers was carried out using an online survey of recent engineering graduates to capture their perceptions on what they spent their time on in a typical working week. Junior engineers, who, following Bailey, we consider to be those with fewer than 7 years of work experience (Bailey & Barley, 2011, p. 267), were selected for logistical convenience because contact details for recent alumni were more readily available than those for the population of engineering professionals at large and we anticipated a higher response rate. Furthermore we wished to compare our findings with those from the Australian study where the population sampled comprised alumni who were followed over the early years of their career. Accordingly, recent graduate alumni from two engineering schools were asked to complete an online questionnaire which was available in both English and Portuguese. Response rate was 15% of alumni contacted. To permit comparison with the results of the UWA alumnus study of Australian engineers an identical survey format was used.

The survey questions were grouped into interactions with people face to face, through documents, and interactions with abstract systems and data, and with hardware, as shown below.

Face to face or telephone interaction with other people:

- with one or two other people face to face
- on site watching and interacting with people doing the work
- meetings

- training sessions or courses
- telephone calls

Interactions with people through text or documents (including reports, spec-ifications, drawings, plans, schedules, procedures, work instructions, operating or maintenance manuals, bills of materials, budgets, tender documents, invoices, contracts etc.):

- text messaging, chat
- e-mail correspondence, queries
- reading or checking formal documents
- writing, preparing formal documents

Interacting with systems or abstract data:

- searching for information on internet, in filing systems, databases, libraries etc.
- calculating, modelling, simulation, data analysis
- designing, drawing, creating software code
- debugging machinery, systems or software code

Interactions with hardware, site work:

- operating, testing, working hardware or systems
- surveying, measuring, inspecting, or observing on site
- maintaining the computers & filing systems you use in your work, installing or updating software etc.
- other hands-on work with equipment hardware, construction etc.
- searching for misplaced items

Respondents were asked to report the time spent each week on each of the activi-ties, choosing from the following intervals: none, <2 hours, 2–5 hours, 5–15 hours, or >15 hours. Following the procedure adopted in the UWA study, these intervals were considered as 0, 1, 3, 7 and 15 respectively for the calculation of the perceived time per week devoted to the activities. The data from 177 respondents is summarised in Table 1.

3.2.2 Results and discussion

Table 1 shows that our survey participants report that they spend around 44% of their working week involved in individual activity (non-bold portion of the right hand column). During the remaining 56% (bold portion of the right hand column), they are engaged in activities involving interaction with others such as meetings, supervision, writing reports and so on. The Australian study of novice engineers in the first years of their career (Tilli & Trevelyan, 2008) noted a similar but more accentuated pattern of 60% of working time involved interacting with other people.

Such findings are surprising given that both studies refer to early-career rather than senior engineers. Historian Rosalind Williams has observed that as early as the 1930s in the US, it made sense for engineers to migrate into management from a

Table 1 Average perceived working time breakdown of Portuguese junior engineers (177 respondents).

	177 Junior Engineers		
Interaction	Hours/week/ respondents	% Time	Cumulative % Time
Interacting with systems or abstract data [searching for information on the internet, in databases, filing system]	773	8	8
Interacting with systems or abstract data [calculating, modeling, simulation, data analysis]	662	7	15
Interacting with systems or abstract data [designing, drawing, creating software code]	844	9	23
Debugging software, machinery or systems	551	6	29
Operating, testing, working hardware or systems	448	5	34
Surveying, measuring, inspecting, observing on site	307	3	37
Maintaining the computers, filing system; installing/ updating software etc.	286	3	40
Other hand-on work with equipment, hardware, construction etc.	298	3	43
Searching for misplaced items	82	1	44
Speaking with one or two other people face to face	874	9	52
Interacting with people or overseeing people on site	536	5	58
Meetings (admin or technical)	547	6	64
Training sessions or courses	226	2	66
Telephone calls/audio conference	477	5	71
SMS, Chat	374	4	75
Email, postal correspondence, queries	1000	10	85
Formal documents: reading or checking	697	7	92
Formal documents: writing or preparing	784	8	100
TOTAL	**9766**	**100**	

career advancement point of view and many did (Williams 2002, pp. 39–40). She comments that this occurred despite the fact that many engineers of the time maintained a disdainful view of managers "who dealt in words ('management bullshit' was a common term) rather than with things and who did not really understand how things work". Accordingly, we might expect that senior engineers, having taken on a more management-oriented role as their career advanced, would be involved in interaction with others relatively often. This is borne out by several earlier research studies with estimates of the time that more senior engineers spend on communication ranging from 40% to 75%, with the majority of estimates around 60% (Trevelyan, 2009, p. 4). The UWA survey findings for novice engineers "were remarkably consistent with the interview and field study data from more experienced engineers" (Trevelyan, 2010, p. 7) and suggested the possible conclusion that there is something intrinsic about engineering practice that requires extensive interaction with other people, rather than the still-common idea that engineers initially focus on largely solitary technical tasks and progress later to tasks that involve a greater extent of interaction with other people. Our survey data appears to support this interpretation and to suggest that the sociotechnical aspects of engineering may have a significant role in the working life of most engineers, including those in the early stages of their career.

3.3   Qualitative study

Although online surveys allow data gathering from relatively large samples in a short space of time, they have the disadvantage of being inflexible and non-interactive. So, given that the findings from our initial survey encouraged us to seek deeper insights into the contours of engineering work and in particular its sociotechnical aspects, we then looked to other techniques to gather supporting data. Qualitative research is frequently used to gather insights of this kind and common data collection techniques include practitioner interviews, documentary analysis and *in situ* participant observation. As we did not encounter relevant documents for analysis and because conducting participant observation is generally accepted to require a significant time investment to produce data (Chism *et al.*, 2008), we chose to adopt the first of these techniques to clarify and enrich our quantitate data. Accordingly the second phase of our study presents data gathered from interviews of practitioners.

3.3.1   Procedure

Qualitative data were collected by interviewing a sample of 18 engineers working in Portugal using a semi-structured interview format. The interviews lasted between 1 and 2 hours and the sample was designed to reflect a range of engineering backgrounds and to include junior (less than 7 years of work experience), mid-level (7–15 years' experience) and senior (greater than 15 years) engineers (Bailey & Barley, 2011, p. 267).

During these encounters, the interviewers, both academic researchers with experience in industry and in the running of SMEs, assumed the role of interested novices with regard to the specialised knowledge of the participants. While a general script based on the 55 questions employed by the Australian researchers (Trevelyan, 2007) provided the overall structure, an empathetic interview format was applied by the interviewers whereby follow-up questions were used to encourage further development of themes which could be seen to be important to the participants. In addition, memoing was used to record instances when participants were particularly passionate in their narratives (Glaser & Holton, 2004).

The interviews were carried out in the language the participant engineers requested, which was normally Portuguese. The recorded interviews were then transcribed in the original language and analysed using the qualitative data analysis software NVivo which in addition to organising interview transcriptions also facilitates pattern recognition and flexible coding of data. The following sections describe the two coding sets that were applied.

3.3.2   Applying UWA codes

The UWA study had followed a common grounded theory procedure (Glaser, 2005) in coding interview responses initially using open coding, followed by selective and theoretical coding until saturation was reached. This process produced a total of 86 codes which were grouped into 10 categories (Trevelyan, 2008). We elected to use the same set of interview questions and also these same codes for our study. The codes are set out in the 10 categories shown in table 2 along with the number of code descriptors for each category.

Table 2 Coding categories from the UWA study.

	Code category	Number of code descriptors
1	Managing self and personal career development	8
2	Technical coordination, working with other people	16
3	Engineering processes, project and operations management	13
4	Financial processes	6
5	Procurement, buying products or services	3
6	Human resource development, training	4
7	Business development or marketing, selling products or services	11
8	Technical: creating new concepts, problem solving, programming	13
9	Technical: reviewing, checking, testing and problem diagnosis	10
10	Hands-on technical work, construction or repairs	2

To exemplify the category and coding process, Table 3 shows the 16 codes clustered to comprise the technical coordination category (category 2 above) while the codes for the other 9 categories (Trevelyan, 2008) can be consulted online[6].

3.3.3 Additional codes

To provide a broad description of the practices being studied it was decided to apply a multiple coding protocol. This meant that a mention by a participant of a particular work aspect could be considered to be categorised by more than one code. Although the 10 UWA coding categories and respective 86 codes were found to be very much applicable to our data, as can be seen in Figure 5, there were other aspects of engineering practice that emerged from our analysis which suggested that further insights could be obtained by assigning additional codes to complement them. Accordingly, during analysis of the first 5 interviews, a process of memoing and theoretical coding was used (Glaser & Holton, 2004) and from this we decided to apply ten further coding categories. Of these, four included sub categories of code descriptors.

The additional categories are set out in Table 4.

These coding categories were applied to all the 18 interview transcripts along with the 10 UWA categories and the following sections will describe the results obtained.

3.3.4 Results

Table 5 shows the number of participant sources and the total number of mentions registered for the original UWA codes and for the additional codes we have applied. We see, for example, that references that were coded as *self and career development* according to the UWA codes were made by all 18 of the engineers interviewed and that the total number of these references was 116.

The results of the application of the UWA code categories are summarised in Figure 5 where the total number of mentions of each code category is expressed as

---

[6]The authors used the codes originally published in Trevelyan (2008); a more recent version can be found available online for Trevelyan (2010) at http://school.mech.uwa.edu.au/~jamest/eng-work/publications/Trevelyan-ES2010-Reconstructing-Engineering-w-appendices.pdf.

Table 3 Coding for references classified in the Technical Coordination category.

Code descriptors	Descriptor examples
Coordinate insiders, mentoring	Coordinate work of peers, subordinates and superiors. Perform technical checks on work, watch for roadblocks, may provide advice and feedback, may review technical competence, may assess training needs or provide informal training when appropriate.
Supervise staff	Supervise staff for which engineer has line management responsibility.
Coordinate outsiders	Coordinate with outside organisations such as other contractors working on same project, community organisations, etc.
Coordinate with client	Liaise with client, expedite solution review and acceptance, coordinate installation, commissioning, monitor acceptance testing.
Advocacy/compromise	Advocacy for a particular technical or commercial view, setting out to change the framework in which a problem is considered. Compromise is probably essential for securing eventual agreement.
Site engineer	Coordinate and supervise work on site: ensure that work is performed according to drawings and specifications, plan site work, coordinate with foremen.
Supervise contractors	Coordinate and supervise work performed by contractors – ensure that work is performed according to specifications and requirements.
Reverse mentoring	Providing mentoring, guidance, coordination, training and supervision to more senior or experienced personnel.
Cross cultural supervision	Coordination, supervision of people from different cultural backgrounds.
Report progress	Report to supervisor, team leaders, peers on project progress, solutions, financial and resource consumption. Verbal, written or in meetings.
Delegate technical work	Allocate responsibility for technical work: balance technical expertise and experience against cost and availability, decide whether to employ additional staff or contractors etc. Select appropriate working methods and tools.
Delegate supervision	Allocate appropriate technical supervision capacity for a given activity to ensure that required performance and quality standards are measured, maintained and recorded.
Review procedures	Review and develop standard organisational procedures or organisational structures.
Build and lead team	Build and lead a project team. Create shared vision, objective, monitor team members, provide care for team members.
Networking	Networking: develop and maintain network of contacts to help with performance of job.
Organise socials	Organise recreational and social activities within organisation.

a percentage of the global number of references in the UWA code categories. In the following sub-sections we shall give more detail on technical coordination, the UWA category which represented the highest percentage in Figure 5, and then go on to discuss the new coding categories we have introduced and exemplify some of the practitioner perceptions they represent.

3.3.5 Technical coordination

In Figure 5 we note that the engineers interviewed, in common with their Australian counterparts, spend a significant amount of their time in coordinating people and information in relatively informal contexts (coded as technical coordination). James Trevelyan has summarised it like this: "While coordination seems to be non-technical,

Practice breakdown for 18 engineers - UWA codes

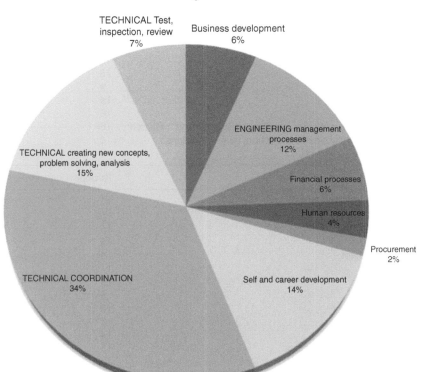

Figure 5 UWA category breakdown applied to our interview data.

Table 4 Additional code categories.

	Code category	Number of code descriptors
1	Team	1
2	Firm	1
3	Reputation	2
4	Professional judgment	1
5	Client relations	2
6	Regulations/codes	1
7	Instruments/technology	1
8	Constraints	2
9	Context networks	3
10	Sub-contracted firms	1

analysis provides evidence supporting the critical importance of technical expertise. Coordination usually involves one-on-one relationships with superiors, clients, peers, subordinates, and outsiders. Coordinating the work of other people seems to be important from the start of an engineering career." In our analysis of the transcripts project

Table 5 Coding analysis of Interviews (18 participants).

UWA Codes	Participant sources	Mentions
Self and career development	18	116
Technical coordination	18	278
Engineering management processes	17	95
Financial processes	18	50
Procurement	9	14
Human resources	14	29
Business development	16	52
Technical: creating new concepts, problem solving, analysis, research	18	118
Technical: testing, measurement, inspection, review, checking	14	55
Hands-on work	0	0

Additional Codes	Participant sources	Mentions
Firm	17	67
Team	16	40
Reputation – Self	16	57
Reputation – Firm	15	48
Professional judgment	14	39
Client – needs	11	21
Client – satisfaction	15	53
Regulation/codes	6	24
Instruments/technology	18	44
Constraints – budget	13	23
Constraints – time	14	31
Context network – market sector	15	54
Context network – national	15	38
Context network – international	16	66
Subcontracted firms	9	51

management, as characterised by formalised written methods, was included in the *engineering management processes* category and was thus classified separately from the non-formalised processes coded as *technical coordination*.

Technical coordination, representing 35% of the activities coded according to the UWA system, was the most prominent in our interviewees' descriptions of their workplace activity. This was also the case for their Australian counterparts in the UWA study where the figure was 27.4% (Trevelyan, 2007). The other activities coded under this system also present broad agreement with those reported from Australia: for example, references coded as *technical: creating new concepts*, a category which includes design and problem solving activities, represent 15% in our data and 13.5% in theirs. Similarly the values for *engineering management processes* were 12% and 19% respectively. Thus, the overall pattern of these Portuguese engineers' perceptions of their workplace practice has contours that are similar to those of the engineers interviewed in Australia[7].

---

[7]Although the Portuguese study employed the interview technique, interview questions and coding system which had been applied by the UWA researchers, it has not yet been feasible to establish inter-rater reliability between the two studies due to the fact that their interviews were carried out in different languages. For this reason, caution is needed in making comparisons between the results of the two.

The following interview extracts from participating engineers illustrate some typical interactions that were coded in the *technical coordination* category. Vicente Chagas[8], a mid-level civil engineer working for an international building rehabilitation firm talks about the range of responsibilities he had on a major job for an overseas client:

"I managed the entire process at this end and this meant launching the tendering process for subcontractors then liaising with them, monitoring the project through the various stages, and being the link between the designers, the architect and our client throughout the project."

Mohsen, a project engineer in his early 30s explains his role in a multinational processor design firm:

"Somewhere in between we have the product engineers, so they try to close the gap between the person that only thinks of creating a prototype and those that make millions of them. Engineering-wise this is what you call product engineering, some one that understands a little bit of the design and the creation of something new and who understands what the 'Fab' is all about."

Alfredo Valdemar, an electrical engineer with 20 years' experience in telecommunications describes how his engineering background is important to underpin the connection between the technical and managerial aspects of his current role:

"Over the course of my career, my work has become gradually less technical and more managerial. Even so, I still keep a strong connection with the technical side. In recent years I have mainly been doing project management which involves a strong management component allied to a strong technical side. In other words, the technical part is necessary to link up the various members of the team, to help the more technical team members design and implement solutions."

3.3.6 Clients

"Our client is the reason for our existence" "Customer feedback has now become a Key Performance Indicator for us"

Variations on these engineers' comments re-echoed during the interviews and, as can be seen in Table 5, this is highlighted by the 74 references to *client needs* and *satisfaction*.

Although engineers working in smaller firms naturally had more direct contact with clients than those in large national or multinational enterprises where there is more specialisation of roles, the multifaceted relations with the clients, either in the form of product end-users or of other firms, in the case of enterprise to enterprise operations were regularly mentioned in our participants' discourse.

---

[8]Names of participants have been changed to preserve anonymity.

A product engineer like Mohsen who has direct dealings with enterprise customers, has needed to cultivate the ability to listen and to explain:

> "In front of me I have the customer, as a product engineer I have to help the customer understand, so talking a lot with the customer means you have to have managerial skill anyway, so you develop that."

On the other hand, João Fanha, a company director and senior engineer who heads the customer experience department handling over 2 million clients of a major telecommunications company, comments that although he rarely has direct contact with end-users, he is, however, responsible 24/7 for their experience with the company's products and expects to be called if there is a major emergency, regardless of the day or hour.

> "If you are involved in customer care then it's part of your DNA to be always on call to solve clients' problems and I guess in parallel with this, one part of your brain is systematically thinking 'If I did this or that maybe it would help to optimise things.'"

3.3.7   Reputation

Both company and self-reputation were also common themes mentioned by interviewees.

Raimundo Pinheiro, a software engineer and CEO of a web-based service consultancy which has gone from a 2-man university start-up to a 32 person sector leader in ten years, sees both of these as being important in maintaining the rapid growth of his start-up consultancy:

> "We have really been working hard on our reputation within the sector both in terms of the company as a brand [uses the English term] and also in the context of personal branding [uses the English term]."

Martina Mendes, a junior project manager in a large construction consortium, believes that the high standards she sets for herself have helped keep her career on track during an economic downturn and major cut-backs in the construction sector:

> "I'm really not the type to put together a so-so piece of work and just hand it over. I like to feels it's impeccable so that when someone picks it up they will say 'that's terrific'. So in that sense I work to my own standards. And I think this has helped my career path. In fact, given the current situation, I probably wouldn't be where I am now if I hadn't had this kind of attitude."

3.3.8   Professional judgment

Although codes in this category were less frequently mentioned than in those previously discussed (mentioned by 14 out of 18 participants), those who did bring it up often described it with some passion; some respondents described this type of decision as being one of the most challenging parts of their work and involving abilities which can only be acquired on the job. In some cases the judgment involved relations with clients while in others it involved taking decisions in the technical domain as exemplified in the following two vignettes.

Sonia Marques, a civil engineer and architect in her 30s who is one of 3 partners in a construction SME, comments on her ethical dilemmas in quoting delivery dates in estimates:

> "Our market, as well as being all about price is also very much about delivery dates, and everybody lies about this – everybody promises to deliver the project within a month, they won't do it but they say they will. So what happens is that if I am going to be completely rigorous in what I am promising, then the client will give the project to the other firm, who will in fact take the same time as we will but they lied right from the start – it's terrible but that's the way it is! [...] I can't accuse my competitor of lying so I have opted for the lesser evil: the work takes the time it needs to take, but people often don't appreciate that giving a bit more time, a month or so, for a home which will probably be an investment for life isn't really a lot to ask."

Arsénio Bento, an experienced freelance construction designer describes having to make judgment calls in structural design while working under pressure:

> "I tend not to refuse work, because basically I'll lose the client ... end of story. The usual thing is that the client company's people are already working all-out on a big job and by the time the work comes to me the deadline has already passed so it's like 2 days to get it all done and that's it. It's two days to check these structures – hardly enough time to go through the results of analyses much less carry out the analyses themselves, but that's how it is [...] you take it or leave it![...] You could obviously produce something better if you had more time. And sometimes you can end up in hot water over this ..." At this point he goes on to recount an anecdote of a new water treatment plant which developed structural cracks.

It has often been commented that new graduates entering technological companies need 2 to 5 years to acquire the competences needed to bring value to their employer[9]. However, our interviews suggest that what is being acquired may go beyond competences and skills to include the kind of professional on-the-job judgment which allows Sonia to decide on what delivery dates will keep her firm's tenders competitive while being compatible with her professional ethics. Similarly, it is specific context-related judgment that enables Arsenio to decide how best to apply the Eurocode specifications

---

[9]See for example Tilli and Trevelyan (2007): "An underlying assumption that has informed our thinking about engineering work is that training and experience is an essential component of the first few years of an engineering career. This assumption is based on data from the framework study interviews in which all participants said that it took between two and five years for a novice to become 'competent'. While each participant had a different interpretation of competence, all identified this early career period as important".

to guarantee structural safety while working under considerable time constraints or that helps a junior engineer in João Fanha's firm to decide if he should contact his head of department over the weekend because customer complaints are peaking at their call centres.

3.3.9   Other codes: team, firm, regulations/codes, instruments/technology, constraints and subcontracted firms

In applying the UWA coding system, references to working in teams were coded by us as *technical coordination* and references to the firm where engineers were employed appeared in utterances coded under *self and career development*. In addition, given that both the team and the firm were often mentioned as actors directly affecting everyday practice, we also decided to code them separately as the multiple coding protocol permits (Table 5).

Similarly, the categories of *instruments/technology*, *sub-contracted firms* and *constraints* were considered necessary for inclusion. Although the theme of *regulations/ codes* was not as universally referred to as the other codes we have employed, we decided there were sufficient references to written codified knowledge (as exemplified in product specifications, tendering regulations, Eurocodes and formal quality or safety procedures) that this warranted a specific coding category.

3.3.10   The workplace as a network

We noted that two of our interviewees resorted to an eco-system metaphor when trying to sum up the challenges they face in their daily work.

Nóe Reis, a young software engineer working for a multi-national market leader in mainframe computing, sees his role as maintaining an eco-system in which the needs of client, his firm and his team members are finely balanced:

> "Of everyone involved, our client is the reason for our existence. Even so, I can't keep the client happy at the expense of one of our people working through the weekend. That's also something I worry about: each one of us who is involved has to feel good about what we are doing and try to maintain this eco-system. We are working for everybody concerned, not only for the client, but also for our team, for the organisation – because L [the company] also has to be happy about it."

CEO Raimundo Pinheiro, while observing that he rarely does hands-on work anymore, also uses an eco-system metaphor to describing his role in a start-up company which has rapidly grown to have a staff of thirty:

> "The eco-system concept makes complete sense for me. In fact, that's pretty much my role in the company. So, neither the client is exploiting us nor are we deceiving the client – and that way I can manage to keep things in equilibrium and provide a fair trade [uses English term] and that's the way we have grown. [...] So if we succeed in getting the balance right, where our client appreciates not only our final deliverable but also the process by which we got to this end product, then that's the way that everyone involved feels good that they have a part in the whole process."

We shall return to this aspect in section 4 when considering ways of representing engineering practice.

3.3.11   Practice embedded in broader contexts: international, national and market sector

Although international dimensions of participants' work context were not specifically addressed in any of the questions asked, they were notably frequent in the interview transcripts. We were initially surprised that this was the case, even for those engineers working in small firms that might have been expected to focus exclusively on local operations. The aspects mentioned included references to international clients and the language and operational challenges involved, engineers' experience working on projects abroad or their possible future emigration plans. Likewise the national context and its effect on the market sector and individual firms *vis-à-vis* job security were common themes as can be seen in Table 5. Given that the country was experiencing a major economic downturn at the time the interviews were being carried out, one that had a particularly strong impact on construction and related sectors (Reuters, 2012), this is perhaps understandable.

3.4   Discussion

3.4.1   The role of technical coordination

It is notable that in both the UWA study and our own, the aspects grouped as *technical coordination* are the most referenced when engineers are describing their workplace practice. We are aware that care is needed in the interpretation of the results obtained by transforming qualitative data into quantitative data in this way (Creswell & Plano Clark, 2011, p. 231) and that modifying the interview questions or technique or the sample could alter the overall number of mentions of a particular item. Nevertheless, the prominence of this range of practices in the samples studied in both Portugal and Australia suggests that this represents a significant dimension of contemporary engineering practice, which merits further study.

3.4.2   Generalising from the data

Karl Weick claims that "the process of theorizing consists of activities like abstracting, generalizing, relating, selecting, explaining, synthesizing, and idealizing" (Weick, 1995). With this in mind, we thought it useful to step back from the detailed coding processes of the qualitative analysis applied to the interview data and ask the question "is there a common generalisable theme in the engineers' descriptions?" A possible answer, and we believe an interesting one, is that a very large part of the practice being described could be characterised as finding workable solutions. Whether we listened to junior, mid-career or senior engineers, from SME startups to national and international firms, with backgrounds from civil, electrical, mechanical or software engineering, a large part of what they related involved an ongoing dynamic process of finding effective and feasible solutions so as to ensure reliable service and commercial success. Such solutions involved coordinating information and people in technical and commercial contexts and implied interaction with a complex mix of factors which include time, budget and regulatory constraints, technological affordance, client needs and demands, working with other colleagues and other firms.

With this in mind, the challenge that then presented itself was to come up with a form of representing these multiple factors in a meaningful way and this is addressed in the next section.

4   REPRESENTING ENGINEERING PRACTICE

Many authors have highlighted the importance of the visual culture of engineering (Henderson, 1999; Vinck, 2003, pp. 113–194; Latour, 2011, chap. 9; Dias Figueiredo, Chapter 1 of this book) and we often noted in our own interviews how engineers would instinctively reach for a sketch pad to help get their ideas across. Accordingly, we believe it useful to give attention to the visual representation of practice and practices.

The representation of engineering practice shown in Figure 6 was proposed by James Trevelyan, based on the UWA research previously mentioned (Trevelyan, 2009). This depiction focuses on an engineering enterprise[10] providing reliable service to its clients based on a number of underlying structural supports that are scaffolded by both formal and informal communication processes.

This may be considered as a meso-scale representation in contrast to more general characterisations of technological knowledge production at the macro-scale level like the Gibbons Mode 2[11] (Gibbons et al., 1994; Nowotny et al., 2001) and the Triple Helix approaches[12] which were proposed in the 90s (Etzkowitz & Leydesdorff, 1998, 1999; Leydesdorff & Etzkowitz, 1996, 1998). At the other end of the scale, we find micro-level characterisations of engineering practice such as the Itabashi-Campbell Integrated Model (Itabashi-Campbell et al., 2011) that adapts existing models from Nonaka and Weick to visually represent one of the processes in the Trevelyan model: technical problem solving[13].

Given that the main focus of the Trevelyan model is on the engineering enterprise as a provider of reliable service, we saw a need for a complementary representation where the focus would be on the role of engineers within the enterprise and we believed that the qualitative data we have been gathering as well as that collected in the UWA study allow us to present a useful visual sense-making representation for this. Although a systems approach to this task was initially appealing from an engineering perspective (Sladovich 1991, pp. 2–22), after examining the data from over 30 hours of interviews we concluded that the complex interaction of technical, social and economic factors

---

[10] As in Chapter 2, we use the term enterprise as the 'unit' or 'locus' of engineering action; this may take place in a firm or a group of firms in consortium or in the context of government or local authority.

[11] Williams and Figueiredo (2010) describe the use of the Gibbons approach in an engineering practice case study.

[12] Shinn (2002) provides a useful comparison of the respective empirical foundations of these two approaches and of their contributions to various fields of inquiry.

[13] See also chapter 6.

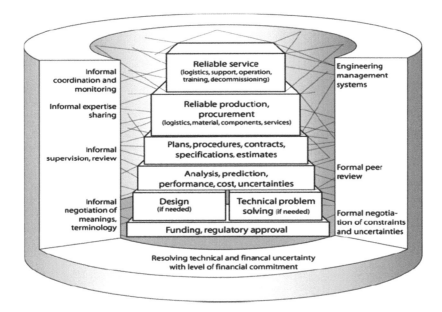

Figure 6 Trevelyan's Unifying Model as a representation of Engineering Practice.

involved would be more effectively represented by an actor network approach based on John Law's work.

4.1 Contemporary Heterogeneous Engineering

As mentioned earlier, our sample was chosen to give a broad range of engineering practitioners in terms of both career experience and engineering training. In practice it was achieved through alumnus liaison and opportunistic contacts with large and small firms based in central Portugal. None of the practitioners we interviewed worked for the government or in laboratories. Moreover, although a minority mentioned having been involved in the development of innovative and cutting-edge technological products and services, the majority of the work they described involved the use or the optimisation of already existing technology.

In chapter 8 Hubert and Vinck describe how they chose John Law's Heterogeneous Engineering concept as a lens with which to explore how work in the field of nano and microtechnologies can lead to the design and introduction of new products and technologies which are likely to affect society. By contrast, our study of engineers in companies, due to its exploratory nature was not guided by a specific framework when we were gathering our survey and interview data. It was only at a relatively advanced stage in the process, when considering how we could visually present the complex web of interactions in the interviewees' descriptions, that we came to see how Law's ideas could rather serendipitously contribute to this sensemaking process for contemporary Portuguese engineers. We were particularly interested in the way his network of actors could relate to the eco-system metaphor used by two of our participants. Likewise his

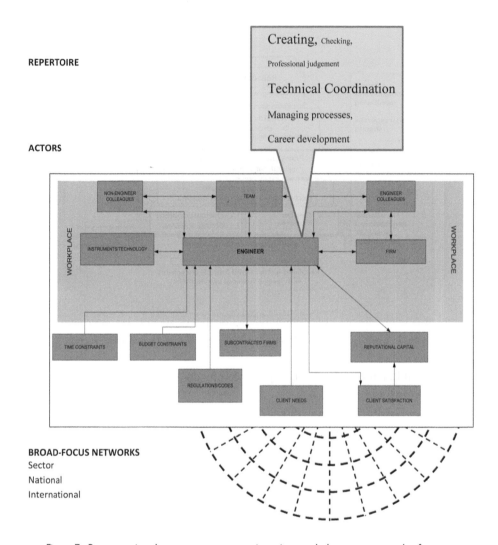

Figure 7 Representing the contemporary engineering workplace as a network of actors.

suggestion that changing the focus[14] in such a system means that a different structure will be seen (Law, 1987, p. 131) was useful in considering how broader contextual circumstances, as in market sector, national and international domains, can have an influence on engineers' work.

Thus, we decided to follow Law's approach by characterising engineering practice as a network involving human and other actors as shown in the middle section of Figure 7. We represent the engineer as an actor interacting with colleagues and

---

[14] Whereas satellite mapping images were not widely available in 1987 when Law originally proposed this analysis, today most of us do have experience of using web-mapping services to visualise complex networks at different scale levels.

with instruments and technology in the immediate workplace, acting in a network which also involves dealing appropriately and effectively with clients, time and budget constraints, reputation issues and with the codes and regulations associated with his or her field. The lines represent the interactions between actors with arrows showing the main directions of influence in these interactions. Interaction with technology is shown as a two-way process, for example, in that some participants described how technological affordance affects what engineers can and cannot do while others related how they had improved or customised technology so as to optimise procedures – by writing a software routine for example. On the other hand, budget and delivery time constraints are represented as one-way interactions: although there may be situations where an engineer can alter them, they are more often a given.

In addition to this network of actors in the engineering workplace, we suggest that when we broaden our focus this can in turn be seen as being embedded in larger networks such as the market sector and national and international commercial spheres, which are shown in the lower part of Figure 7.

The figure is completed by what we describe as the repertoire of the engineers within this network of actors, which is made up of the activities identified by the UWA study with the addition of an extra factor we have called professional judgment. To visually represent the relative time perceived by engineers as devoted to the repertoire of activities shown in Table 5 we have used a word cloud convention where the font sizes represent relative frequencies[15].

It is important to clarify that what we present in Figure 7 is not a model in the sense of a representation of a process or system which enables us to make predictions. Rather we propose it as a visual aid based on empirical data, which helps to make sense of the complex contours of contemporary engineering practice by providing a framework for engineering students (and those educating them) to help negotiate the complex transition from student to practitioner.

5 CONCLUSION

The quantitative data from surveying 177 junior engineers that we present here suggests that the sociotechnical aspects of engineering may have a significant role in the working life of most engineers, including those in the early stages of their career. In addition, our analysis of the qualitative data from interviews with 18 practitioners at various career stages provides a portrait of engineers in companies finding workable solutions by coordinating information and people in technical contexts.

These solutions involve successfully interacting with a heterogeneous mix of actors in a complex network system so as provide reliable service and achieve commercial viability. This mix includes both human actors such as team-members, colleagues and

---

[15]To aid legibility we have simplified the codes shown in the repertoire section of Figure 7. Thus the UWA codes in section 5.2. are represented as follows:

Technical Coordination = Technical coordination, working with other people;
Career development = Managing self and personal career development;
Creating = Technical: creating new concepts, problem solving, programming;
Checking = Technical: reviewing, checking, testing and problem diagnosis; and
Managing Processes = Engineering processes, project and operations management.

clients and other actors such as technology, regulations, deadlines, budgets, reputation and sub-contracted firms as we represent in the actor-network diagram in Figure 7. This representation is derived from the actor-network analysis proposed by John Law to characterise the Portuguese navigators in the 15th century who gained a competitive advantage through their successful deployment of technology to achieve solutions that were "forged in situations of conflict" (Law, 1987, p. 111).

Engineers frequently emerge from their formal education with a belief that the principal factors that determine what is possible and not possible in engineering are almost entirely associated with engineering science, largely based on physical sciences and expressed in the language of mathematics. The data that we and other researchers have gathered from engineering professionals suggest that the reality faced by young engineers is much more complex, calling upon social and communication skills and powers of judgment with which their academic and technical training has frequently not equipped them. Identifying these skills and competences and studying the paths by which junior engineers can successfully acquire them can be expected to benefit engineering graduates in making the transition from student to professional and to suggest ways of providing value to those who employ them. We present the findings from our study in the belief that they can play a part in this important process.

ACKNOWLEDGEMENTS

This work is supported by a grant from the Portuguese Fundação para a Ciência e a Tecnologia (FCT) of the Portuguese Ministry for Education and Science as part of the project *What Engineers Do* PTDC/CPE-PEC/112042/2009.

REFERENCES

Bailey, D. E. & Barley, S. R. (2011) Teaching-Learning Ecologies: Mapping the Environment to Structure Through Action. *Organization Science*, 22 (1), 262–285.

Bucciarelli, L. L. (1994) *"Designing Engineers," Inside Technology*. W.E. Bijker, W.B. Carlson, and T.J. Pinch (eds.), Cambridge, MA, MIT Press.

Chism, N. V. N., Douglas, E. & Hilson Jr, W. J. (2008) *"Qualitative Research Basics: A Guide for Engineering Educators."* National Science Foundation.

Creswell J. W., and Plano Clark, V. L. (2011) *Designing and conducting mixed methods research*. Thousand Oaks, CA, Sage Publications, Inc.

Downey, G. L. (1998) *The machine in me: An anthropologist sits among computer engineers*. New York, Routledge.

Diffie, B. W. & Winius, G. D. (1977) *Foundations of the Portuguese Empire: 1415–1580*, Europe and the World in the Age of Expansion Series, Volume 1, University of Minnesota Press.

Eça, V. A. (1894) *O Infante D. Henrique e a arte de navegar dos Portuguezes*. Lisbon, Imprensa Nacional.

Etzkowitz, H. & Leydesdorff, L. (1998) The Endless Transition: A Triple Helix of University-Industry-Government Relations. *Minerva*, 36 (3), 203–18.

Etzkowitz, H. & Leydesdorff, L. (1999) The Future Location of Research and Technology Transfer, *Journal of Technology Transfer*. 26 (2–3), 111–123.

Faulkner, W. (2007) Nuts and Bolts and People. *Social Studies of Science*, 37 (3), 331–356.

Fernández-Armesto, F. (2007) Portuguese Expansion in a Global Context. In: Bethencourt F. & Curto, D.R. (eds.) Portuguese Oceanic Expansion, 1400–1800. Cambridge University Press.

Gago Coutinho, C.V., 1951, A náutica dos descobrimentos, Lisbon, Agência Geral do Ultramar, Divisão de Publicações e Bibioteca.

Gibbons, M., Limoges, C., Nowotny, H., Schwartzman, S., Scott, P. & Trow, M. (1994) The new production of knowledge: The Dynamics of Science and Research in Contemporary Societies. London, Sage Publications.

Glaser, B. G. & Holton, J. (2004) Remodeling grounded theory. *Forum: Qualitative Social Research*. [Online] 5 (2), 1–17. Available from: http://www.qualitative-research.net/index. php/fqs/article/view/607/1315 [Accessed 26th March 2013].

Glaser, B. G. (2005) *The grounded theory perspective III: Theoretical coding*. Mill Valley, CA, Sociology Press.

Henderson, K. (1999) *On Line and On Paper: Visual Representations, Visual Culture and Computer Graphics in Design Engineering*. M.I.T. Press Inside Technology Series.

Itabashi-Campbell, R., Perelli, S. & Gluesing, J. (2011) Engineering problem solving and knowledge creation: An epistemological perspective, Proceedings of Technology Management Conference (ITMC) IEEE International, 27–30 June 2011, San José, California, [Online] Available from: doi:10.1109/ITMC.2011.5996057.

Law, J. (1987) Technology and Heterogeneous Engineering: The Case of Portuguese Expansion. In: Bijker, W.E., Hughes, T.P. & Pinch, T.J. (eds.) *The Social Construction of Technological Systems: New Directions in the Sociology and History of Technology*. Cambridge, MA, MIT Press.

Law, J. (2003) On the Methods of Long Distance Control: Vessels, Navigation, and the Portuguese Route to India. Centre for Science Studies, Lancaster University, Lancaster LA1 4YN.

Latour, B. (1987) *Science in Action: How to Follow Scientists and Engineers Through Society*. Milton Keynes, Open University Press. Latour, B. (2011) Drawing Things Together. In: Dodge, M., Kitchin, R. & Perkins, C. (eds.) *The Map Reader: Theories of Mapping Practice and Cartographic Representation.*, Chichester, UK, John Wiley & Sons Ltd [Online] Available from: doi:10.1002/9780470979587.ch9.

Leydesdorff, L. & Etzkowitz, H. (1996) Emergence of a Triple Helix of University-Industry-Government Relations. *Science and Public Policy*, 23, 279–86.

Leydesdorff, L. & Etzkowitz, H. (1998) Triple Helix of Innovation: Introduction. *Science and Public Policy*, 20 (6), 358–64.

Magee, C. L. & Devezas, T. C. (2011) How many singularities are near and how will they disrupt human history? *Technological Forecasting & Social Change*. [Online] 78 (8), 1365–1378. Available from: doi:10.1016/j.techfore.2011.07.013 [Accessed 26th March 2013].

Nowotny, H., Scott, P. & Gibbons, M. (2001) *Re-Thinking Science: Knowledge and the Public in an Age of Uncertainty*. London, Polity Press with Blackwell Publishers.

Reuters, June 28 2012, "Portugal construction bleeds jobs, threatens banks", retrieved on 1 August 2012 from http://www.reuters.com/article/2012/06/28/ us-portugal-construction-debt-idUSBRE85R18T20120628.

Shinn, T. (2002) The Triple Helix and New Production of Knowledge: Prepackaged Thinking on Science and Technology. Social Studies of Science, 32 (4), 599–614.

Shavelson, R. J. & Towne, L. (eds.) (2002) *Scientific Research in Education, National Research Council*. Washington, DC, National Academy Press.

Sladovich, H., (ed.) (1991) *Engineering as a social enterprise*. Washington, D.C., National Academy Press.

Strauss, E. (1994) Dictionary of European Proverbs. Routledge, Taylor & Francis Group.

Tilli, S. & Trevelyan, J. P. (2008) Longitudinal Study of Australian Engineering Graduates: Preliminary Results. *Proceedings of the American Society for Engineering Education (ASEE) Annual Conference & Expositio. 22–26 June 2008, Pittsburgh, PA.*

Trevelyan, J. P. (2007) Technical Coordination in Engineering Practice. *Journal of Engineering Education*, 96 (3), 191–204.

Trevelyan, J. P. (2008) A Framework for Understanding Engineering Practice. *Proceedings of the American Society for Engineering Education (ASEE) Annual Conference & Expositio. 22–26 June 2008, Pittsburgh, PA, [Online] Available from:* http://www.asee.org/search/proceedings?search=session_title%3A%22Professional+Skills+and+the+Workplace%22+AND+conference%3A%222008+Annual+Conference+%26+ Exposition%22 [Accessed on 1st August 2012].

Trevelyan, J. P. (2009) Engineering Education Requires a Better Model of Engineering Practice. Proceedings of the Research in Engineering Education Symposium, REES 2009, 20–23 July 2009, Cairns, Australia.

Trevelyan, J. P. (2010) Reconstructing engineering from practice. *Engineering Studies*, Special Issue: Situated Engineering in the Workplace, 2 (3), 1–21.

Vinck, D. (2003) *Everyday Engineering: An Ethnography of Design and Innovation*. Inside Technology, Bijker, W.E., Carlson, W.B. & Pinch, T.J (eds.) Boston, MA, MIT Press.

Weick, K. (1995) What Theory is Not, Theorizing Is. *Administrative Science Quarterly*, [Online] 40 (3), 385–390. Available from: http://www.jstor.org/stable/2393789 [Accessed 26th March 2013].

Williams, B. & Figueiredo, J. (2010) Engineers and their practice: A case study. Proceedings of the 2010 IEEE EDUCON Conference on Education Engineering, 14–16 April 2010, Madrid, Spain. [Online] Available from: doi:10.1109/EDUCON.2010.5492533.

Williams, R. (2002) Retooling: *A Historian Confronts Technological Change*. Cambridge, MA, USA, MIT Press.

Zacuto A. (1496) *Almanach Perpetuum de Abraão Zacuto*, p. 1, stored in the Portuguese National Library, Lisbon.

Chapter 8

Going back to heterogeneous engineering: The case of micro and nanotechnologies

Matthieu Hubert[1] & Dominique Vinck[2]

[1]*Department of Social Sciences, University of Quilmes, Bernal, Argentina*
[2]*Institut des Sciences Sociales, University of Lausanne, Lausanne, Switzerland*

I INTRODUCTION: HETEROGENEOUS ENGINEERING (RECALLING LAW'S STATEMENTS)

Based upon John Law's concept of 'heterogeneous engineering', we set out to explore engineering work in the field of Micro and Nano Technologies (MNT) visiting a variety of places where technology and society are shaped; international committees where roadmaps are defined, research labs where new devices are designed, platforms where technological exploration and transfer are performed, big collaborative research programs where researchers and companies shape sociotechnical infrastructures, and places where societal inscription of novelties is regulated.

The concept of 'heterogeneous engineering' was first proposed by John Law (1989). It emerges from a theoretical discussion about the ways in which a technological object gets stabilised. This discussion opposes two alternative approaches to the social study of technology. The first considers that the interests and resources of social groups which are able to impose their views explain the stabilisation process (social constructivism). The second approach considers that the stabilisation process is better explained by accounting for the way in which an innovator relates an artefact to social, economic, political, and scientific factors and variables.

Analysing technological change in terms of heterogeneous engineering is part of the second approach. More precisely, it considers the innovator as a seamless web (Hughes, 1983), a network or a system builder (Law, 1989, pp. 111–112), and it emphasises three main characteristics of innovation processes: "(1) the heterogeneity of the elements involved in technological problem-solving, (2) the complexity and contingency of the ways in which these elements interrelate, and (3) the way in which solutions are forged in situations of conflict" (Law, 1989, p. 111). Consequently, according to Law, the ability to articulate heterogeneous elements explains technological success and power relationships.

The approach in terms of heterogeneous engineering uses the actor-network theory of Michel Callon (1986) that rejects any privileged social explanation, proposes to recognise the agency of non-humans, and applies a generalised symmetry between human and non-human in order to describe the 'heterogeneous network' or the actor-network which corresponds to a specific technology. A new technology is seen as coming into being and being performed anew through engagements with mediators of all kind and the construction of translations between entities. In order for an actor-network to become stable, many entities have to be engaged and enrolled.

Shaping, building, controlling and taming these entities and relations are all heterogeneous engineering.

Considering technological practice as heterogeneous engineering also relates to the thesis of 'multiple constraints' that has been proposed by the science historian Peter Galison (1997). As John Law did for the social study of technology, Galison (1997) argues that taking into account the 'multiple constraints' of the material culture of microphysics is necessary to understand its long-term dynamics: indeed, neither scientific factors ('internalist' thesis), nor socio-economic elements ('externalist' thesis), nor technical constraints ('technological determinism') may explain the dynamics of material culture separately.

However, beyond the similarities with Law, Galison takes a stand regarding Actor-Network Theory (ANT):

"their work emphasises the effectiveness and elasticity of the leveraged power that comes of creating alliances and enlisting other groups toward a predetermined end. My work tends to focus on the material (and non material) obstacles that shape and delimit action in the sphere of science over time" (Galison, 1997, pp. 75–78).

In other words, whereas Galison emphasises optimisation work under a fixed set of given 'obstacles', ANT authors and Law's heterogeneous engineering perceive innovators as system builders who may *a priori* shape reality and displace technoscientific or socioeconomic 'constraints'. In fact, ANT and Law's heterogeneous engineering define this shaping as the capacity to enrol and to translate heterogeneous entities (whatever they are, human or not) which conversely will be paid by an enrolment and a displacement of the innovators. Furthermore, enrolling an entity has a cost, which depends on the attachments of this entity to other entities. To translate an entity, the heterogeneous engineer needs to engage in dissociation and detachment mechanisms, subtracting the entity from its previous heterogeneous network (Goulet & Vinck, 2012).

The concept of heterogeneous engineering is useful to show the sociotechnical 'embeddedness' of the construction of technological orientation, artefacts and systems. It sheds light on the 'articulation work' that enables the construction of 'do-able' technoscientific problems and solutions (Fujimura, 1987). It also shows that technology does not successfully establish itself because of its intrinsic characteristics alone.

Taking the case of Micro and Nano Technologies (MNT), this chapter shows that heterogeneous engineering is the fate of everyday engineering practices (Vinck, 2003), at whatever place or 'level' engineers work: defining technological priorities and roadmaps, designing new technological devices, organising technological platforms, developing sociotechnical infrastructures inside public-private partnerships, or regulating the use of technology in order to inscribe it in society. With the concept of heterogeneous engineering, it makes sense to emphasise the mix of constraints in the definition of MNT. What is going on cannot be understood without articulating the variety of places and engineering practices which are shaping the field. From the design of small compounds to the regulation of invisible infrastructures, the concept helps to shed light on engineering practices, which rarely are 'pure' technological work. Circulating in various places where engineers are key actors of the MNT (international committees, research labs, international cooperative projects on pervasive devices, etc.), we will show how they are engaged in the negotiation, translation and articulation work of heterogeneous and interrelated elements. The success of MNT (i.e. the shaping of both the technology *and* of the society) depends on the engineering ability to articulate these elements.

2 HETEROGENEOUS ENGINEERING OF TECHNOLOGICAL ORIENTATION: ROADMAPS

A part of engineering work deals with the making of new technological orientations. Defining such orientations is particularly crucial in technology-driven sectors such as microelectronics, where technical functions and potential markets are often deduced from technological innovation (Devalan, 2006). That is why MNT engineers and researchers use prospective instruments to make important technical choices: one of them is the 'roadmap'[1]. Understanding the making of roadmaps in terms of heterogeneous engineering leads them to ask questions such as: how are roadmaps made? What is materialised through the process of collective decision? Which arguments support the technological choices that will then orient daily engineering practice? Let's consider an example to investigate these questions deeply; the *International Technology Roadmap for Semiconductors* (ITRS)[2].

The ITRS is the main roadmap of microelectronics research and industry. In concrete terms, it is a document of several hundred pages of text and tables. It identifies the main technological problems that have to be solved in order to achieve the next microchip 'generations' also called 'technological nodes'. It is the result of debates and negotiations in committees of experts. Such committees gather engineers and researchers from the main industrial and research centres of the sector. They meet twice a year in order to actualise, renew or readjust their recommendations about the main technological and scientific challenges for the next ten years.

At first glance, the making of the IRTS roadmap seems to be a purely technological activity materialising the consensus of the scientific and industrial community. Indeed, expert' committees give three types of information. First, they identify the relevant problems that have to be treated by engineers and researchers of the sector (depending on the application). For instance, during a scientific conference, an ITRS expert claims: "there are three challenges to be taken up: we want the device to work at room temperature (. . .) control the threshold voltage (. . .) control interference effects". Second, the ITRS gives some recommendations about the 'good' way to solve these problems: it could be new technological solutions to be explored (new material, process or architecture such as double gate transistor, molecular electronics or Silicon-on-Insulator technology), or effective compromises between two correlated parameters (e.g. "we have to reduce performance in order to avoid electronic leaks", or "the necessary replacement of the silicon dioxide leads to new interface problems", says the same expert). Third, the ITRS gives quantified objectives ("we must go under

---

[1] The roadmap as a prospective instrument for technology management was first proposed by Motorola (Willyard & Mclees, 1987).
[2] For more details, see the ITRS website: www.public.itrs.net.

one volt of supply"), and lists the values to be reached for various parameters (critical dimension, leak current, power, threshold voltage, etc.).

However, although all these engineering activities have a strong technological background, the making of the ITRS has important economic and strategic dimensions. In particular, the main challenge is that "everybody pulls in the same direction", says an ITRS expert, "because we are all together in the same boat". The collective goal is to organise a combined process of competition and cooperation ('coopetition') between all industrial and academic actors of the sector: "it is in the well-known interest of everybody", says a technologist-researcher, according to whom the roadmap is useful to know "when we will need what". In particular, expert' committees agree about recommendations for equipment suppliers who need time to develop new machines: "it is an advantage for the suppliers" that they may limit financial risks by developing the machines in relation to the future needs of microchips manufacturers, as explains the same expert:

> "Today, it is sure that nobody might stay alone at home . . . it means that equipment suppliers have accepted to play the collective game. It is a challenge that they have accepted to take up. Well, there is a competition between them: 'may the best man win'. But there is also a lot of collaborative work with universities and manufacturers."

Beyond the combined cooperative and competitive logics that are materialised through the roadmap, the ITRS produces a number of injunctions whose effects on technological orientation depend on its effective use by MNT engineers and researchers. Even from public researchers' point of view, reference to the roadmap constitutes a good reason to address a technological problem: "it is urgent to deal with it; that is what it means", says an engineering researcher. In the official documents of scientific institutions, arguing about the roadmap is often a way to justify the choices of technological orientations that have been made. For instance, we have studied the strategy of some French researchers and engineers developing a new technology of plasma etching that reduces the critical dimension of the transistors. In particular, they justify the interest of their technological orientation by arguing about the objectives of the ITRS:

> "The ITRS put forward a significant gap between the resolution of a given generation of lithography and the dimensions that must be reached in the silicon for the equivalent generation of microchips. In 2016, transistors with a gate of 9 nanometres must be realized, whereas the lithography that will be used will have a resolution of 13 nanometres" (Activity report of a public research laboratory).

Thus, these researchers use the ITRS roadmap as an argument to show the relevance of their orientation toward a new technology of plasma etching that could compensate for the future failure of another technology (lithography). In this context, the roadmap is perceived less as a constraint than an opportunity to seek new financial resources and to interest potential partners (especially industrial ones). More generally, public researchers and engineers make strategic use of the ITRS.

However, beyond declared consensus and strategic uses, MNT engineers and researchers evaluate variously the consequences of the ITRS. In particular, they repeatedly criticise the adverse effects that are produced by the roadmap. According to criticism, the ITRS fosters defensive strategy instead of stimulating innovation: low-creative solutions are encouraged because "everybody is following the same way". The 'cautiousness' is increased by the predominant influence of industrial actors in the process of decision-making: "some try to go faster than the roadmap or to influence it", says a researcher, making notable reference to the largest multinational firms such as Intel, that have a strong influence on the decision process; however, none of the protagonists seems to be able to challenge the mainstream of the ITRS.

For instance, during an MNT conference gathering of industrial and scientific actors, a researcher in microelectronics denounces "the extreme conservatism of the microelectronics industry". Then he calls for a dialogue between electronics and optics scientists and engineers in order to break such 'conservatism' and to solve the 'bottle-neck' of 'interconnection problems' – and he suggests, with a certain irony, that laser sources must be installed far from transistors "for the microelectronics engineers not to be afraid". Similarly, in a more informal meeting, a former manager of microelectronics industry admits that "the roadmap is an innovation killer".

Finally, the inscription of technological research and engineering in the roadmap of microelectronics research and industry is not something neutral, because it defines scientific and technological objectives for a whole community of academic and industrial actors. In that sense, the roadmap is a powerful coordination tool. Although it is strongly criticised for the mainstream perspective that it promotes, the argumentative and effective use of the ITRS is a common strategy for MNT engineers and researchers who often consider it necessary to follow the roadmap in order to interest industrial partners, to convince financing agencies, or to mobilise potential users who could be associated with the optimisation of new devices.

In doing so, the strategic orientations that are made and adopted by engineers and researchers of the field are not purely guided by scientific and technical considerations. On one side, negotiations, compromises and power relationships play a central role in the creation of the ITRS. Accounting for that dimension, the concept of heterogeneous engineering is useful to show the social 'embeddedness' of the construction of technological orientation. On the other side, MNT researchers and engineers have to take into account the objectives and priorities that are inscribed in the roadmap. Using the concept of heterogeneous engineering, it makes sense to emphasise the mix of external and internal constraints in the definition of technological orientation: it sheds light on the 'articulation work' that enables the construction of a 'do-able' technoscientific problem (Fujimura, 1987). Furthermore, as we will present in the next section of the text, the concept is also useful to show that a technology does not successfully establish itself because of its intrinsic characteristics alone.

3 HETEROGENEOUS ENGINEERING OF DEVICES: LAB ARTICULATION WORK

Beyond orientation choices, technological devices are also shaped by the research, design, and experimental practices that occur inside public and private laboratories. How do technology orientations translate into daily laboratory practices? And how does the concept of heterogeneous engineering help to explain what is at stake in the laboratory? Let's take an example to answer these questions, the one of "nanoimprint lithography".

From the beginning of the fabrication of transistor-based integrated circuits, lithography technology had a crucial role (Sydow *et al.*, 2012), because its 'solution' determined the critical dimension of the transistors (i.e. 'the technological node')[3]. Since 2003, nanoimprint lithography has been included in the ITRS: it is part of "the family of advanced lithography technologies that should address the 32 nanometres [transistors] in 2013", as emphasised by a researcher during a conference in 2005 gathering together industrial and scientific actors in the field. In doing so, nanoimprint lithography establishes itself as an alternative technology for future MNT. "So, the deadline is there", adds the same researcher. But the story is not as simple as it seems to be.

As we have demonstrated previously, the inscription in the ITRS is not something neutral, because it has implications for the scientific and technological objectives to be pursued. What are the consequences at the level of laboratory practice? Beyond the general orientation of the investigation, it is important to understand that the inscription in the ITRS also modifies the *technical* characteristics that have to be considered in the research activity. More precisely, it means that crucial properties of the nanoimprint lithography are therefore 're-qualified', and that the requalification process constitutes an important form of heterogeneous engineering that occurs in the MNT laboratory.

Thus, if nanoimprint lithography does not have the same characteristics after its inscription in the ITRS, what are its new characteristics? First, as for other industrialised lithography technologies used in order to design microelectronics devices, the 'resolution' becomes a crucial parameter. More precisely, the important change impacts *the resolution that can be achieved on a standardised microelectronics sample*, whose form and size are fixed for the whole microelectronics industry which leads the same researcher to raise key questions: has it been calculated on "large industrial samples" (i.e. a disc with a diameter of 150, 200 or 300 millimetres), or has it been calculated on "samples which have been specially made for publication" (i.e. samples that are smaller and less challenging in terms of engineering capabilities)?

In addition to the resolution, the inscription of nanoimprint lithography in the microelectronics roadmap means that 'speed' becomes a 'fundamental point', as she explains:

"The speed is the second very interesting characteristics in the perspective to integrate these processes in microelectronics industry. It's excellent, which may be surprising for you, it's excellent; especially comparing it to the use of electronic lithography, which means direct inscription with an electronic beam. Electronic lithography processes in a linear way, microchip by microchip, whereas our replication technology process in a collective way; this is the point, the second fundamental point."

---

[3]In fact, transistors may be significantly smaller than the resolution of the lithography machine thanks to processes (such as oxygen plasma) that have been recently industrialised, and that reduce the critical dimension during the etching step.

Finally, in addition to its capacity to collectively process all the chips that are gathered on a single sample, 'reproducibility' is a third essential characteristic of nanoimprint lithography in order to be of interest to industrial manufacturers. According to the researcher, its importance depends strongly on "technological application" and "engineering optimisation": "concerning the reproducibility, I would say: it has to be worked. It's mould engineering; it's process engineering".

Thus, after its inscription in the ITRS, what is relevant to characterise nanoimprint lithography is heavily transformed. Before its inscription, nanoimprint lithography was a relatively confidential technology that was destined for a few marginal applications: it was considered as a laboratory technology to be tinkered with – a 'waffle mould', says another researcher from the same laboratory. However, after its inscription in the roadmap, nanoimprint lithography became a benchmark technology that circulates easily from one research centre to another, and from research to industry. It circulates better not only because it is made credible by the ITRS, but also because its relevant technical characteristics have been 're-qualified': the parameters that have to be taken into account (resolution, speed, reproducibility) come within the competence of industrial and technology management.

The requalification process – i.e. the reformulation and transformation of what are the relevant technical characteristics – may be analysed in terms of heterogeneous engineering. In that other meaning, heterogeneous engineering enhances the technical qualities of nanoimprint lithography by linking them to the conditions of mass production, and to the standards of the microelectronics industry. The concept emphasises the 'multiple constraints' (Galison, 1997) that have to be taken into account concurrently to develop an experimental device, and possibly transfer it to industry. Furthermore, it shows that a technology does not successfully establish itself only because of some intrinsic technical characteristics, but rather through a specific requalification and articulation process.

4 HETEROGENEOUS ENGINEERING OF RESEARCH ENVIRONMENT: PLATFORMS

Thus, at the laboratory level as well as for the construction of technological orientation, heterogeneous engineering is a useful concept to account for technology research and engineering practice. It is also true if we widen the focus, and we consider the broader research environment of MNT activities. Let's take the case of the setting up of technological platforms.

Platforms are devices organising experimental activities for external users who may come from research, education or industry (Merz & Biniok, 2010). In order to carry out the day-to-day work of the platform, engineers, researchers, technicians, and managers accomplish a great variety of heterogeneous activities such as: mobilising financial support, negotiating with suppliers, following the setting up of new tools and machines, interacting with users, managing maintenance contracts and their effective realisation, redacting some access procedures, teaching the use of instruments, etc.

The organisational work of an experimental environment as a platform is an important part of technology research and engineering practice in MNT, and sociotechnical skills are necessary to perform it. In particular, communication skills play a key role. For instance, presenting a new platform to a public of potential users (researchers, industrialists or students) requires a capability for promotion. Indeed, one of the objectives of giving a talk to present a new platform is to interest future users: in doing so, the platform spokesperson promotes a global technological offer (concretely, through the projection of a list and some pictures of tools and machines) that users will pay for in order to realise their own experimental activities.

Furthermore, beyond the communication work, some other forms of heterogeneous engineering are even more crucial when we observe the effective process that leads to the setting up of a new platform. In particular, technology research and engineering practice go through some processes of negotiations and decision-making regarding the sharing of instruments and the regulation of experimental activities. Such interactive processes deal with: the institutional and operational governance of the structure; the possible differentiation between experimental work spaces (e.g. for education, research, or technology transfer to industry); the concrete modalities of project evaluation; the conditions of sustainable financing and the possible contribution of users (depending on users' institutional belonging); the rules (material compatibility) and procedures of administrative and physical access to the platform; the effective role of users on the platform (doing the experimental work, or giving instructions to some local technicians who perform it); the respective roles of local technicians and assistance from equipment suppliers in the set-up and maintenance of machines; the rules of intellectual property protection and secrecy versus sharing knowledge and experiences (Hubert, 2013).

In these kinds of negotiations, many stakeholders of the platform may be involved: technicians and engineers, managers, researchers, students, equipment suppliers, etc. They confront arguments about innovation strategy (e.g. knowledge production versus technological advancement), collaborative tradition (e.g. cooperative versus competitive, or scientific community versus public-private partnership), evaluation criteria (e.g. financial versus scientific, or publication versus patent), or, more generally, professional culture (e.g. academic versus industrial, teaching versus research, or microelectronics versus microsystems). For instance, technicians, engineers and researchers may confront the managers of the platform regarding the purchase of instruments: does the choice rely on an economic logic or a more technological one? Another example of confrontation deals with the possible modification of experimental devices: some researchers want to explore new ways of experimentation (through the tinkering or unusual tuning of instruments), whereas some technicians and engineers (and equipment suppliers) stick, as much as possible, to the standard uses of machines (in order to avoid unexpected technological problems on the machines that would lead to corrective maintenance). In the same way, those performing exploratory research want to work on low cost wafers (100 millimetre), just to do a quick and dirty experiment in order to have a preliminary idea of what could be expected and done with a given technological process, whereas those involved in technology transfer want to work on industrial standards which requires costly 300 millimetre-wafers[4].

---

[4]Both standards (100 or 300 millimetres) are the standardised sizes of samples that have been discussed in the previous section.

Finally, in the case of platforms, heterogeneous engineering is a relevant concept to account for the great variety of practice through which the 'right' research environment is performed: communicating, negotiating, regulating, organising, and coordinating the experimental work of others are some the activities that technicians, engineers, researchers, and platforms managers do to produce and maintain their experimental environment. In doing so, heterogeneous engineering refers not only to what is pointed at by the concept of 'technical coordination' (Trevelyan, 2007)[5], but also includes the management of tensions, disagreements, and conflicts that are created or revealed by a new or changing regulation, or by a reconfiguration of power relationships (inside the organisation or in its relationships with external partners).

5 HETEROGENEOUS ENGINEERING OF PROBLEMS AND TECHNOLOGICAL SOLUTIONS: A COLLABORATIVE RESEARCH PROGRAM

The shaping of a technological orientation, of a laboratory device, or of a R&D environment can be considered as heterogeneous engineering because it implies sociotechnical embeddedness, re-qualification, articulation, and a variety of practices like organising, coordinating, communicating, and negotiating (sometimes in conflictive situations). Moreover, if we widen the focus once more and consider a technological research program, the heterogeneity of the engineering involved is even more evident. Let's take the case of the design of a digital fingertip.

5.1 Articulating research interests: setting up the problem of naturalness

Among the emerging MNT projects was the idea of developing tactile sensors. A few engineers, always looking for new applications, anticipated that the field of the 'touch' was not explored enough and there would probably be a new frontier for sensors; it would be at the crossroads between fundamental unresolved questions and many potential industrial applications. Thus, two engineers at a microsystems lab decided to engage in an exploratory project on a tactile sensor. The aim was to design a concept and a first demonstrator which means a single sensor to design, to produce and to check its effective performance. As the project was relatively far from direct industrial application and expressed needs, they negotiated internal funds for this basic technological research. An electrical engineer on the team, having more than 10 years experience in the field of sensors (piezoelectric material, micro pressure sensors, force sensors and accelerometers for vibration measures), was in charge of the project. With the assistance of a technician and the support of the technological platform, they recycled a concept they had previously developed for a pneumatic industry: a force sensor dedicated to the measure of deformation and pneumatic adherence. They translated the specifications and designs from the earlier project made in collaboration with the

---

[5]According to James Trevelyan's definition, technical coordination "means working with and influencing other people so they consciously perform some necessary work in accordance with a mutually agreed schedule" (Trevelyan, 2007, p. 191).

automotive pneumatic industry towards new developments in touch perception. They got interesting results for the new tactile sensor and then engaged in cooperation with a research lab in physics working on friction. The sensor was then used for a PhD thesis exploring the way to differentiate various materials thanks to this sensor. The aim of the microsystems lab was to develop the sensor; that of the partner lab was to use it as an investigative tool. Their respective objectives and interests were articulated and translated from one to the other.

Further, this collaboration was involved in the setting up a European network wishing to write a research proposal for the "Measuring the Impossible" initiative. This initiative relates to measuring human responses and feelings: individual susceptibility to diseases, determinants of music emotions, aesthetic sound qualities of artefacts, human experience of media enjoyment (the fun of gaming), perceived air quality, eyewitness memory, feelings and expectations associated with texture, etc. The program relates also to "Emerging Metrology", a section of the International Metrology Congress in 2007.

The design objectives of the microsystems lab were then translated into a new set of arguments related to human perception and cognitive processes. Under the leadership of the National Physical Laboratory, a British leading centre in the development of measurement techniques acting as the UK's national standards laboratory, six research institutes are networked in Europe with an industrial research lab belonging to a multinational care products company. The project involves heterogeneous teams with expertise in the areas of physical measurement, instrumentation and microsystems, cognitive neuroscience, psychology and mathematical modelling. Some are close to industry while others are closer to basic science.

The range of their cognitive interests is also, *a priori*, large. A lab specialising in metrology tries to ensure consistency and traceability of measurements, even in subjective aspects: appearance and visual perception, roughness, smoothness, hardness, and tactile and thermal perception. It is looking for new methods which are more appropriate for predicting the perception of naturalness and for applications to novel situations. Insofar as this lab has also an extensive experience in the development of standards and reference materials, it shows its interest in developing test materials and measurement instrumentation or methods. In the same way, insofar as it has also some expertise in mathematics, modelling and data mining, it engages in analysing and modelling the results of experimental investigations carried out by other partners. It also establishes a link between physical measurement and human perception. As it was involved in another project using a model of the human visual response to light, it suggests using that and building on past experience with visual perception. Furthermore, having experience in coordinating collaborative research projects and a certified ISO 9000 quality management system, it takes charge of the management of the project.

The shape of both the network and the content of the project also relate to the presence of a company's research facility interest in the delivery of new home and personal care products. Two of its research teams, already involved in cognitive neuroscience and sensory physics, have expertise in cognition and in metrology. They also have experimental facilities customised for exploring material sensory properties and cognitive responses. Thus, they are interested in developing test materials.

A university research group has its interests in cognition (exploration of the neural and cognitive bases of the human brain) and is equipped with electroencephalography

(EEG), neuro-imaging techniques (scanner), and sound attenuated rooms for behavioural investigations. It is interested in human visual and tactile senses and cross-modal interactions between these senses. It is also involved in studies of the effect of cortical damage on tactile and visual perception in, and the recognition of objects by, the blind. Another university research group works in experimental psychology and cognitive neurosciences and is interested in the perceptual and brain mechanisms associated with human sensory processing (among others, the interactions between touch and vision). Its expertise in psychophysical measurement of perceptual processes leads it to involve itself in the establishment of psychophysical procedures and to interface between the physical measurements and the neuro-imaging. Another university research group is specialised in image analysis for food applications (colour measurements of flowers and the biochemistry of plant material).

In France, the two labs had already collaborated around a tactile sensor: one has an interest in the physics of surfaces (mechanics of adhesion and friction) and has equipment to produce substrates and to characterise them. It is interested in designing a bio-mimetic tactile sensor and in understanding the phenomena occurring between a finger and a substrate. The microsystems laboratory is interested in developing new sensors and in transferring technologies to companies. It relates also to engineers interested by microsystems for motion capture and prepared to take into account usability in the design of new technologies.

Through the articulation of the experience, resources and expectations of these labs, their network is progressively shaping the content of the project. They involve one another in the project according to their expressed research interest, looking for complementarities in order that some of them can supply inputs for the others. For instance, they may choose the subject of naturalness because the perception of whether or not a material is natural – which is different from the perception of its beauty, for instance – shows good reproducibility from one person to another. This is important for neuroscientists and psychologists because this means the underlying neural processes will have a good repeatability and the knowledge they could publish would have a high impact. On the other hand, this choice of the subject of naturalness joins with the interest of the industry because it is important in a wide range of high-value products. So, naturalness is a "boundary object" (Star & Griesemer, 1989) between two subsets of actors involved. Furthermore, others, mainly those who are interested in the differentiation between materials, translated their interest for substrates variability; naturalness is one characteristic among others. As for the microsystems lab, it is not particularly concerned as to whether the problem was naturalness or not. What is of interest to it is the design of a haptic (tactile) sensor according to specifications defined by the others.

For shaping the object, they also define the specific contribution and competence of each of the partners inside the network. Conversely, defining work packages, they negotiate and set up a common objective and its breakdown into sub-objectives and tasks. Progressively, a hierarchy of priorities emerges with some main goals: understanding the neural process involved in the perception of 'naturalness'; identification of the extent to which the material properties drive this perception; determination of the most appropriate measures for this perceptual process and modeling of the relationships between the physical attributes, the sensory networks, and the cognitive functions, in order to make some predictions of the human perception of naturalness in new situations. In this shaping of priorities, the aim of the microsystems lab (developing new sensors) is not a priority for the whole project, but a useful sub-task. When the project was accepted, the requested budget was significantly reduced causing the main research groups (who define the core priorities) to abandon the sensor development. The engineers of the microsystems lab had to fight and negotiate to maintain their position, their contribution and their part of the grant. The microsystems lab, anticipating that the domain of touch is a growing industrial preoccupation, needed funds for resolving some technological problems regarding the sensor fabrication processes. As engineers, they had to defend their position in a research community dominated by neuroscientists and cognitivists.

5.2 Articulating conceptual approaches: setting up the human perception

Microsystems engineers, physicists, psychologists and neuroscientists apparently share the evidence that: "We instinctively know whether something is natural or a synthetic mimic". They black box this statement and base upon it their arguments to industrial partners and public authorities. Founded on their previous experience and on the literature, they shape the perception of naturalness as dominated by visual appearance and touch. Basic visual sensory information would be sufficient to differentiate between natural and synthetic materials; touching the material would serve to reinforce the initial visual perception. Touching is decomposed into various mediations: skin pressure and thermal perception. The project aims to establish the "chain of perception" (from the physical attributes of the material and sensory inputs to the cognitive processes) for naturalness. This information would be used for mathematical modelling and prediction of the perception of naturalness for a range of materials.

Shaping the goals leads to shaping the work. According to the hierarchy of goals, sub-objectives are defined, such as the development of stimulus materials, of methods for characterising the physical attributes of the materials (among them a digital finger) and of techniques and models relating the physical attributes to the perception of naturalness. Progressively, a new configuration of materials, techniques, procedures and concepts draws what seems to be the perception of naturalness for a human being: a re-engineering of the human perception and cognition.

Entering into the conceptual frameworks of the involved disciplines, they draw articulations and complementarities. For instance, cognitive neuroscience employs a structure/function relationships approach for the various stages of sensory-perceptual-emotional processing. Neuroscientists have access to it through neuro-imaging techniques (identification of neural pathways of the brain activity). They also build upon knowledge developed within the sensory science relating to the properties of the relevant sensory mediators (among them touch). Psychologists, for their part, look into the way in which consumers react to products, through panel tests for marketing purposes. They determine which sensory attributes elicit particular subjective responses in a given situation. The articulation with neuroscience goes through the connection with the underlying neuropsychological processes, and empirically through coupling psychological methods, measurements of the physical parameters and neurological and cognitive processes for these same materials. Then enter the metrologists with a dual approach: centred on both materials and human sensing. They would generate robust objective measures of the physical properties, but taking into account the fact that real products are not large samples with uniform bulk characteristics.

They also need to develop systems for force measurement with high resolution to measure the friction between the skin and, for instance, a soft material (textile, flower petal, cream). The challenge is to produce measures that mimic a human sensory system. Such an approach already exists for visual perception, but needs to be extended to the measurement of the mechanical properties through force measurement sensors. Thus, the specifications of the sensor that the microsystems lab has to develop are related to this challenge: to mimic human skin mechano-sensors to provide a direct window on those material aspects a human sensory system can actually perceive. A bio-mimetic sensor would be developed to mimic the tactile sensitivity of the human fingertip. This 'digital finger' will use micro-technologies. It would be used to demonstrate the capability to discriminate natural from synthetic surfaces within the range of materials. Thanks to such an instrument – labelled as "cutting-edge MEMS[6] technology" – a new area of science and technology, the measurement of subjective parameters would be developed with many potential applications related to the quality of high added value goods, their desirability and consumer expectations in terms of comfort. What is at stake, mainly for industrial companies, is to predict human reactions to new materials and objects – but also fighting against fraud by detecting fake materials and counterfeit goods. This is related to the statement the researchers make that natural products (silk, leather, walnut, rosewood, ivory) are perceived as highly desirable and associated with exclusivity but they are also limited natural resources. Thus, the project would open up new opportunities for the development of improved artificial materials and for product design in fashion, cosmetics, and the food industry. In the longer term, the challenge is to replace the panel testing approaches currently used in various industrial sectors – thus reducing time-to-market and development costs. Thus, some relations are expected with international organisations for standardisation. Furthermore, researchers relate the project to a general analysis of the European economy qualifying it as dependent on the ability to produce high quality products and to differentiate these from lower cost and poor quality imitations imported from outside Europe. Setting up a digital fingertip is a way to protect itself from new competing countries – just as CAD/CAM[7] was a way to save the US nation (Downey, 1992). They also relate the project to worldwide challenges like sustainable development (designing artificial products in order to protect natural resources), improvement of well-being and health (often associated to naturalness), and productivity (thanks to the design of comfortable workplace environments). The researchers also relate the project to a heterogeneous set of applications like remote surgery and surgical training, and other virtual reality situations in which having confidence in virtual systems seems to be essential.

5.3 Articulating the device: setting up a digital fingertip

The project is divided into work packages. One is dedicated to the identification of the key attributes of a material that influence the perception of naturalness; it is supposed to set the boundaries between the psychophysical and neuroscience investigations and

---

[6]Micro Electro-Mechanical System.
[7]Computer-Aided Design/Computer-Aided Manufacturing.

the design of measurement devices, among them the fingertip-mimetic sensor. The work package dedicated to the development of the tactile sensor then relates to the physiological and mechanical characteristics of a human fingertip and the friction mechanisms between the skin surface and the material being touched.

The design and fabrication of the bio-mimetic tactile measurement device represents 1/5 of the whole project in terms of person-months budget. It is under the leadership of the physics laboratory of which the microsystems lab is a 'provider'. These two labs cooperate in the realisation of the fingertip-mimetic sensor in order to match the characteristics of the human fingertip but the engineers have no leadership in the project.

Physicists design the device. They consider the human fingertip as a tribological instrument in which mechanical distortions of the skin are converted into nervous signals. This human instrument is represented as a set of hundreds of mechanosensitive cells embedded within the first two millimetres below the skin surface. Thus, the device should also be a matrix of sensors with a specific density of sensors. Furthermore, touch is modelled as a complex of friction mechanisms between the skin surface and the material being touched. Therefore, the device should reproduce these friction mechanisms. The digital finger takes the form of a friction measurement device. The principle is to rub down the substrate with the embedded sensor. The substrate is stuck on a cylinder, which comes to press – under controlled conditions – on the sensor put on a flat surface. Thus, it simulates the material either caressing or pressing on the finger.

The design and construction of the digital finger started from an existing prototype already developed by the physics lab and the microsystems lab. It consists of a single micro-force sensor, a micro electro-mechanical system simulating a biological mechanosensitive cell, embedded in a textured elastic membrane simulating the skin. The microsystems lab designed and constructed the first one – following a previous prototype development for pneumatic adhesion measure – and the laboratory of physics inserted it into the whole device. In the new project, the novelty is to use an array of micro-force sensors that could replicate the multiplicity of mechanosensitive cells in a human fingertip.

From the engineers' point of view, moving from a single sensor toward a matrix of sensors is not so much becoming closer to a human fingertip as achieving a technological breakthrough. Being expected to align themselves with the industrial trend to increase the ability to design single sensors, their challenge is to advance in some lesser-known directions, among them designing an integrated matrix of sensors. Thus, the move is, for them, considered as 'a logical continuation'.

In fact, two types of arrays of micro-force sensors would be developed. The first one is a basic version obtained by assembling existing micro-force sensors to produce a demonstrator system. To do it, the engineers start from specifications defined by the physicists but they start the design and realisation for internal reasons: the opportunity to still work on 100 millimetres equipment before the technological platform moves towards new industrial standards[8]. They expect to deliver five calibrated

---

[8]The standards (100, 150, 200, or 300 millimetres) are the standardised sizes of samples that have been discussed in the previous sections 3 and 4.

demonstrators of fingertip-mimetic tactile sensors (one-dimensional arrays with the associated electronics). These demonstrators are not considered as a good representation of the fingertip, regarding their *a priori* sensory performances, but they are needed for immediate use in the first phase of experimental investigations, and it will provide guidance for the design of the second array which will be built as a fully integrated micro-system, composed of many sensors in a single chip. The performance would then be closer to the human fingertip, particularly in terms of pitch and sensitivity. They plan to deliver ten calibrated prototypes of fingertip-mimetic tactile sensors with performance optimised to match the sensory performance of the human fingertip.

In this case, we observe that heterogeneous engineering is a collective endeavour, which partly escapes engineers. They are involved in a network of actors, with their research and industrial strategies, previous knowledge and experience. They have to negotiate and to translate their know-how, existing devices, and strategy in order not to be excluded from this broad heterogeneous network shaping the new device.

6 HETEROGENEOUS ENGINEERING OF A SOCIOTECHNICAL INFRASTRUCTURE: EUROPEAN SENSORS' NETWORKS

The making of MNT refers to heterogeneous engineering when orientations are defined and translated into R&D projects, instrumental infrastructures, and cooperative research programs. The previous example, even if it involves a broad heterogeneous network, still stays inside the R&D world around a specific device and its local applications. Now, let's take the case of big European programs leading to the shaping of a sociotechnical infrastructure made of sensors' networks. It engages potential major changes in the society for which heterogeneous engineering appears in full light.

The engineers' dream and challenge here is to merge physical, digital and virtual worlds, putting sensors in everything, such as cars, clothes, keys, books, homes, and consumer goods. A leading engineer of the program, involving twenty research labs and companies, makes public promises saying, for instance: "this would make our daily lives easier, safer and more efficient". To build such 'ambient intelligence' for smart homes, cities or countries, he says, they need to bridge the gap between physical and virtual worlds. The project is to set up smart cities all over Europe with imperceptible and autonomous technology helping everyone with everyday tasks like controlling the temperature at home remotely or getting instant offers based on one's location and interests. The system needs to be able to handle big data (vast amounts of heterogeneous data): conservation, circulation, treatment, visualisation and intuitive ways to interact with it. It refers to the 'Internet of things' where, beside web servers, the requests will search for information from a large amount of disseminated sensors that are integrated in everyday goods all over the world. The engineering challenge is just how best to connect the virtual world to the physical world. It is problematised as a problem of articulation between goods, parameters, sensors and actuators, data, servers and software, IT infrastructure, interfaces and users for which most of these things must be 'transparent', which means invisible.

In doing so, engineers contribute to the re-shaping of society. The network articulating physical and virtual worlds, in fact, would connect and transform various sociotechnical networks. The interface, for instance, would take the form of services (e.g. free service on Internet, services offered by private companies, or public services) giving understandable information, leading to automatic action (e.g. opening a door) and preserving security, trust, and privacy. They also imagine and shape the potential user within this development as a person asking questions of the machines in their natural language, e.g. "what is the temperature at home and in the village where my grandmother lives alone?" The technology would decode the question, look for the relevant sensors' networks and access to the temperature sensors in the village (e.g. in the streets and bus stations, embedded in the cars circulating in the village, or distributed in the grandmother's clothes), checks the data from various sources of information, structures an answer (ensuring data privacy), and returns it to the user who can decide with whom the information can be shared.

The project has already engaged a test bed within a Spanish city where 12,000 sensors are deployed to help "to smooth the flow of traffic in the city" through the monitoring of parking places and the information to drivers about the available places. It is expected to provide benefits for the people, to transform Santander into a smart city, and to generate experimental data for researchers. A specific sociotechnical network was set up there with local public authorities, companies and researchers. It still pursues its development extending connections to other networks of sensors and actuators for the control of street lighting according to the presence or absence of people on the streets. The system would turn light up when increased activity or incident are detected. This would be connected with public (police, emergency medical service, etc.) and private services and with energy distribution, management and regulation.

Similar implementations are in train for a set of other cities in Europe. In Aarhus, the aim is to have intelligent and autonomous management of water and sewage infrastructures. In Berlin, the smart city is going through the development of a network of 'intelligent waste baskets' connected to waste management services. In Birmingham, it is shaped through transport infrastructure and services (trams, buses, cycle paths, walkways, and roads) in order to have a streamlined transition between modes, saving people's time. These projects involve public authorities and services but also major industrial companies like Ericsson, SAP, Thales, or Telefónica, which integrate the infrastructure solutions into their business plans and offers of services to cities – like Ericsson with a fleet tracking system for the Belgrade transportation system. In fact, the programs develop just like burgeoning heterogeneous networks, which progressively are re-shaping society and its sociotechnical infrastructures.

Wireless engineers say that in a few years, personal networks will have connections between more than a thousand devices and networks (telephones and TVs, laptops, sensors, embodied intelligent prosthesis, sensors monitoring vital signs, smart clothes, cars, machines, homes, and environments). To connect all these devices in a 'network of everything', IT engineers shape smart personal network tools (the 'fourth generation') where many heterogeneous devices would work together, providing customised services, and ubiquitous connectivity. The engineers, therefore, write reports, R&D proposals and industrial plans, and give demonstration and public presentations, trying to convince a variety of people to share the same conviction and vision for the future of the society. They convince public authorities to put the money into projects like the "personal adaptive global network", which translates the concept of 'personal area networks' into devices, infrastructures and tools. They state that the number of personal devices will multiply considerably (a thousand for every person on the planet)[9], even if it is hard to know which ones. They pave the way for a 'federation' of personal networks, translating friendship, family and professional relations. They also project (at least in their public argumentation) a user who would be able to control which devices and information other people can access, through a tailored security, a trust framework, and the management of identity and credentials. The ESNA (European Sensor Network Architecture) project is preparing for the setting up of the support for wireless sensor networks allowing users to connect things and people within an 'Internet of whatever'. Based on this infrastructure, engineers project that things could communicate between themselves so that clothes could self-colour-coordinate, refrigerator and kitchen cabinets create a shopping list, foods at home could self-meal-plan. With e-tags equipped with micro or nanosensors that could measure various parameters (moisture, light, temperature, pressure, motion, etc.), goods, cars, people, and places would generate information on what is present, moving, damaged and so on. For that, the ESNA project is developing a multifunctional platform supporting a variety of applications to meet a huge diversity of uses.

In fact, this heterogeneous engineering is invested in by companies which shape a business-oriented new world and by 'geeks' and social movements fighting for opening or democratising the future of the sociotechnical networks. As in the case of the Portuguese maritime expansion (Law, 1989), heterogeneous engineering is also the shaping of power relations and its analysis details mapping the pattern of forces revealed through the adversities met by IT engineers.

7   CONCLUSION: ARTICULATING THE DESIGN OF SMALL DEVICES AND REGULATED INVISIBLE INFRASTRUCTURES

Through the five cases presented in this chapter, we give an idea of the variety of places where engineers are nowadays engaged. The heterogeneity of the work situations shows the heterogeneity of the connections and of the engineering practices. Whatever the places or the 'level' (road mapping, laboratory articulation work, technological environment building, shaping of a new device, or setting up a new sociotechnical infrastructure), the notion of heterogeneous engineering helps to shed light on a variety of processes of which much concerns negotiation with a variety of partners, materials and constraints, sociotechnical embeddedness, displacement and re-qualification, articulation with existing sociotechnical networks, and fighting to convince, to organise, to coordinate and to connect and transform these networks.

So far as the design and introduction of new products and technologies could affect society, it is not a surprise to encounter a variety of actors enrolling scientists and engineers in the making of new objects and infrastructures. But we saw that engineering practice engages also the shaping of the society, identities, and power relations: engineering work is also the construction of rules and frameworks, supporting the societal inscription of the novelties into the society. This leads to another set of practices in which engineers are heavily engaged: regulation. Indeed, through the making

---

[9]The World Wireless Research Forum (WWRF) predicts of 7 trillion devices for 7 billion people by 2017.

of norms and standards, of metrology and data treatment, of monitoring systems and infrastructures, of new technical constraints or possibilities, engineers also extend their engineering to the shaping of the rules, the functioning and the evolution of society.

In the case studies, we have chosen to focus on the many practices that deal with the shaping of engineering artefacts. In doing so, we have not been so much concerned about engineers' individual and collective motivations (except the most strategic ones, that *directly* shape technological development). However, it is crucial to understand how engineers valorise their work, how they qualify the importance of the project in which they are engaged, and also to know how they appreciate and feel the heterogeneous nature of their activity. This would help to analyse how individual motivational factors affect heterogeneous engineering. Furthermore, the collective nature of engineering work would be better understood, allowing us to analyse how engineers collectively defend what they value, and how heterogeneous engineering has shaped their professional culture. This could set the stage for future inquiry.

REFERENCES

Callon, M. (1986) Some Elements of a Sociology of Translation: Domestication of the Scallops and the Fishermen of St Brieuc Bay. In: Law, J. (ed.) *Power, Action and Belief: A New Sociology of Knowledge*. London, Routledge & Kegan Paul. pp. 196–233

Devalan, P. (2006) *L'innovation de rupture. Clé de la compétitivité*. Paris, Lavoisier.

Downey, G. (1992) CAD/CAM saves the nation? Towards an anthropology of technology. In: Hess, D. & Layne, L. (eds). *Knowledge and society: the anthropology of science and technology*. London, JAI Press Ltd. pp. 142–168.

Fujimura, J. (1987) Constructing "Do-able" Problems in Cancer Research. Articulating alignment. *Social Studies of Science*, 17 (2), 257–293.

Galison, P. (1997) *Image and logic. Material culture of microphysics*. Chicago, The University of Chicago Press.

Goulet, F. & Vinck, D. (2012) Innovation through withdrawal. Contribution to a sociology of detachment. *Revue Française de Sociologie*. [Online] 53 (2), 117–146. Available from: http://www.cairn.info/resume.php?ID_ARTICLE=RFSEN_532_0117 [Accessed 26th December 2012].

Hubert, M. (2013) *Partager des expériences de laboratoire. La recherche à l'épreuve des réorganisations*. Paris, Editions des Archives Contemporaines.

Hughes, T. P. (1983) *Networks of power: Electrification in Western society, 1880–1930*. Baltimore, Johns Hopkins University Press.

Law, J. (1989) Technology and heterogeneous engineering: the case of Portuguese expansion, In: Bijker, W., Hughes, T. & Pinch, T. (eds) *The social construction of technological systems. New directions in the sociology and history of technology*. Massachusetts, MIT Press. pp. 111–134.

Merz, M. & Biniok, P. (2010) How Technological Platforms Reconfigure Science-Industry Relations: The Case of Micro- and Nanotechnology. *Minerva*. (48), 105–124.

Star, S. L. & Griesemer, J. (1989) Institutionnal ecology, 'translations' and boundary objects: amateurs and professionals on Berkeley's museum of vertrebate zoology. *Social Studies of Science*. 19, 387–420.

Sydow, J., Windeler, A., Schubert, C. & Möllering, G. (2012) Organizing R&D Consortia for Path Creation and Extension: The Case of Semiconductor Manufacturing Technologies. *Organization Studies*, 33 (7), 907–936.

Trevelyan, J. (2007) Technical Coordination in Engineering Practice. *Journal of Engineering Education.* (July), 191–204.

Vinck, D. (ed.) (2003) *Everyday engineering. An ethnography of design and innovation.* Inside Technology series. Cambridge, MA, MIT press.

Willyard, C. & Mclees, C. (1987) Motorola's technology roadmap process. *Research Management.* (September–October), 13–19.

Chapter 9

Professional lock-in: Structural engineers, architects and the disconnect between discourse and practice

Andrew Chilvers & Sarah Bell
Civil, Environmental and Geomatic Engineering, University College London, London, UK

I INTRODUCTION

Recent challenges such as sustainable development and climate change, along with high profile technological controversies and failures, have brought ethical and normative elements of engineering practice into focus. Particular attention has been drawn to the need for engineers to actively engage with normative goals such as sustainability. This focuses attention on the agency of engineers within sociotechnical networks where these same networks can both enhance and constrain engineers' capacities to contribute to positive change.

This chapter presents ethnography of a structural engineering project undertaken by the firm Arup. Arup have long engaged with issues of holistic design and the role of engineering in achieving societal goals. The chapter summarises key themes from the writings of the firm's founder, Sir Ove Arup, relating to the need for more integrated approaches within the built environment sector, as exemplified in his proposition of 'Total Design' (detailed in section 3.4). The implications of Total Design in addressing the challenge of sustainability have been taken up by the firm in recent years yet, as the case study of structural design shows, providing leadership to clients in this field is highly constrained in practice. The professional arrangements within the project to deliver a public building constrain efforts to deliver a more efficient, sustainable structure. Instead, the architect takes a central role, mediating relationships within the actor-network of the design team such that the structural engineer is dissociated from the building users and owners, as well as the construction contractors. In this way the project conforms to the divisions in the industry that Ove Arup was explicitly trying to overcome when he set up the Arup firm in 1946. The persistence of these divisions points to 'professional lock-in' which may be just as pervasive and immutable as the 'technical lock-in' that has been identified as a key barrier in the transition to more sustainable sociotechnical systems.

This chapter reminds us that engineers rarely operate in isolation; working instead in networks of relationships which include clients, other professionals, colleagues and team mates, legislation and standards, finance, contracts, technologies, production facilities, the environment and communities. As such, drawing the boundaries of responsibility for engineering action is often fraught and close attention to the real-life constraints on engineers, manifest from the actor-networks in which they must operate in practice, is vital. The ethnography we present shows how the structural engineer negotiates trade-offs and conflicts to meet the client needs and, in doing so, is confronted with dissonance between Arup's two strategic goals of supporting high-end architectural innovation and leading clients towards sustainability.

We first draw on Actor-Network Theory (ANT) to highlight the networks of relationships that constrain the agency of the engineer. We then outline the culture and discourses of Arup, from the vision of its founder to its current corporate strategy. The case study of the structural design project is then presented. This shows the emergence of dissonance between the firm's strategic goals as everyday practice is constrained by long standing relationships between the architecture and engineering professions which persist in the built environment sector despite Arup's efforts at reform.

2   ACTOR-NETWORK THEORY

Actor-Network Theory (ANT) deals with the networks of relations which define and enable entities to present themselves in the world (Latour, 1993). It rejects the human/nonhuman distinction; all phenomena possibly encountered in the world are part of networks of relations between actants. Both humans and nonhumans have agency in processes that 'translate' interests and build or disrupt material relationships. In this way actants are defined by their existence within networks, rather than by essential properties intrinsic to themselves. The power held by any one actant is not a property inherent to them but is determined by the control or influence they have over the networks in which they are enlisted. Humans and nonhumans are intricately bound up and defined within networks of relations and the way things are experienced and the way we come to know reality is manifest from these networks of sociotechnical relations (Latour, 1987; 2005).

According to ANT (e.g. Latour, 2005), faithfully tracing actants and the processes of translation avoids flawed use of pre-defined and often isolated sociological categories. There are many examples of such studies (e.g. Callon 1986a; 1986b; 1989). Studies of actors who mediate between producers and consumers, including journalists, public sector agencies, policy makers and patient advocacy groups, have shown their power to constrain designers both within their own organisation and without, and thereby challenge simplistic designer-to-user models of design and technological development (Oudshoorn & Pinch, 2008).

Demonstrating the value of close attention to roles and relationships in the professions contributing to our built environment, Anique Hommels (2005) deploys ANT alongside concepts from theories of Large Technical Systems and the Social Construction of Technology to explain the obduracy of the built environment. Buildings and infrastructure are resistant to change as a result of the complex interaction between factors including the framing of urban design problems by professionals and other interested social groups, the stability of relationships which are embedded in technologies and artefacts, and the longer term development of technical and economic structures which lead to the persistence or 'lock-in' of particular technologies and built forms.

ANT's methodological principles of 'following the actors' through the processes of building and disrupting material relationships, and of describing human and nonhuman actants in the same terms have also influenced ethnographies of engineering such as those by Bucciarelli (1984, 1988, 1996) and Vinck et al. (detailed in Vinck

2009a). Bucciarelli's subject engineers were design engineers at two engineering design firms – one making photovoltaic cells, the other x-ray equipment. His focus was:

"... on the designing of the artefact, on the design as it evolves. I study participants' images and ideas about the artefact, the qualities it ought to have, and how they design to meet these goals. I record what participants say, sketch, write, or imply while they are at work designing. I allow for, and gather ample evidence of, variations in their thoughts and perceptions over time ... (1984, p. 185)."

Bucciarelli's work sought to move away from the mechanistic, often linear, views of design that miss the uncertainty, under-determination and ambiguity of the social process. In his view, part of this was attending to the object of design as icon within the culture of the firm (Bucciarelli, 1988).

"... participants in design come to their different design responsibilities with different perspectives, interests and expertise yet, at the same time, they share certain norms and common goals. I study how their individual views, beliefs and interests, grounded in their different technical competencies, contend, are resolved and reflected in the design ... (1988, p. 161)."

It is based on this groundedness of participants in their technical competencies that Bucciarelli develops a key conceptual contribution; what he terms the engineer's 'object world' around which thought on an object is structured, and esoteric language and concepts exist:

"The mechanical engineer, designing a structure to hold the plates used to collimate an X-ray beam, moves within an object world of beams, of steel, of geometric constraints, of stress levels, of close tolerances, of bearing surfaces, of positioning errors, of fasteners, and of metal machining practice. The electrical engineer designing a photovoltaic module works in terms of voltage potentials, and of current flows. He sketches networks with special symbols for diodes and current sources, resistive elements, all within a meaningful topology. These are two different worlds ... These frames are constructed on the job, within the firm, as well as in the schools. A mechanical engineer working in the steel-making industry designs within a different world from his classmate who joined IBM (1988, p. 162)."

Thus, there is a sociotechnical interplay within situated contexts – the object infiltrates thought and thought configures the object, and contexts, including organisational ones, matter. This becomes even more important because ultimately these object worlds need to come together where design objects are subcomponents that have ambiguity in the ways that they might work and fit together. This notion is taken up in Vinck's (2009b) study of a mechanical engineer that illustrates in fine-grained detail that the technical complexity of engineering tasks is often matched by and entangled with organisational complexity and that, within their different object worlds, the design interests of different actors often conflict. Furthermore, design interests are not always objectively presented and require an understanding of how they are put together both technically and socially.

This chapter presents some of the outcomes of a wider study, undertaken between 2007 and 2012, on normative issues in engineering practice (Chilvers, 2013). An overview of organisational discourses on aims and means is given first. This was developed through historical analysis on the work and reflections of Sir Ove Arup, the founder of the Arup firm, and analysis of contemporary strategy documents and interviews with staff. An ethnographic case study of a structural engineering project, conducted in one of the firm's regional offices, then places organizational discourses in contrast with the realities faced by the engineer at the level of practice. This ethnography involved shadowing the structural engineer throughout the project, including attendance at key project meetings, day-to-day observation, interviews with key participants and analysis of drawings and other documentation. The overall outcome of this mixed methods approach allows the analysis of the formation and disruption of networks of relationships within and beyond the firm and the identification of key points of dissonance and alignment between strategic goals and everyday practice, especially in relation to the achievement of goals associated with sustainability.

3   OVE ARUP AND TOTAL DESIGN

Arup is an international engineering consultancy, founded in the UK by Sir Ove Arup in 1946. As founder of the firm, Sir Ove remains a key figure in understanding the culture and structure of Arup today. He believed in a moral purpose for engineering and, from this standpoint, pushed for innovation and integration across professional disciplines and a bridging of the divide between clients, architects, engineering consultants and contractors in the construction industry. His own work in structural engineering established the firm's reputation in delivering high quality engineering services to support architectural innovation, including structural design for buildings such as the Sydney Opera House[1]. Sir Ove's vision of engineering included engaging with society in order to contribute to human development and welfare. From this moral standpoint, he pursued three critical themes within the building industry: the architect-engineer divide; divisions between briefing, design and construction; and the limits to the specialisation of knowledge. These ultimately led him to devise his particular design philosophy and approach to managing the Arup firm.

3.1   The architect-engineer divide

Closely aligning with the artistic and functional ideals of modernist architecture, Ove Arup saw the longstanding division between architect and engineer as outmoded.

---

[1]More recent examples of high-profile projects the Arup firm have contributed engineering design services towards include the award-winning Beijing National ('Birdsnest') Stadium and the National Aquatic Centre ('Watercube') in Beijing as well as Beijing's Capital International Airport Terminal 3 and Beijing South Railway Station (which will eventually cater for over 100 million passengers a year), 30 St Mary Axe (London's 'Gherkin'), the Metropol Parasol in Seville, the Centre Pompidou Metz, Paris, and the China Central Television Headquarters, Beijing. Bridges to which Arup have contributed engineering solutions include the Øresund link joining Denmark and Sweden, the Helix Bridge, Singapore (the first double-helix structure used for a bridge), the Hong Kong-Shenzhen Western Corridor, and Stonecutters Bridge (the world's second longest spanning cable stayed bridge).

He places his emphasis on a balanced synthesis of the architect's concern with human reactions to form and space, and the engineer's emphasis on conquering natural forces in a rational way with the aid of science and technology. His treatment of this problem focused on trade-offs; an architectural understanding without engineering conceives of buildings and spaces without regard for any implied trade-offs in efficiency of structure or in methods of construction. Conversely, optimised efficiency does not appropriately prioritise human goals of architectural delight and humane design (Arup, 1972).

3.2 Divisions between briefing, designing and construction

Ove Arup objected to the rules and norms surrounding the persistent division between design (assigned to the architect and consulting engineers) and construction (the domain of the contractor who was absent from design). He argued that constraints on design undermined efficiency and quality:

> "You cannot create designs for which the technical and constructional facilities do not exist, yet on the other hand no contractor is interested in creating facilities which are not yet called for by design ... The architectural design is very largely the special interpretation of the client's wishes. The client himself does not really know what he wants before the architect has put pencil to paper and has shown the client what could be done ... wise decisions can only be based on a knowledge of facts, and this means that the technical adviser should be brought into the business ... at an early stage. It is essential for economy that the design takes into account the method of construction as well as the final structure (Arup, 1956, p. 2)."

For Ove Arup, design needed to become an interactive process involving all the above parties from the earliest possible stage. The client should formulate their brief alongside an exploration of design possibilities with the designer, and the designer should be closely informed by the contractor's knowledge of construction possibilities, processes and costs.

3.3 Specialisation and the limits to knowledge

A further barrier to the synthesis of design-pertinent knowledge across the industry was the specialisation resulting from scientific and technological advances. The ever broadening body of knowledge and technique was causing ever greater specialisation in all areas of the industry with no one group covering a wide enough field to discern all design information from often bewildering possibilities. Specialisation was necessary to deal with problems in a manageable way, but for Ove Arup the danger was to forget the connections "so ruthlessly severed" (Arup, 1970a, p. 391). Ove Arup's characterisation of specialised views on any design correspond well with Bucciarelli's (2002) 'object worlds' which explain the different knowledge, values and languages of specialists in the design process. He identified these as a barrier to the 'synthesis' he sought between quality, form, and safe and efficient functionality.

3.4 Total design

Sir Ove's focus on synthesis and his concerns with the key problems of fragmentation within the design and construction industry led him to formulate the concept of 'Total Architecture' for the built environment, or 'Total Design' more widely:

> "The term 'Total Architecture' implies that all relevant design decisions have been considered together and have been integrated into a whole by a well organised team empowered to fix priorities (Arup, 1970b, p. 1)."

Sir Ove described a design as the sum of all the decisions recorded and communicated in the form of drawings, sketches, models, prototypes and so on, covering all the facts that needed to be known and processes that needed to be gone through to achieve the aims that had been collaboratively explored. In line with his criticisms of current practices, this had to occur across:

– design perspectives (between both architectural and engineering disciplines and emerging sub-specialisms therein), and;
– client/designer and designer/builder boundaries.

Arup freely recognised that integrated planning and design of this sort for the whole human environment was sufficiently lofty an aim never to be achieved, nevertheless he still explicitly stated this as the Total Design ideal (Arup, 1970a). In any case, if he and his colleagues strived to find what was needed for the best possible result in any single case, then what applied to one entity might well apply to most, as the need for proper integration of parts is a feature of all design (Arup, 1970a).

4 ARUP AND SUSTAINABILITY

The modern-day Arup firm's organisational strategy document for the 2010–2015 period reflects a continuation of the firm's traditional objective of offering Total Design or, more broadly speaking, 'holistic design solutions'. The provision of client value through creative and sustainable solutions and 'joined up' approaches is placed at the heart of their business and ongoing brand of engineering.

The challenge of sustainability in design and engineering is seen within Arup as a continuation of Sir Ove's early commitment to delivering integrated design solutions to improve society. According to the 2007 Arup Sustainability Statement:

> "Since 1946, Arup has been helping clients create a more sustainable future. Our founder, Sir Ove Arup, was an engineer and philosopher who believed in integrating environmentalism and social purpose into our projects. This commitment to sustainability influences the way we conduct our business, the way we treat our staff and the way we interact with the community and society as a whole. Our mission to 'shape a better world' defines our purpose ... Our approach to business has always aimed to deliver solutions based on the cornerstones of sustainability: environmental integrity, economic viability, social welfare and efficient

resource utilisation ... This journey is now entering a new phase as we formalise our commitment to a published Sustainability Policy (Arup, 2007, p. 1)."

Working for Arup is thus presented not only as professional practice but as 'living by values' and Arup self-identify themselves as thought-leaders in sustainability.

Significant as both a constitutive part of internal discourse and as an attempt at external 'thought-leadership' in sustainability is the work 'Entering the ecological age: the engineer's role', a document written by Peter Head CBE for the prominent Brunel Lecture Series of the Institution of Civil Engineers (ICE) and ultimately presented by him in lectures around the world. A civil and structural engineer, Head was a director at Arup and led their planning and integrated urbanism team until 2011. The paper resulting from this work was widely circulated and publicised. It aims to bring together and build on disciplines such as sustainable systems engineering, life-cycle analysis, industrial ecology and earth systems engineering. As the title suggests, it also claims to be "primarily aimed at highlighting the role of engineers". It states that:

"Engineers have global experience, are adept at multidisciplinary team working, which will be essential for success and can design and deliver these new infrastructure systems. However we recognise that resource levels are limited to undertake such an unprecedented challenge in a very urgent timescale of no more than 50 years and so we need to train and motivate young people to join this challenge and be the Brunels of the 21st Century (Head, 2008, p. 73)."

It is clear that staff members identify with the 'ecological age' as part of the Arup brand. For members in interviews it also comes down to leadership of external parties, encouraging clients to align themselves with the agenda in their adoption of design solutions:

"We are charged, everybody in the company, with encouraging clients to take on board sustainability. So we have an obligation within Arup to do that. Which is a very positive thing I think, and it raises the profile and makes people aware of it ... that's really how companies make things happen; I mean it's by people at the top saying, 'Look, this is what I want, this is what we as a company ...' and this isn't Peter Head going off on a tangent, this is the company policy, he's the driver, he's the owner if you like ... and Arup now are in a position to influence clients."

5 STRUCTURAL DESIGN IN PRACTICE

The above presents a view 'from within' of the organisational discourses of Arup, the narratives on aims and means that are on-going within the firm and which, most recently, have taken a positive turn towards tackling challenges of sustainability. We now present the detailed ethnography of a structural design project in order to explore the extent to which these principles and strategies influence the everyday engineering practices of Arup. The case studied was a project providing structural design services to a US$4.9 million development, consisting of a new building combining local government offices with public amenity space, at a tourist site situated within the major cultural district of one of the region's main tourist destinations. Contractually,

Arup's remit was to work with 'Paul Smith Architects' (a pseudonym), to develop their concepts through detailed design to tender-stage drawings. Key moments in the project are described – the development of the scheme design and the detailing of an element of the roof structure.

As is common in the work of this team, the architect had already been awarded the design contract. In this case the architect's scheme was based on two individual building spaces separated by a covered promenade. Over both building sections and this promenade section was to be an ambitious and elaborate 'mirrored steel-clad' roof. The roof was to be of a complex 'wavy' form with unique, facetted extrusions in all three dimensions. One of the two distinct building spaces was to be an elongated office space for local council workers with a decking area for them on the far side. For the latter the architect's scheme designated a reception and foyer area, a shop and a separate coffee shop, kitchen, toilets and a large interpretative/multi-use exhibition space. Directly outside and roughly to the North-East of the multi-use space was to be a landscaped grass amphitheatre over which the complex, 'wavy' mirrored-steel-clad roof was to extend. A seating area outside the coffee shop was also to be overhung by the roof structure.

The roles and pseudonyms of the project participants active in the forthcoming account are summarised in Table 1. Note that the following account concentrates on observations of the work of the main structural engineer assigned to this project within the Arup office, Adam. Unless otherwise stated all observations are of him operating at his desk within the working area of the office's 'Structures Team'. For the duration of the work observed, the researcher occupied the desk next to Adam's in this office space.

5.1 Scheme design

Following the award of contracts to the design consultants detailed above, an initial design meeting is held. Before this, however, having looked over the architect's

Table 1 List of project participants.

Employer Organisation	Pseudonym used	Formal role(s) within employer organisation	Formal project role
Paul Smith Architects	Paul Smith	Architect and Founder	Lead Architect and Project Manager for end-client
	Simon	Architect's Assistant	Architect's Assistant
Arup	Adam	Structural Engineer	Lead Structural Engineer
	Aaron	Senior Structural Engineer	Project Manager
	Ben	Associate and Structural Engineer	None, but acts as a senior mentor to Adam
	Ryan	Senior Associate, Senior Structural Engineer, Structural Team Leader and Office Leader	Project Director
	Leo	Civil Engineer	Civil Engineer
	Mike	Drafter	Structural Drafter
Consultancies external to Arup	David	Mechanical and Electrical ('MechElec') Engineer	
	Ian	Hydraulic Engineer	
	Colin	Landscape Architect	

scheme design, Adam requests a meeting with Aaron. They discuss concerns around the lack of uniformity and repetition in the architectural scheme. Repetition simplifies the structural engineer's design task and reduces the structural analysis and design effort. Currently, Adam and Aaron are concerned that the amount of analysis and design work likely to be entailed in the project is over and above the scope of works they had anticipated. The onus of their discussion is on the need for simplification of the scheme.

The initial design meeting is held at the Arup office. At the 'head' of the table at the far end of the room from the glass doors sits the architect, Paul, whilst around the table sit all the project participants detailed above (with the exception of Mike and Ryan) as well as a landscape architect, the researcher and a representative of the local government (the project's end-client). Paul chairs the meeting throughout, initially stating that he is the only person with a full overview of the project. He stipulates that the meeting be structured such that each contributing consultant presents their current state of thinking and level of design, and then others can comment and question. Before asking Adam to kick this off, timeframes are covered – the design is set to go to tender in four weeks (revised to six weeks a few days later).

Adam tells the group that he hasn't "crunched numbers" yet and, once the architectural scheme is "locked in"[2], he can start detailing the fascia[3] and sizing structural members[4]. Everything in the design seems achievable though there are various details that need to be worked through, requiring detailed discussion. Paul notes there will be a detailed "sit down" over the structural design after this main session. In keeping with this, after the other project consultants speak and make similar requests for information or 'design lock-in' from Paul, the discussion turns to structural issues.

Firstly Ian, the hydraulic engineer, voices a concern that anyone reviewing the architectural scheme will "see curves and think expensive"; the other consultants all seem to indicate shared concern over this with nods or murmurs of agreement. Paul states that their output documentation will seek to obscure the extreme extent of the wavy form and pursues a discussion on roof drainage including 'funnels' (to allow water to free-fall at two places through the promenade roof, forming water features). To paraphrase consultant responses; various arrangements are possible but all of these will determine drainage flows, the need and location of drainage pits, etc. As such, the need to have concrete decisions on these details in order for work to proceed is the main focus of discussion.

Adam then expresses to Paul the concerns that he had discussed with Ben and Aaron over the lack of uniformity and repetition throughout the architectural scheme

---

[2] The term 'lock-in' of the architectural scheme or of individual design decisions, when used in this account, refers to the need for design decisions to be finalised. That is, all details pertaining to a design decision to be agreed upon and thus made stable.

[3] 'Fascia' is a term used in structural design referring to the band that runs in a horizontal direction along the perimeter of the roof, situated under the roof edge.

[4] The terms 'structural members' or 'structural steel elements' are, in this account, used to refer to the range of structural components used in structural engineering to construct structures. In this case, the 'primary' roof members/elements are composed of steel beams and trusses providing the main support to the roof structure. 'Secondary' elements are those which are supported by the primary elements and that achieve all other structural features.

for the building. In response, Paul encourages Adam to consider an approach that he (Paul) has used on previous designs where:

> "Support and bracing was achieved on a completely opportunistic basis based on opportunities available for vertical supports, rather than a fully rationalised and repetitive grid."

Though Adam doesn't state it in the meeting, this is precisely the approach to achieving structural support that he had discussed trying to avoid with Ben and Aaron at the outset, due to the amount of analysis and design work likely to be entailed. He frames his expression of concern with another reference to timeframe and the scope of works entailed by this approach but then accepts that this is the scheme that he will work towards.

Following this, an internal Arup 'Design Inception Meeting' is held in the same conference room. Participants, clockwise around the table from the researcher, are; Leo, Mike, Ryan, Aaron and Adam. The meeting, chaired by Aaron, follows a standard, itemised pro-forma developed for such meetings by the team[5] (the office's 'Structures Team' who work regularly together on such projects). At meeting-close Ryan summarises:

> "He [Paul] is out to win an award with this one and we have to support him in that because we are his consultants but we've also got to come up with a practical design that's going to come in within budget. We don't want the scenario where it goes out to tender and comes back over budget and we end up going back over it for no extra fee and take money out [the build][6]."

Some days later, after developing the design further, Adam receives a phone call from Paul; he has been working with a quantity surveyor on findings ways to economise in the design. For costing purposes, Paul asks for information on the number and sizes of structural members. Adam agrees to send through his preliminary work on this, stressing that he is being restrained from progressing by floor plans and elevations that are still not locked in. Following this discussion, Adam expresses relief to Ryan that the Architect's aspirations seem to have been "reined in" by a quantity surveyor.

Later that afternoon, Adam meets with Paul and his assistant, Simon, in the cafe across the street from the Arup office (where both meeting rooms are fully booked). Adam starts to talk through his proposals for framing out the fascia. However, upon

---

[5]The items for discussion stipulated by the team's pro-forma for project design inception meetings were 'Delivery Approach/Lessons Learnt from Previous Projects'; 'Technical Approach'; 'Managing Change'; 'Resourcing and Deliverables'; 'Major Project Risks'; 'Safety In Design'; 'Communications' and 'Opportunities for Innovation'.

[6]Here the phrase "take money out of [the build]" refers to the situation the Arup team wish to avoid whereby a design for the building is finalised but is such that resulting tenders from contractors for construction come in over the amount budgeted for construction phase. Associated with this outcome is the likelihood that Arup, as consulting structural engineers, would have to go back over the building design with the architect and make modifications that would bring the cost of construction down to within budget. This would likely be for no extra fee to their client.

hearing that it utilises hot-rolled steel elements[7] for fascia construction, Paul refuses to agree to Adam's proposed method. As Paul and Simon start to brainstorm alternative ideas for Adam's consideration, Adam is still trying to present his reasoning, he stresses:

> "It will be great if we stick to hot-rolled elements because they'll be easy to model, easy to mould, easy to do and it gives us a certain continuity across the roof perimeter."

Paul is adamant that the method cannot involve hot-rolled steel elements and encourages Adam to focus on an alternative based on horizontal roof beams with vertical 'droppers'[8] on their ends, Adam counters with the following problems:

– Achieving both convex and concave curves between the vertical drop-downs; and
– The wavy nature of the fascia meaning horizontal bracing will be seen at the points where the curve rises above the horizontal supports within a 'span' between droppers.

However the unacceptability of hot-rolled elements is a fundamental sticking point for Paul. He explicitly states that he recognises that Adam's proposal is the "best technical solution" and that he appreciates the work he has done to develop this solution. However, his concern is that contractors tendering for the construction work will see lots of hot-rolled steel and pre-fabricated steel elements and their prices will escalate five-fold:

> "I know contractors around here, alright. It's a case of the C.B.F's, the 'Can't Be (expletive)'ed's', if they see stuff like that they bump up the price on principle."

Eventually Adam accepts that he must work on an alternative that does not involve hot-rolled steel elements[9].

First thing next morning, Paul phones Adam at his desk in the Arup office to say that he has been looking at screening the large condenser units to be located on the roof by extending the parapet further upwards in certain areas. In a cost saving

---

[7]'Hot-rolled steel' is steel that has been processed when hot and easier to manipulate. Hot-rolled steel elements are thus steel elements that have been manipulated at the point of production into bespoke forms to suit a specific purpose. In the case referred to here, Adam had agreed with Ryan that the most structurally efficient means of constructing Paul's proposed wavy form of fascia was with hot-rolled steel elements as this is the only way to produce steel elements that follow directly the unique curves of the outer edge of the roof perimeter as defined by Paul.
[8]The term 'droppers' refers to steel members that are orientated vertically, fixed to and extending downwards from the ends of the primary horizontal members in the roof structure (i.e. around the roof perimeter). These are being used as supports onto which to fix the fascia.
[9]On the use of hot-rolled elements in fascia construction, Adam retrospectively said the following in interview: "[Framing using hot-rolled elements] was the best solution in terms of structure, in terms of the weight of material . . . but there would have been more labour costs involved in the job in terms of preparing, cutting and welding these panels . . . sometimes the best technological solution is not the best solution overall and I guess that's the key point here. Sometimes something works on a technical level and doesn't work on account of, in this case, a contractor going; 'Hmmm, I'm not familiar with that' and bumping their prices up."

exercise, he also wants to reduce the size of the parapet to a minimum where screening is not required. Adam and Paul run through sight-lines on the plans identifying where the parapet will need to be extended upwards or can be reduced. Adam stresses that increased parapets will mean more weight at roof perimeter which will be largely borne by purlins. This means using more structural steel throughout the design and aborting design work that he has already completed.

Paul: "Well look that's alright, design is an iterative process. Add 300 mm to the parapet height to make it an average of 450 mm. Do you think that's enough?"

Adam: "Look sorry, that's not my call. I've told you what it will probably mean for the purlins and I can do a check on how it changes steel sizes..."

Paul: "Look screening is an issue and if you could do that that'd be great."

Paul then explains that he and his quantity surveyor want to keep open the option of using a cheaper but heavier cladding system – one made from condensed cement fibre board – as an alternative to the more expensive mirrored steel or aluminium options already specified, in case savings need to be made in the build. Adam is quick to point out that he has already done work on this. Paul insists that this won't change the layout but rather only the sizing of steel elements in order to be able to accommodate a heavier alternative if this becomes a budgetary requirement. Adam agrees to work with this.

To clarify the material and practical implications of designing to keep open the option of building with the cheaper, heavier cladding system, it is worth stressing that the cladding is to cover the whole of the protruding roof structure including the extremities and underside of its sizeable overhangs. The support of this cladding exterior is the main function of the primary steel. Thus an increase in any support requirements (weight) of cladding material is a profound change in the weight of material to be supported by the whole roof structure and thus, as acknowledged by Paul, a significant up-sizing of all primary steel. Adam has already modelled the weight of the lighter cladding system and designed the primary steel members to this architectural stipulation. This work will now have to be aborted and repeated to accommodate the heavier system.

Adam phones the cladding supplier for the weight and support arrangement for the new option. His reaction amounts to an exclamation at the increased weight of the new cladding option: "Wow! Big difference!" He seems troubled and later explains his concern: the increased weight will increase the overall sizes of structural members and this might be to the point where the design, even with savings made from cheaper cladding, becomes unaffordable to the end client.

5.2 Detailed design

When Adam turns his attention to developing the structural design for the 'Transitional Roof' which bridges the two building sections and covers the promenade section in between, he expresses two difficulties:

1   Ian, the hydraulic engineer, has highlighted some areas of the roof which, due to its unusual shape, do not drain into the guttering system in its current position, and;

2   Since the roof cladding system for the roof attaches directly to the roof purlins[10], Adam cannot easily visualise a system of support to the guttering given that:

   –   Its lip needs to be flush with the roof cladding surface, and
   –   The gutter depth will drop below the level of the roof purlins (which are currently its only means of support).

Adam puts forward a proposal to Paul that solves both of these issues and is economic in its use of steel. It does, however, entail modification to the wavy form of the transitional roof. Paul rejects this on the basis that he does not "like the shape" this gives the transitional roof. He instructs Adam to follow a more complex arrangement of guttering in order to retain the original form of the roof. Upon receipt of this, Adam discusses the new scheme with the researcher as follows:

> "Just so that you know, we've received a bunch of updated plans and, looking at them, there's a couple of points where I don't see how it's going to work – areas that seem to have been complicated, overly so."

Again, Adam makes a number of suggestions for simplification, which Paul rejects in favour of the retention of a more complex form through the introduction of additional steel. As Adam works under these demands he requests a meeting with Aaron. Adam confesses that trying to find an arrangement which works sensibly is proving very difficult. He shows great frustration. There are numerous problems. In addition to the previous challenges he is also struggling to achieve the points where 'droppers' are required for fascia framing. Together Adam and Aaron examine the options for steelwork in detail, concluding as follows:

Aaron: "Whilst I recognise it's a mess and far from ideal, we should be able to come close to what he [Paul] wants to see."

Adam: "When you've got a minute have a look at [Mike's 3-dimensional model being developed from Adam's designs for the detailing of final drawings], it gives you some idea of the amount of steel being used up in this job".

Aaron: "Yeah, I do wonder by how much this roof is going to go over budget [refers not to Arup's fee but the actual build budget], it'll be interesting."

These frustrations carry through when Adam comes to interpret the design codes for the sizing of the additional steelwork introduced with Paul's preferred solution. There are ambiguities regarding how to characterise the loads, given the unique structural form. He consults Ben and talks of how he has had to be very conservative in his treatment of the structure to date due to its unusual, non-uniform form and that all the primary structural elements have become "heavily designed" i.e. that there is a high amount of steel designed in due to conservative applications and readings of design codes. Vigorously expressing frustration, Adam impresses on Ben:

Adam: "There is a lot of steel in this design!"

Ben: "You didn't come to Arup to design nice rectangular boxes, did you? If you did, then you were misinformed!"

---

[10]'Purlin' is a term used in structural design for structural members in a roof that support the main roof surface. In the typical system they are supported by and lie over beams or rafters.

It is around this time that Adam goes to speak to Mike, who is working on developing the 3-dimensional model of the design, to discuss the progress relative to the schedule for the project. Adam also looks up some details regarding their [Arup's] fee for this job and relays this to Mike who responds:

Mike: "(expletive) we're going to lose money"
Adam: "We certainly are."

6   DISCUSSION

Having presented the above a account, we can now turn our attention to drawing out the findings salient for assessing the achievement of desired outcomes from the Arup-organisational perspective presented earlier. When we do so, we find significant dissonance between the organisational aims and the self-identity of Arup, and the realities of the project events and outcomes in practice.

A primary feature of the account is that the Architect, Paul, can be viewed, in ANT terms, as an 'Obligatory Passage Point' (OPP), an actor through whom all interests in the network are translated and who mediates interactions between actors in the network (Gonçalves & Figueiredo, 2010). As 'the client' to the engineer, Paul subsumes the interests of those actors both up and downstream in the design process – the building owners and users on the one hand and building contractors on the other. Adam configures his own networks of actors, including humans such as his colleagues and other engineering consultants, and non-humans such as design codes, computer models, steel and drawings. However, ultimately, all his representations are translated through the architectural interests of Paul, as gate-keeper to the building.

As Ben reminds him, Adam's technical ability to translate the creative demands of the Architect into a workable structure is a central element of the Arup identity as an innovative firm delivering high quality design solutions for unusual and iconic buildings. Yet throughout the project Adam is also constrained by the pre-eminence of Paul, as the Architect, when pursuing other Arup goals of sustainability and profitability. Despite his efforts to renegotiate the demands of the structure, Adam delivered a design that was materially inefficient and required a design effort beyond that needed for Arup to turn a profit for the work.

None of this is new. These are precisely the problems that Ove Arup identified in the 1940s and sought to overcome in creating and expanding his own firm. The collaborative distance between client, architect, engineer and building contractor, and the associated division between practices of briefing, designing and construction, were two of the central barriers he identified to the realisation of normative social goals and which drove the way he sought to develop his firm. Yet, while Arup have created a multidisciplinary firm and have been able to deliver more integrated outcomes in high profile, large scale projects, the case above indicates that they continue to be constrained in the everyday world of structural consultancy by the persistence of conventional professional arrangements in the sector. The initial multidisciplinary design meetings and late 'lock-in' of the building scheme resonate with some elements of Total Design, but the project hierarchy is far from Arup's ideal collaborative partnership of clients, architects engineers and contractors working together from the inception of the project in order to bring to the table and openly negotiate their varying perspectives, aims and concerns throughout the design. Whilst they maintain loyalty to the Total Design ideal, in the world of everyday practice Arup's engineers find themselves locked-in to professional networks constraining their agency and capacity to pursue normative goals such as sustainability and efficiency.

This example of practice challenges the idea that Arup successfully assert their desired approach in everyday structural engineering projects and, whatever the legitimacy of the criticisms made by Sir Ove Arup throughout his career, points to professional lock-in to the Architect-led arrangement for the building industry that does not support the collaborative sharing of design-pertinent knowledge and perspectives in support of normative goals. In the above, the consequence was that sustainability, manifesting in the work of Adam as concerns over structural efficiency and material consumption within the structure, was reduced to a muted participant in the social and technical interactions of the project. We struggle to hear it coming through as a legitimate issue of concern in the project account, despite its prominence in the stated aims and organisational discourses of Adam's employer organisation. This dissonance between Arup's organisational discourses and the engineering practices displayed in this example is fully recognised by Adam in interview[11].

7 CONCLUSIONS

Adam's thwarted efforts to renegotiate the structure, as a spokesperson for material efficiency and sustainability and as an engineer within a firm that holds leadership towards sustainability to be central to its mission, are prescient to the problem of engineering agency within sociotechnical networks. The key constraints to pursuing sustainability and material efficiency in this design are not economic, technological or institutional, as is often assumed. Adam's capacity to achieve a sustainable design is constrained by the obduracy of the professional hierarchy within the building project. The centrality of the Architect, as 'client' to the engineer and as gatekeeper to the building owners and construction contractors, prevents more integrated design practices that may have resulted in more sustainable outcomes.

---

[11] In retrospective reflections on the project Adam said the following: "On a professional level we [Arup] have a client and we have to service their needs as best we can. He had a very clear vision in terms of the shape of the building and we had to work with that. Personally, I thought, and perhaps a few others thought, that the aspirations for sustainability didn't really match, or rather let me rephrase that: the design did not match the aspiration in terms of sustainability and it probably could not match those aspirations because of the pretty – what's the word? – adventurous nature of the building, the form of the building. Or rather what particular decisions required in material terms based on an adventurous form for the building. The form of the building dictated to a large extent (structural solutions)." Asked whether there was a contradiction between Arup's aspirations for providing leadership towards sustainability in all their practices and the building design conceived by the architect in this case, Adam answered "Yeah, yeah. Absolutely, I think so."

Adam's plight highlights the need to consider the role of other actors in sociotechnical networks which translate and constrain engineering agency. Reforming engineering education and practice to more explicitly address normative concerns such as sustainability is necessary but not sufficient. Wider reconfiguration of the sociotechnical networks of design and construction are required if engineers are to contribute more fully to addressing these challenges. Whilst high level strategic thinking, such as demonstrated in Arup's Total Design and Sustainability Statement, are important in outlining the nature of such change, it is in the details of engineering practice that it will ultimately be achieved.

REFERENCES

Arup, (2007) Sustainability Statement 2007. [Online] Available at: <http://www.arup.com/_assets/_download/165DCD97-19BB-316E-408765B8431A520D.pdf> [Accessed 6th September 2012].

Arup, O. (1956) The importance of design. Speech delivered in Nairobi on October 26th. Arup Library Archive, OA51B. London: Arup Library, 8 Fitzroy Street.

Arup, O. (1970a) Architects, Engineers and Builders; The Alfred Bossom Lecture By Ove Arup. Journal of the Royal Society of Arts, 118, 390–401.

Arup, O. (1970b) The Key Speech. [Online] Available at: http://www.arup.com/Publications/The_Key_Speech.aspx [Accessed 6th September 2012].

Arup, O. (1972) The built environment. The Arup Journal, 7 (4), 2–7.

Bucciarelli, L. (1984) Reflective Practice in Engineering Design. Design Studies, 2 (3), 185–190.

Bucciarelli, L. (1988) An ethnographic perspective on engineering design. Design Studies, 9 (3), 159–168.

Bucciarelli, L. (1996) Designing Engineers. Cambridge, MA, The MIT Press.

Bucciarelli, L. (2002) Between thought and object in engineering design. Design Studies, 23(3), 219–231.

Callon, M. (1986a) The Sociology of an Actor-Network: the Case of the Electric Vehicle. In: Callon, M., Law, J. & Rip A. (eds.) Mapping the Dynamics of Science and Technology: Sociology of Science in the Real World. London, Macmillan. pp. 19–34.

Callon, M. (1986b) Some elements of a sociology of translation: domestication of the scallops and the fishermen. In: Law, J. (ed.) Power, action and belief: a new sociology of knowledge? London, Routledge. pp. 196–223.

Callon, M. (1989) Society in the Making: The Study of Technology as a Tool for Sociological Analysis. In: Bijker, W. E., Hughes, T. P. & Pinch, T. J. (eds.) The Social Construction of Technical Systems: New Directions in the Sociology and History of Technology. London, MIT Press. pp. 83–103.

Chilvers, A. (2013) Engineers and Values: Ethnographic studies of the normative shaping of engineering practice. EngD thesis. [Online] Available at <http://discovery.ucl.ac.uk/>.

Gonçalves, F. A. & Figueiredo, J. (2010) How to recognize an Immutable Mobile when you find one: Translations on innovation and design. International Journal of Actor Network Theory and Technological Innovation, 2 (2), 39–53.

Head, P. (2008) Entering the Ecological Age: The Engineer's Role. The Brunel Lecture 2008 [Online] Available at <http://www.arup.com/Publications/Entering_the_Ecological_Age.aspx> [Accessed 22nd July 2011].

Hommels A. (2005) Unbuilding Cities. Cambridge, MIT Press.

Latour, B. (1987) Science In Action: How to follow Scientists and Engineers through Society. London, Open University Press.

Latour, B. (1993) We Have Never Been Modern. Cambridge, Harvard University Press.
Latour, B. (2005) Reassembling the Social: An Introduction to Actor-Network Theory. Oxford, Oxford University Press.
Oudshoorn, N. & Pinch, T. (2008) User-Technology Relationships: Some Recent Developments. In: Hackett. E. J., Amsterdamska, O., Lynch. M. & Wajcman, J. (eds.) The Handbook of Science and Technology Studies. London, The MIT Press. pp. 542–565.
Vinck, D. (2009a) Everyday Engineering. London, The MIT Press.
Vinck, D. (2009b) Sociotechnical Complexity: Redesigning a Shielding Wall. In: Vinck, D. (ed.) Everyday Engineering. London, The MIT Press, pp. 13–27.

Chapter 10

Observations of South Asian engineering practice

James Trevelyan

School of Mechanical and Chemical Engineering, The University of Western Australia, Perth, Australia

1 INTRODUCTION

This chapter argues that engineering can be seen as a lever that amplifies the effect of a given economic investment, yielding human productivity improvements. Engineering reduces the human effort, energy and material resources needed to provide a given level of benefits. Largely seen as the application of scientific and mathematical principles, there is a tacit assumption that a given engineering investment will yield much the same benefits, wherever it is applied. Observations of engineering practice in South Asia and Australia reveal that this assumption is questionable. The social heterogeneity of engineering practice and the political economy in which it is situated influence the performance of engineering investments, causing large differences in benefits.

The chapter briefly reviews observations of engineering practice in South Asia and Australia that allow some comparisons in three types of settings: water supply utilities, metal manufacturing, and telecommunications. These observations reveal the mediating influences of social hierarchy, application of engineering management systems, system knowledge representation, community education and skill development, and the entanglement of engineering networks with political power structures. These influences might explain part of the large performance differences observed in a comparison between Australia and South Asia. The performance of water supply utilities and manufacturing enterprises appear to be less in South Asia than Australia.

Recent developments in mobile telecommunications, however, demonstrate that certain socio-technical choices might reduce or even eliminate factors that interfere with engineering practice in South Asia.

By understanding these factors, engineers could make more reliable outcome predictions. They would need to understand the influences of social, economic and political factors that are seldom if ever mentioned in education or management discourses. Further research could help provide improved education resources for engineers. Improved practice resulting from education could eliminate much of the poverty and misery that characterises the lives of billions of people on the planet.

2 BACKGROUND

If there is one issue that gnaws at the conscience of most people in the industrialised world, it is the enormous gap between rich and poor in the world today, particularly the gap between wealthy nations and what are now called low income countries (LICs). While some countries, such as Korea and Malaysia, have successfully transitioned from exploited colonial states to wealthy nations, many LICs are making slow progress, if any, towards improving the quality of life of their people. Contemporary development economics attributes the wealth disparity to differences in governance, economic policy and health so it is not surprising that most interventions have focused on capacity development in these aspects.

Through a series of chance encounters, I came face to face with another possible explanation for poverty in LICs. I was employing engineers in Pakistan to design and construct prototype equipment for landmine clearance. These engineers were unable to match expectations based on my experience with similarly experienced mechanical and electrical engineers in Australia. Neither the level of qualification (bachelor's or master's degree) nor the country in which it was gained (Pakistan, UK, USA, or Australia) appeared to make any difference. Visiting local industrial firms led me to the realisation that one has to adopt a different level of expectation in order to make realistic performance predictions for engineering work in South Asia.[1]

However, it was very difficult to explain the differences when compared with Australia. There seemed to be little difference in the understanding of fundamental technical issues. There was a significant difference in the ability of local firms to supply many specialised engineering components and materials. Few South Asian technicians worked from drawings in the way that their Australian counterparts would, yet they were just as skilled in fitting and machining. The most significant qualitative difference between Pakistani and Australian engineers was in aspects that one might describe as 'practical skills' but it was very difficult to articulate exactly what this term meant.

Around 2001, I turned my attention to the provision of water supplies and sanitation for government schools in the urban backstreets on the fringes of large South Asian cities.

It was in these backstreets that I began to appreciate the real significance and social value of effective engineering because it was conspicuously absent. We take good engineering for granted in Australia where copious quantities of potable water are available 24 hours a day from kitchen, bathroom and garden taps at around US$2 per tonne including connection charges. Families typically spend less than 2% of their economic resources on water and sanitation.

Even though most people in South Asian cities have access to a piped water service, water only flows from pipes for an hour or so every other day and it is almost certainly unsafe to drink.[2] Many consumers have installed their own pumps in an effort to extract more when water flows, leaving others with less. Potable water has to be prepared or often carried, with the result that the real economic cost is many times higher than in Australia in equivalent currency terms per ton. The economic cost

---

[1]South Asia includes India, Pakistan, Bangladesh, Sri Lanka, Nepal, Sikkim, Bhutan, and Maldives.
[2]Progress towards the Millennium goals is measured in terms of improved water sources, that by the nature of their construction or active intervention, are protected from outside contamination, particularly faecal matter. Supply quality and continuity is not currently a performance indicator (World Health Organization, 2012).

for a minimal 10 liters daily per person represents more than 10% of GDP (Gross Domestic Product) per capita, whereas in Australia this requirement represents a total cost less than 0.01% of GDP per capita.[3] Sanitation mostly relies on open drains that are often blocked by garbage. Heavy rains bring sewerage water onto the streets and into domestic compounds. For locals, most of the cost is invisible: water charges are significant but small in comparison. There is no charge for water from public purification plants and women's labour is regarded as a free household commodity. Indeed, the notion that basic goods could be far cheaper in a wealthy country seems beyond comprehension. Compared with industrialised countries, the performance of large engineering investments in urban water supply is poor, both for the owning public and end-users.

The cost of other services that rely on engineering (such as electricity supply and construction), taking into account total costs and comparable end-user service quality, tend to be higher than in Australia, though the differences are not as great as for potable water which is heavy to transport. Electricity utilities have insufficient generation or transmission capacity to meet the demand, and may not collect sufficient revenue to pay for fuel, particularly in summer weather when wealthy customers use air-conditioners. A generator is essential for anybody who needs continuous power, increasing the average cost of electric power by a factor of three or four.[4] Even people who do not have use of a generator are affected by indirect costs. Without being able to rely on refrigeration, uneaten food has to be discarded: it cannot be safely kept for eating later. Fresh food has to be brought in daily: it cannot easily be stored. Commodities purchased through shops and markets are more expensive because the shops and their suppliers have to run generators. Then there is opportunity cost when productive work has to be suspended for the lack of lighting or electric power.

Many would argue that these difficulties reflect political and economic challenges: they would assert that there are no "technical engineering" issues without readily available solutions. However, electricity and water utilities are predominantly engineering enterprises: the core staff comprises professional engineers with university training. Transport, communications and construction also rely on engineering to a similar extent though seldom with monolithic enterprises that characterise water and electricity distribution. All the engineers in these enterprises who participated in this research saw their roles in terms of contributions that influenced economic performance, and the products or services received by end users, even though they still perceived much of their work as "not real engineering" (Trevelyan, 2010).

What could explain the difficulties with water and electricity distribution compared with industrialised countries? Engineering is seen by many as simply 'applied science' with principles that apply equally in Perth, Pune and Peshawar. The cost and service quality difference is ascribed by some to "lack of resources", or "subversion

---

[3] The equivalent economic cost can be calculated from the shadow priced cost of unpaid domestic and child labour for carrying water. Other sources come at similar cost, including 20 litre plastic bottles from water sellers, private or government water carriers, including cost of queuing time and necessary 'facilitation payments' (Trevelyan, 2005). Depending on local market conditions, the cost can range from $30 to $70 per tonne.

[4] This issue has been covered extensively in the media, for example (Qizilbash, 2009).

by many of the self-interested social elite".[5] I struggled to understand why basic engineering clearly was not, and still is not, working in South Asia. Electricity and water utilities rely on substantial contributions from engineers and there is little need for advanced technology beyond the reach of LICs.

Is it possible, therefore, that the practical skill shortcomings that I encountered first-hand are being replicated on a grand scale in these large enterprises, influencing the relative level of benefits from a given engineering investment? Is it possible to identify specific skill shortcomings?

The research originated from these challenging questions, and led to a series of mostly qualitative studies on engineering practice in Australia and South Asia informed by a combination of interviews, field studies and research visits.

As explained elsewhere, it turned out that there were remarkably few published accounts of systematic research examining the work of ordinary 'everyday' engineering in industrialised countries[6] and none from the developing world at the time. Therefore, much of this study has been based on interviews and fieldwork in Australia as an example of the industrialised world to provide a reasonable 'baseline' for comparison with South Asia. Several focused studies by the author and students contributed data from particular engineering settings.

Gradually, the objective became clearer: a conceptual and detailed framework for engineering practice that could explain observations on performance differences, in different settings and disciplines. Traditional caution would have suggested a more restricted focus, perhaps on a single setting such as automotive manufacturing engineering that exists in both Australia and South Asia. However, by obtaining data from a variety of different settings and locations, it was possible to develop a framework that is more likely to be valid in a wide variety of settings and engineering disciplines.

3 RESEARCH METHODS

Constructing a conceptual framework for engineering practice required rich data sources provided by qualitative research. All recent conceptual investigations of engineering practice have been based on qualitative methods.[7]

This work is based on several overlapping qualitative research studies, and also draws on the author's first-hand experience of engineering practice in Australia (20+ years) and Pakistan (several periods of one-two months each). Data includes transcripts from more than 120 semi-structured interviews lasting 90–120 minutes exploring participants' engineering careers, all aspects of their current work, and perceptions related to job challenges and achievement satisfaction. Some interviews included questions on dishonest behaviour (of others), checking and mistakes. Follow-up questions (probes) elicited more detail and specific examples when the respondent generalised. In some instances, circumstances required small focus group discussions. Data has

---

[5]For detailed accounts of water distribution systems in South Asian cities, see (Anand, 2011; Coelho, 2004) For metal manufacturing, see (Domal, 2010).
[6]Several reviews have appeared: (Barley, 2005; Trevelyan, 2010; Trevelyan & Tilli, 2007) (Stevens et al., 2012 – in press). Most of the relatively scarce empirical studies have focused on design in advanced technology contexts.
[7]A useful and detailed example: Zussman (1985).

also come from several field studies: a limited number of subjects were shadowed (with their consent) for 1–2 days to triangulate interview data. Transcripts were prepared from recordings (obtained with participants' consent) or notes. In the latter case the participants checked transcripts for accuracy. The author's students, several with substantial first-hand engineering experience, performed the field studies and many of the later interviews to examine specific aspects of practice using the same interview protocol with minor variations in questions to suit their research. Training, joint interviews, observations and reviews of the recordings helped ensure consistent data. Earlier accounts have provided extensive details on the interview methods and participants.[8]

The sampling was partly opportunistic and partly purposeful for maximum variation to include engineers in all major disciplines, experience levels and types of business (except defence). Both companies and individuals were approached. Some companies responded with a cross section of their engineers and others with individuals: none declined our requests. While most were mechanical engineers, the sample included telecommunications, civil, electrical, mechatronics, control and instrumentation, petroleum, offshore, subsea and several other engineering specialisations. Six were female and almost all had engineering degree qualifications.

First-hand experience of local engineering practice and field studies were essential to make sense of interview data from the Indian subcontinent. Language skills were often important: many participants spoke English, code switching with Hindi or Urdu and other languages. Some of the senior engineers had an understandable tendency to present an over-optimistic view of operations for which they were responsible. On-site observations were also valuable for improving the trustworthiness of the data even when it was not possible to shadow engineers.

Analysis followed standard ethnographic analysis techniques.[9] Some published data together with independent field observations provided additional triangulation evidence.[10]

Re-reading the original transcripts, searching for missing aspects of practice, reflecting on the data, the language and meanings, and discussing the main themes emerging from the analysis with some of the Australian study participants helped to validate or refute aspects of the analysis and distil the essential features.

This chapter draws on the data to compare engineering practice in Australia and South Asia with reference to three different settings – water utilities, metal manufacturing and telecommunications.

4   UNREGULATED SOCIAL INFLUENCES

In this chapter it is only possible to expose a small part of the evidence collected in this research, limiting the discussion to some key factors.

---

[8]Further details appeared in the text and on-line appendices of recent publications: (Trevelyan, 2007; Trevelyan, 2010).

[9]A combination of codes was used, partly with emergent themes (grounded theory) and pre-assigned aspects of practice based on first-hand experience (Huberman & Miles, 2002; Miles & Huberman, 1994; Patton, 1990; Strauss, 1987).

[10]Some detailed field observations have appeared in reports of major studies (e.g. in Bucciarelli, 1994; Darr, 2000; Orr, 1996; Vinck, 2003; Winch & Kelsey, 2005).

The conceptual framework for engineering practice that emerged from this research portrays a human social performance in which technical knowledge is distributed unevenly among both the human actors[11] and also inanimate 'actors' such as procedures, standards, documents and information systems.[12] Engineers play a critical role as they are largely responsible for arranging the presence of the necessary actors and also coordinating their performance to achieve predicted technical and commercial outcomes. Therefore the results depend not only on their combined technical expertise but also on the social interactions between the actors by which the expertise is accessed and contributed through skilled and coordinated individual performances.[13]

The data from South Asia revealed how the prevailing social culture permeates engineering practice, inhibiting social interactions crucial for enacting a performance that relies on distributed knowledge and cognition.

At a superficial level, the daily work of engineers in South Asia often resembles a pattern observed in Australia. Most engineers check their e-mail on arrival at the office and briefly inspect work in their areas of technical responsibility to identify issues that need immediate attention. They often attend coordination meetings, meet clients, and inspect work done by other people or outside contractors. They display similar dedication and gain as much personal satisfaction as their Australian counterparts. They often work long hours into the evenings. While it is specific to their domain of practice, the technical expertise of experienced engineers in South Asia covers similar depth to their Australian counterparts.

4.1   Social hierarchy and autonomy

At a deeper level, significant differences start to emerge. First, engineers in South Asia still report a high degree of autonomy but not as high as their Australian colleagues. The hierarchy exerts significant control over their daily work. A production engineer at a manufacturing plant described how this can interfere with a more organic approach.

"We started a new idea in our assembly plant called a cellular manufacturing system. We organised small cells and the participants were from all levels of the company including executives from different departments like maintenance. [And did it work out?] No, it didn't last long. [Why not?] Because I would say that here the culture is such that people do not like change so they were not receptive to it and secondly not all departments were involved in it that you needed to execute a job. For example if there was a problem and the maintenance department had to fix it they would say 'Okay we have indented the parts and when we get

---

[11]The details appeared in the main text and on-line appendices of (Trevelyan, 2010). For an historical perspective on distributed knowledge and cognition, see Mukerji (2009).

[12]Actor network theory provides a way of describing social interactions in a technical context in which actors can be people or inanimate objects created to influence human behaviour (Latour, 2005).

[13]Distributed expertise in engineering is coordinated by engaging participants in relationships that promote willing and conscientious collaboration. Social capital is also an essential component (Brookes et al., 2007; Friesen, 2011; Larsson, 2007; Levin & Cross, 2004; Trevelyan, 2007).

them you would have the thing fixed'. Of course, the indent would have to be processed through the administration department and they would take their own time. Of course, if the cell had been given both responsibility and authority it would have worked but the authority was not given to them, only the responsibility. That's why it did not work."

This quotation reveals the impact of the firm's hierarchy on the day-to-day activities of engineers. In Australia, informal networks across the firm allow engineers to coordinate work across and even outside an enterprise, but this is much more difficult in South Asia. Engineers in South Asia occupy a privileged niche, but they are still some distance below the top of the social hierarchy. Nearly everyone maintains subservience to the social hierarchy, reflected in the firm's organisational structure, and also to a large personal network of extended family and relatives, friends, and acquaintances. Together, the hierarchy and one's personal network provide the only social safety nets that can protect an individual from a personal catastrophe and loss of earning capacity. Getting a job invariably requires access to someone with influence, often a significant financial payment as well. Most people are only a short step away from destitution. Therefore, loyalty to the social hierarchy and one's personal network takes precedence over everything else, including production work.

This is the first of Domal's five 'dimensions' that characterise the differences between Australian and Indian engineering practice.[14]

It is important to understand that subservience to the social hierarchy that we see in South Asia is also present to a degree in Australia. While class, caste and the firm's organisational chart highlight hierarchy in South Asia, in Australia the hierarchy is less visible, less obvious. It is not uncommon, for example, for senior engineers in an Australian firm to wear clothing and drive cars almost indistinguishable from technicians and tradespeople. Most Australians address each other by their first names and may interrupt while the other person is speaking even if they are more senior. The difference that we see between South Asia and Australia is one of degree and the degree of difference changes between different settings.

Nevertheless, addressing more senior staff by their first name, even speaking out of turn is rare in South Asia. It is not uncommon in meetings that involve staff at different levels of the hierarchy for junior staff not to speak at all unless specifically asked to by one of the senior staff. Even when this happens, the response will be respectful, and minimal. It is often considered disrespectful, for example, to ask a more senior person to repeat an explanation or ask a clarifying question. Silence is preferable to an admission that one was not listening carefully in the first place. Without the opportunity to clarify and resolve uncertainty, action is usually deferred in the absence of a supervisor who can lift the responsibility for mistakes from one's shoulders. This helps to explain the need for constant supervision, another of Domal's dimensions of difference.

---

[14] 1 – Social hierarchy disrupts informal coordination and access to distributed expertise; 2 – Weak systems and procedures; 3 – Need for constant supervision; 4 – Differences in production supervision skills; 5 – Differences in labour cost perception. Only the first two are discussed in this chapter (Domal, 2010, pp. 103–167).

The prevailing social requirement that one demonstrates respect for one's seniors often blocks informal coordination and sharing of expertise. Knowledgeable trades-people and technicians keep quiet rather than point out misunderstandings by even the most junior engineers. Yet, in Australia, firms rely on experienced foreman to help junior engineers acquire practical knowledge, as explained by a senior engineer:

"A wise construction boss will make sure that a young engineer has a highly experienced foreman on his site. The experienced foreman will be teaching the engineer how to do the project management, a kind of reverse mentoring. Older foremen with the right skills get a tremendous thrill from doing this kind of thing especially when they're explicitly asked to do it."

The effect of the hierarchy, apart from disrupting access to distributed knowledge, is to centralise decision-making at the most senior levels of the enterprise. Even a factory manager may have negligible financial authority in South Asia. At one manufacturing plant turning over millions of dollars, the managing director had signing authority for about US$3,000. At a cement plant with similar turnover the managing director had authority to spend only US$200. In Australia, even relatively junior engineers have much greater financial responsibility. Many of the larger engineering firms use a risk-based decision hierarchy that allows for practically all decisions to be delegated, freeing up the time of the most senior staff to focus only on the most critical decisions with the greatest financial consequences for the firm.

In an Australian metal manufacturing company, an electrical engineer explained how he made most decisions for himself and only informed his boss later:

"My boss wants me to bring solutions, rather than problems."[15]

Centralising decisions introduces delays and the risk of misunderstanding, as another Australian senior engineer explained:

"If you work through the traditional lines of authority it's going to take too much time and may even get forgotten."

In comparison to a commercial manufacturing plant, hierarchy and power rela-tions in the context of a South Asian water utility are infinitely more complex. While engineers work within an elaborate bureaucratic hierarchy within the utility, they are also directly responsible to local agencies of political power. Describing a young depot engineer investigating an apparently illegal connection, Coelho wrote:

"What was interesting to me in this incident was the intense dilemma the engineer was thrown into by the seemingly straightforward problem of an illegal connec-tion. She had to negotiate a labyrinth of plots constituted by rumours, illicit acts and transgressive collaborations in order to enact or exert her own agendas of personal survival, responsibility to her workers and colleagues, and a wider official

---

[15](Domal, 2010, p. 118).

accountability. She was also caught in a classic bureaucratic conundrum where, as head of the unit, she was also the newest kid on the block with at best a shallow grasp of local geographies and histories of power and collaboration. All these needed to be unravelled in order to act effectively, or at least safely."[16]

In this discussion excerpt, two assistant district water supply engineers (AE1, AE2) explained why they receive up to 200 mobile phone calls a day as they coordinate a host of manual valve operators who direct water to different neighbourhoods for an hour or two every other day. Each house and office has a rooftop storage tank that overflows when full. Directing water pressure for too long results in unnecessary wastage.

AE1: "Allotted time will be given to particular areas or a posh colony … they will wait for 5 min and they have our mobile numbers"

AE2: "Almost after five minutes water has not come … we receive such type of call also …"

AE1: "Also from top bureaucrats also we receive calls"

AE2: "Senior citizen also"

AE1: "Also the slum people … we can (get calls from) below poverty line people"

With a host of illegal and ad-hoc connections in the network, and official records that can be decades out of date, both knowledge and authority are diffuse, uncertain, and subject to a myriad of intersecting political and commercial influences. Engineers directing work gangs searching for underground pipes, both legal and illegal, will often have to rely on nearby residents who have more accurate information than they can obtain, even from their depot staff. In some of the more marginalised settlements, local political alliances may even organise their own parallel water supply systems using groundwater. Alternatively they organise their own plumbers to create large scale illegal bypass connections in the public water supply system, especially if they cannot exert enough pressure on either engineers or central city politicians to provide a more reliable supply using the central distribution network.[17]

Engineers find they have to use their technical and social authority to make up for the lack of it among clerical staff, as one of their direct responsibilities is revenue collection. Householders will often ignore payment requests from meter readers. The engineers explained that minimising 'non-revenue water' in their district is one of their performance indicators, directly affecting their salaries, so they need to devote part of their time personally applying pressure to recalcitrant consumers. Beyond regular phone calls to chase consumers, they often have to take firmer action:

AE2: "people will be paying electricity bill, cable bill everything but water they won't be paying … until and unless we go there … people will be … sometimes we block the sewage connection also … some of the customer they don't pay even if

---

[16]Coelho's work provided extensive supporting evidence, complementing focus group discussions with utility engineers in an Indian city which has one of the best urban water supplies (Coelho, 2006, p. 501). Kayaga and Franceys provided complementary evidence from Africa, revealing similar patterns of social behaviour (Kayaga & Franceys, 2007).

[17]Anand (2011, p. 559) provides an analysis of the ways that political and organisational power impinges on the work of engineers and the water system as a whole.

you disconnect the water supply line they won't pay ... so ultimately we have to block their sewage system"

AE1: "we use coarse methods and we use all types of arm twisting techniques"

AE2: "see if we block water connection ... he gets the water from next houses or he gets the tanker so after 3 or 4 days ... we again go to his house ... we ultimately disconnect ... we block his sewage system ... then thereby he will come for part of the payment".

This excerpt reveals how social hierarchy requires pressure from engineers to persuade many consumers but even this is not always successful. Ultimately, engineers can use technical sanctions but in doing so compromise the integrity of the pipes. With re-made joints in the sewage pipe and water pipe in close proximity, cross contamination from sewage is inevitable.[18]

Australian engineers, particularly those in smaller businesses, are also exposed to the necessity to apply personal pressure in order to recover outstanding payments. The difference is one of degree, as explained before.

Still another aspect of the social hierarchy is the issue of language. Engineers will have received at least their tertiary education in English, and most will have studied in English through secondary schooling. Jobs in the public service are tightly restricted by local quota requirements so it is likely that engineers have a working knowledge of local dialects spoken by labourers and production workers with dirty hands, a mark of low social status. Anyone who can afford to study in English can acquire an occupation that does not require hands-on labour in a dirty environment. For engineers, the lack of hands-on work experience does not significantly differentiate them from their Australian colleagues who do not have time to engage in hands-on work.

The language issue surfaces on the shop floor, particularly in a commercial enterprise that may hire engineers without fluency in the local dialect. While the engineers can speak their national language (Hindi, Bengali or Urdu) and English, they will usually adopt their mother tongue (often Punjabi or the other major languages common around centres with quality education) and code switch with other languages.[19] Code switching not only complicates transcription for the researcher but also makes comprehension difficult, even for locals, because words don't necessarily have corresponding meanings. The local language may subsume the national in many settings. Production workers converse in local dialects and may not even be able to speak or comprehend more than a few words of the national language. This means that engineers may not have a common language with many of the people working on the shop floor, particularly tradespeople, technicians and casual day labourers. Instead they have to work through supervisors from a similar background who can traverse the languages, though not necessarily with the meaning or intent intact.

---

[18] For most of the time there is little or even negative pressure in water pipes and so sewerage seeps into the scheme water pipes from thousands of leaking or corroded connections.

[19] Code switching: the speaker will switch without warning to another language and back again, without any apparent reason, by habit. This habit is so deeply ingrained that anyone who does not switch between three languages is marked out as being distinctly different.

Once again, the difference between Australia and South Asia is one of degree: language issues surface in Australian workplaces but for different reasons. Many technicians and skilled workers have migrated from other countries and in a workplace with compatriots they will often prefer to speak their mother tongue. However, all will be fluent in English to a greater or lesser extent.

In order to better appreciate the ways in which social interactions mediate the results of engineering practice, it is useful to review some research findings on distributed knowledge and cognition.

4.2   Procedures and documents: embodied expertise

Much of the distributed knowledge that supports an engineering enterprise is embodied in processes and procedures that formalise interactions between engineers and other actors to ensure that interactions take place that might otherwise be overlooked. It is not uncommon for engineers to regard much of the work associated with procedures as administration, "not real engineering", yet their importance is readily appreciated. Relatively weak management systems, procedures and processes were another of the distinguishing aspects of difference that Domal observed.[20]

An Australian control systems design engineer working for a multinational firm explained some of his firm's procedures:

> "We have very formalised document review systems. For the critical design phases we have actual site meetings and we have checklists of what has to be done at those meetings. We have found that they are very effective. In this project we have probably gone further than just about anyone else has in the world in terms of formalising what previously has been a very informal process. We have check lists for all the major meetings, we have a formal checklist for document review, we have templates for all the major documents, so I think we are very structured in that regard. The client has to check all our conceptual design documents."

He went on to explain how the firm's engineers would work through the documents, sitting with the plant engineers, line by line, page by page, following checklists to make sure that every aspect of every detail is discussed and checked with the engineers on site. This is time-consuming, tedious work with the aim of ensuring that the control system designers can systematically access the practical contextual knowledge of the plant engineers. However, the results pay off. The firm was able to charge a much higher rate per hour of work for its engineers (more than twice as high as smaller local firms) because the way they worked with such meticulous detail resulted in fewer mistakes being made and a more reliable implementation of the control system that had been designed for the plant. In the end, the client was happier, and the plant was much safer to operate, more reliable, and more profitable for the owner.

This is an instance of distributed cognition, aided by formal procedures that embody knowledge accumulated within the firm, to ensure that knowledge is exchanged through social face-to-face interaction.

---

[20]The second of Domal's dimensions of difference (Domal, 2010, pp. 103–167).

The engineers that we interviewed in South Asia reported few organisational procedures, and often they reported making them up as needed, such as this refinery
operations engineer:[21]

> "I don't have fixed procedures ... each person in this position will institute
> different procedures to suit himself ... (he) will design the procedures to suit the
> requirements as he sees fit."

Gradually, I realised that engineers in South Asia make much less use of 'embodied knowledge' in the form of procedures and documents than their counterparts in
Australia. The tendency to focus on aspects of their work associated with their technical identity is common to both.[22] Even large firms seemed to impose fewer formal
processes on engineers in the form of procedures. The notion of building on embodied
knowledge and experience from other people as an explanation seldom surfaced. The
invisibility of such embodied knowledge is one possible explanation for difficulties
expressed by many engineers interviewed in South Asia in understanding the apparent
superiority of engineering firms from industrialised countries. This issue surfaced in
an interview with a senior consulting engineer:

> "Private sector clients often are reluctant to pay extra for a consultant ...
> Interestingly, foreign institutions such as the World Bank are prepared to pay 6%
> to a foreign consultant but only 2% or 3% to a local consultant for infrastructure
> projects."

After the interview, I spent some time touring the design office with junior engineers
in the firm. It seemed to me to be dark but relatively clean, with piles of rolled drawings
and box files heaped in corners. Most of the staff had their own computer workstations
and access to at least shared telephones. Individual directors had their own offices and
male secretaries sitting just outside. There were one or two female engineers. Compared
with many similar offices it was relatively spacious and uncrowded. There was no
documentation storage and indexing system so much of the useful documentation was
relatively inaccessible: access relied on somebody's memory of where they might last
have left it. While documentation and configuration management systems are perceived
by Australian engineers as an administrative overhead, a non-technical issue, the access
they provide to embodied knowledge underpins a significant part of their competitive
advantage.

Lack of formal documentation is common on the shop floor, even in a well-
organised firm. A production superintendent explained that in his factory, the manufacturing engineering department provides the tooling needed for his production line:

> "their only input is a drawing of the original component that we want to make:
> they have skilled toolmakers. They don't make any drawings of the tooling itself.

---

[21] See also Domal & Trevelyan (2008, pp. 6–10).
[22] Engineers have a tendency to hide the social aspects of their work behind a superficial technical
description, and also to relegate to subsidiary status those aspects of their work not directly
associated with their technical identity (Trevelyan, 2010, pp. 8–11).

It is a largely undocumented process. Of course, it makes it difficult to maintain quality and consistency in the product."

Much of the manufacturing is outsourced to street side 'vendors' who perform basic cutting and machining work on materials that they either purchase themselves or are supplied by the enterprise. Vendors are often selected by a 'purchasing department' on the lowest priced tender with an agreed rate per piece produced. Their working methods are even less formal than the larger factory enterprises that hire them. Drawings and documentation are often ignored: most of the vendors cannot reliably read drawings and specification documents. As a result, quality control for engineers in the more organised enterprise tends to be a constant fire-fighting occupation.[23]

A quality assurance engineer described his experience working with vendor-supplied components this way:

"It is a very stressful job. All day long I am faced with different problems, from vendors or the assembly section. I never have the time to find myself in a relaxed condition."

A senior development engineer echoed these comments:

"there is a constant stream of 'crises' concerning part quality through the company. This seems to be an accepted state of being."

The respective purchasing departments do not have any method or data to take into account the real economic consequences in terms of engineering time cost, production disruptions and low quality products. They take their decisions to award vendor contracts on price and firmly resist informal 'meddling' by engineers who, in their opinion, would send the company bankrupt if they were given a free hand in purchasing. In a wide variety of firms, this is 'an accepted state of being'.

The end result is a loss of accumulated knowledge that can only be represented with written documents or computer records. I have observed many other instances, far too numerous to describe in this chapter. While South Asian engineers are aware that well-organised firms tend to use more documentation, none described contemporary informality in terms of knowledge or information loss.

In the public sector utility, documentation takes on quite a different significance. A senior utility engineer explained his experience with household connections:

[You were doing household connections?][24]
    "House connections ... we used to submit as a plan, how they are going to connect it which mean we have to connect it. We used to have plans from state department, but nowadays we don't have plans for house connections."

---

[23]Domal provided more details and photographs of street-side vendor workshops from India and described familiar scenes in any South Asian engineering-oriented community (Domal, 2010, pp. 122–124).
[24]Questions posed by the interviewer in square brackets.

Gradually, over time, the water distribution utility had moved from formal plans and documentation to an informal system based on outdated plans and *ad-hoc* adaptations, bypasses and connections arranged verbally with depot engineers with varying degrees of documentation (Coelho, 2006, p. 506).

In terms of enterprise performance, the South Asian manufacturing companies employing the engineers participating in this research were domestic leaders, but not yet internationally competitive at the time (2008–2010). With labour costs around 7–10% of expenses, their performance was largely determined by plant availability, throughput and product quality, all of which were compromised by factors related to, among others, the issues discussed above. While labour was a higher proportion of expenses for Australian firms, engineers were more effective in maintaining high quality, availability and throughput. The manufacturing settings we observed appear to be representative and the issues may lie behind continuing difficulties in that sector (e.g. Anon, 2012).

5 MOBILE TELECOMMUNICATIONS – A NEW START?

At this stage, a reader could be forgiven for suspecting that this chapter is arguing that certain social cultures can perform engineering better than others, and that I have analysed South Asian engineering practice through a critical ethno-centric lens, particularly in pointing to 'disruptive influences'. The author, naturally, has far more experience of engineering practice in Australia than South Asia, so the reader could be forgiven for this perception. I would ask the reader to consider some particular aspects of the evidence presented in this study before reaching his or her interpretation of this aspect of the chapter. First, my knowledge of South Asian practice and culture is based on extensive engagement over two decades. Second, the findings of this chapter have been discussed with engineers native to the region (as it was not possible to meet most of the participants when the findings were finally ready for publication). Third, much of the evidence was collected by an engineer of South Asian origin, in his home district and Australia. There was no major qualitative difference between that evidence and the other evidence collected by myself or between our respective interpretations. Last, the propositional knowledge base, technologies, tools and materials used for metal manufacturing, water utilities and electric power utilities have been widely dispersed and available for enterprises in South Asia and Australia for several decades. It is therefore difficult to argue that the pronounced differences in the economic effectiveness of these enterprises can be explained other than by local social and cultural differences. Certainly, few South Asian people regard the performance of their manufacturing, water and electric power enterprises as being at a generally satisfactory standard. It is only by examining the engineering performance differences closely, particularly how engineers perform their practices, that we can begin to understand what is currently a significant economic performance penalty for South Asian citizens.

There is widely available evidence that individual engineers from South Asia are equal to the best when they are employed by engineering enterprises in Europe, America and other industrialised societies. It would be easy to conclude, therefore, that the evidence presented so far suggests that there are social or cultural factors in their home environments that prevent South Asian engineers from being effective. In the research that contributed to this chapter I found (by chance) a small number of South Asian engineers with an outstanding record of domestic success, contradicting this hypothesis. However, in the present chapter it would be more instructive to describe a class of South Asian engineering enterprises that have been outstandingly successful.

Originally introduced in the mid-late 1990s, mobile telephone networks have become well established in South Asia with access costs well below the rates in industrialised countries. At the time of writing, calling rates vary between US$0.05 per hour and $0.02 per minute of talk time in Pakistan, whereas in Australia the cost is between $0.05 and $0.10 per minute. Most adults and many younger people carry internet-enabled phones with them at all times, even in some of the remotest areas of the Himalayan Mountains. Mobile phone networks have eclipsed public landline telephone monopolies and have such a reliable reputation that mobile phone credits often serve as an informal payment settlement system.

In contrast to water and electricity utilities, mobile telephone networks provide high quality service at extraordinarily low cost, and also provide profits for investors. What insights can we gain into engineering practice in South Asia that might help to explain such large and apparently sustainable beneficial outcomes?

Mobile phone networks in South Asia have adopted technology designed originally for industrialised countries. Some of the choices seem to have provided critical differentiating factors that enabled these privately owned utilities to become commercial successes in developing countries, in contrast to the publicly owned landline monopolies they have displaced. These factors include:

1   Mobile telephones provide lasting economic benefits for end users, mainly in the form of time savings. For example, a tenant farmer situated many kilometres from the nearest economic centre can send a text message to his seed and fertiliser suppliers to find out when his consignments will be ready for collection. He can also, without having to leave his farm, negotiate advance sales of produce with processors who are several hours away by bicycle or donkey cart. A few text messages and phone calls replace entire days of exhausting travel. Even though the hourly value of time for a tenant farmer is small, the savings are significant and more than outweigh the cost of mobile phone access.

2   The technology permits incremental extensions of network capacity once the initial installation of antenna towers, exchanges, microwave and optical fibre links has been completed. Although the upfront capital investment is still large, it is within the capacity of financial institutions and investors in developing countries. (As demonstrated by the Indian example, the size of capital investment is usually beyond the reach of government. Initially the Indian government wanted to retain a controlling share of ownership in critical infrastructure but has since allowed a much greater proportion of private and foreign ownership.)

3   Prepaid scratch cards with encrypted access code numbers enable the efficient collection of large numbers of small payments by end users.

4   Centralised payment processing enables efficient sales tax revenue collection for government.

5   The service is not accessible by users unless they have paid in advance or they have provided a reliable line of credit. In South Asia, the vast majority of users access

networks with prepaid cards. Unlike water and electricity, users cannot simply tap into the service without paying.

6   Information technology has so far enabled the implementation of effective barriers that raise the cost of fraudulent intervention by individual users. Even though mobile phones are an extension of the Internet, they are protected with special access controls that have restricted unauthorised access to an acceptably low level. In the case of water and electricity utilities, the cost of influencing meter readers, even engineers, is well within the resources available to medium and large-scale businesses.

In summary, technological choices appropriate for industrialised countries have served to protect both investors and end users in developing country environments.

Interviews and observations of telecommunication engineers have revealed further significant differentiating factors.

In water and electricity utilities, even in manufacturing, hands-on aspects of production and service delivery work are performed by people with limited education who have little or no influence in the social hierarchy. Having to get one's hands dirty at work or engaging in physical labour is considered to be an indicator of low social standing.

In the telecommunications sector, on the other hand, apart from the initial installation of towers, antennas, standby generators and connecting cables, hands-on service delivery, production and maintenance work is performed by highly skilled technicians, most of whom have obtained degree qualifications in an English-speaking environment. In Pakistan, for example, there was a surge of demand for private tertiary education in electrical and computer engineering and computer science in the second half of the 1990s and early 2000s. Initially this demand was driven by the prospect of migration to Europe and or America where IT skills seemed to be in strong demand at the time. In many cases, what passed as a degree in computer science was little more than a practical hands-on introduction to the use of spread sheets, word processors, e-mail and databases. As a result, people with good computer operating skills and reasonable English were readily available in the employment market and were paid only a moderately higher salary compared with other clerical staff.

This reservoir of people with 'applied' IT skills provided an important skills base for mobile communications.

Much of the work in the mobile telecommunications sector requires technicians to configure digital exchanges, apply software patches and upgrades, and configure network routers. Online certifications from companies such as Microsoft and Cisco have provided additional skills.

The mobile phone systems that they were supporting simultaneously provided technicians with on-the-job access to a wide circle of friends. When confronted with a fault that they could not solve on their own, they could always resort to this network: someone would know the answer.

Level 1 field service engineers explained some aspects of their technical responsibilities:

"It is hard to diagnose a fault. I have to think of the justified logical evidence and locate the fault with precision."

"The procedures for fault tracing are set down in international standards for the company. I have to trace fault events in hexadecimal code (a computer code for representing numbers). Where the exchange sends an invalid message then I have to trace through the software to find out the problem for myself."

At level 2, the engineers also have supervision responsibilities, as this excerpt demonstrates.

"First I sit with my computer (PC) and carry out patch administration. These are software corrections that have to be applied to the systems operating in the various digital exchanges. I receive the patches from Germany and I issue them to the field technicians. I then check on the final status of the patches once they have been installed in the exchanges."

More extensive studies in Brunei and Australia (Tang, 2012) have confirmed that technical aspects of telecommunications engineering practice largely involves incremental network capacity expansion through equipment procurement, installation of cables and equipment in buildings, and using software to configure network switching. High level mathematical analysis and modelling expertise, which comprise a large proportion of a telecommunications engineering degree course in industrialised countries, play only an indirect role.

Another engineer explained how he was required to comply with company procedures and standards laid down in Europe. In addition he had travelled to Europe for training courses.

[Tell us about fixed procedures you are required to follow.]
"Yes, as laid down in the company standards. The company software centres decide very detailed procedures to be followed for every aspect of the work. This is done in Germany and the software development centre in Ljubjana in Slovenia."

In contrast with the other utilities and manufacturing, there was little if any disparity between the engineers and technicians performing hands-on work in terms of social hierarchy. Both had acquired tertiary engineering qualifications in English. Their informal social networks blurred the distinctions between them and they freely exchanged technical expertise when needed. The social barriers that we observed interfering with the successful enactment of distributed knowledge in water utilities and manufacturing did not appear to the same extent in telecommunications work. Fluency in technical English coupled with comprehensive implementation of standards and procedures, as well as e-mail access to specialist international communities of practice, enabled engineers to draw on similar embodied expertise to their European counterparts:

"I use internet, intra-net, specifications, procedures. The internet is helpful for 'personal grooming', improving my background knowledge. The MultiTel[25]

---

[25]Multinational telecommunication equipment supplier pseudonym.

intranet (we call it sharenet) is really useful. We post technical questions on the "MultiTel Everywhere" net and someone who knows the answer will respond. We study the draft international standards for interfaces between telephone systems, you know, between us and companies like Siemens and Alcatel etc. Also we have sales presentations using Netmeeting, and we listen on the phone using a conference call. We use that for some sales training courses."

In contrast to the water utilities, mobile telecommunication engineers had little or no contact with end users of the service other than through their own social network and family. Operational knowledge of the system is encoded in software. Engineers seemed have no idea of the credit of individual customers, or any ability to influence the level of service that an individual customer received. Unyielding financial control is maintained through software, allowing little if any space for political and other social influences, thus protecting both end users who have contributed advance payment for service and investors who have paid for the infrastructure. In the water utilities, on the other hand, operational knowledge of the system is diffused in outdated documents and the memories and histories of power and influence within and around the enterprise, far beyond the reach of reasonable technical and financial restraint. Individual engineers and technicians can readily influence the service level provided for powerful individuals or groups of customers. There is little or no protection from socially disruptive manipulation beyond the reach of governance.[26] These differences seem to contribute significantly to the extent to which end users gain real economic and social value from the services provided by their engineers.

6 CONCLUDING REMARKS

One way to view engineering is as a lever that amplifies the effectiveness of investments in enterprises such as water and electricity utilities, manufacturing and communications, reducing the human effort, energy and material resources needed to provide given benefits. In the case of South Asian water and electricity utilities, the leverage is currently not nearly as effective as it is in the industrialised world. Large investments have been made, but the resulting costs for end-users are high and service quality is poor in comparison. In South Asian metal manufacturing, the leverage is observable, but still not as effective as in the industrialised world. In mobile phones, on the other hand, the leverage has been at least as effective, if not more so.

It is interesting to reflect that the engineering for water distribution, sanitation, electricity distribution, telecommunications, concrete construction, and motorised transport have all been imported into South Asia from industrialised societies. Among these, mobile telecommunications stands out as an extraordinary success, not only in South Asia but also in every other developing country.

---

[26]The perception of social disruption that decreases the quality of service for end-users has to be distinguished from social disruption by informal local power structures that improve service quality for their local users. Power relations in water supply are explained by (Anand, 2011).

This success demonstrates that engineering enterprises can work in low income developing societies. However, this success has been dependent on certain technological and sociotechnical factors. It is interesting to speculate on whether these success factors could be transferred into other engineering enterprises in developing societies such as water, sanitation and electricity distribution. If this could be achieved, the current economies in most of these societies might be transformed and much of the present poverty might be eliminated.

However, it seems unlikely that the current levels of energy and material resource intensity required for the present "third world" to achieve an equivalent economy to the industrialised world can be sustained within the constraints of current and foreseeable energy and material supplies. Achieving this desirable objective will take considerable ingenuity beyond our current levels of achievement. Engineers are likely to be able to succeed more often if they can recognise factors that enable engineering enterprises to succeed in their own societies.

This has been only an exploratory study. More research is needed to improve confidence in these findings and to provide a foundation for education resources for future generations of engineers.

The ultimate test for this hypothesis is successful intervention; changing the way that engineering is performed and the ways that engineers see themselves in South Asia so that they can begin to regulate the social interactions on which their practice depends. Most of today's engineers regard this aspect of their work entirely as 'a management issue' and restrict themselves to a technical discourse in which their authority is largely unquestioned, where the predictable laws of physics and mathematics hold sway rather than the insecure vagaries of human nature.

Engineering in developing societies would be greatly helped by a deeper understanding that the results of engineering investment depend on local culture and social interactions, as much as the universal laws of physics. Effective solutions for global poverty may depend on understanding that interventions need to go far beyond governance and economic policy, and take engineering practice issues into account.

The comparisons between South Asia and Australia demonstrate that there are patterns of social behavior that support the enactment of distributed knowledge that seems essential for successful engineering in European societies. In most South Asian enterprises, the patterns of normal social interactions can impose barriers that interfere with access to distributed knowledge, arguably with significant effect on the level engineering success measured in terms of both economic and social benefits. One can hope that with appropriate insights from this and similar research, future engineers might be able to devise effective ways to enact distributed knowledge in their own social cultures.

ACKNOWLEDGEMENTS

The author would like to acknowledge numerous contributions from his colleagues and students, particularly Vinay Domal through whose work much of the detailed substantiation for this argument has come. Countless engineers and companies donated their time, patience and enthusiasm to enable the author to understand their work. The author is grateful for family financial support that enabled this research.

REFERENCES

Anand, N. (2011). Pressure: The PoliTechnics of Water Supply in Mumbai. *Cultural Anthropology, 26*(4), 542–564. doi: 10.1111/j.1548-1360.2011.01111.x.

Anon. (2012, August 11). Manufacturing in India: the marsala mittelstand. *The Economist.*

Barley, S. R. (2005). What we know (and mostly don't know) about technical work. In S. Ackroyd, R. Batt, P. Thompson & P. S. Tolbert (Eds.), *The Oxford Handbook of Work and Organization* (pp. 376–403). Oxford: Oxford University Press.

Brookes, N. J., Morton, S. C., Grossman, S., Joesbury, P., & Varnes, D. (2007). Analyzing Social Capital to Improve Product Development Team Performance: Action-Research Investigations in the Aerospace Industry With TRW and GKN. *IEEE Transactions on Engineering Management, 54*(4), 814–830. doi: 10.1109/TEM.2007.906859.

Bucciarelli, L. L. (1994). *Designing Engineers.* Cambridge, Massachusetts: MIT Press.

Coelho, K. (2004). *Of Engineers, Rationalities and Rule: and Ethnography of Neoliberal Reform in and Urban Water Utility in South India.* PhD, University of Arizona, Tucson.

Coelho, K. (2006). Leaky sovereignties and Engineered (Dis)Order in and Urban Water System. In M. Narula, S. Sengupta, R. Sundaram, A. Sharan, J. Bagchi & G. Lovink (Eds.), *Sarai Reader 06: Turbulence* (pp. 497–509). New Delhi: Centre for the Study of Developing Societies.

Darr, A. (2000). Technical Labour in an Engineering Boutique: Interpretive Frameworks of Sales and R&D Engineers. *Work, Employment and Society, 14*(2), 205–222.

Domal, V. (2010). *Comparing Engineering Practice in South Asia and Australia.* PhD, The University of Western Australia, Perth.

Domal, V. K., & Trevelyan, J. P. (2008, June 20–22). *Comparing Engineering Practice in South Asia with Australia.* Paper presented at the American Society for Engineering Education Annual Conference, Pittsburgh.

Friesen, M. R. (2011). Immigrants' integration and career development in the professional engineering workplace in the context of social and cultural capital. *Engineering Studies, 3*(2), 79–100. doi: 10.1080/19378629.2011.613571.

Huberman, A. M., & Miles, M. B. (Eds.). (2002). *The Qualitative Researcher's Companion.* Thousand Oaks, California: Sage Publications.

Kayaga, S., & Franceys, R. (2007). Costs of urban utility water connections: Excessive burden to the poor. *Urban Policy, 15*, 270–277.

Larsson, A. (2007). Banking on social capital: toward social connectedness in distributed engineering design teams. *Design Studies, 28*(6), 605–622. doi: 10.1016/j.destud.2007.06.001.

Latour, B. (2005). *Reassembling the Social: an Introduction to Actor Network Theory.* Oxford: Oxford University Press.

Levin, D. Z., & Cross, R. (2004). The Strength of Weak Ties You Can Trust: The Mediating Role of Trust in Effective Knowledge Transfer. *Management Science, 50*(11), 1477–1490. doi: 10.1287/mnsc.1030.0136.

Miles, M., & Huberman, A. (1994). *Qualitative Data Analysis: An Expanded Sourcebook* (2nd ed.). Thousand Oaks, California: Sage Publications Inc.

Mukerji, C. (2009). *Impossible Engineering: Technology and Territoriality on the Canal du Midi.* Princeton: Princeton University Press.

Orr, J. (1996). *Talking About Machines: An Ethnography of a Modern Job.* Ithaca, New York: Cornell University Press.

Patton, M. Q. (1990). *Qualitative Evaluation and Research* (2nd ed.). Newbury Park, California: Sage.

Qizilbash, T. (2009, August). Powerless. *Newsline, 22*, 18–52.

Stevens, R., Johri, A., & O'Connor, K. (2012 – in press). Professional Engineering Work. In A. Johri & B. M. Olds (Eds.), *Cambridge Handbook of Engineering Education Research*. Cambridge: Cambridge University Press.

Strauss, A. (1987). *Qualitative Analysis for Social Scientists*. Cambridge: Cambridge University Press.

Tang, S. S. (2012). *An Empirical Investigation of Telecommunication Engineering in Brunei Darussalam*. PhD PhD, The University of Western Australia, Perth.

Trevelyan, J. P. (2005). Drinking Water Costs in Pakistan, from http://www.mech.uwa. edu.au/jpt/pes.html.

Trevelyan, J. P. (2007). Technical Coordination in Engineering Practice. *Journal of Engineering Education, 96*(3), 191–204.

Trevelyan, J. P. (2010). Reconstructing Engineering from Practice. *Engineering Studies, 2*(3), 175–195.

Trevelyan, J. P., & Tilli, S. (2007). Published Research on Engineering Work. *Journal of Professional Issues in Engineering Education and Practice, 133*(4), 300–307.

Vinck, D. (Ed.). (2003). *Everyday Engineering: An Ethnography of Design and Innovation*. Boston: MIT Press.

Winch, G. M., & Kelsey, J. (2005). What do construction project planners do? *International Journal of Project Management, 23*, 141–149.

World Health Organization. (2012). *GLAAS 2012 Report: UN-Water Global Analysis and Assessment of Sanitation and Drinking Water*. Geneva, Switzerland: World Health Organization.

Zussman, R. (1985). *Mechanics of the Middle Class: Work and Politics Among American Engineers*. Berkeley: University of California Press.

Chapter 11

Mathematics in engineering practice: Tacit trumps tangible

Eileen Goold[1] *& Frank Devitt*[2]

[1]*Department of Electronic Engineering, Institute of Technology, Tallaght, Dublin, Ireland*
[2]*Department of Design Innovation, National University of Ireland, Maynooth, Ireland*

1 INTRODUCTION

In the introductory chapter James Trevelyan notes the scarcity of systematic research on engineering practice and he maintains that a theory of engineering practice would benefit engineers, researchers and students. This chapter presents the findings of a study of professional engineers practising in Ireland. More specifically this chapter investigates the role of mathematics in engineering practice. The study was inspired by the observation that there is a lacuna in the scholarly literature concerning the nature of mathematics' role, if any, as a significant cause of the declining number of students entering professional engineering courses and the absence of any broad picture of the mathematical expertise required or used by practising engineers. The research shows that it is use of broader mathematical thinking – tacit mathematics, rather than the more explicit 'syllabus' mathematics – that is of most value to all engineers in their workplace.

2 BACKGROUND

While engineering expertise is key to sustaining a modern economy and to the advancement of civilisation, the interest of young people to pursue careers as engineers has diminished, in Western Europe and the USA in particular (Elliott, 2009; Forfás, 2008; King, 2008; McKinsey, 2011; Organisation for Economic Co-Operation and Development, 2010). It is widely thought that mathematics is the "the key academic hurdle" in the supply of engineering graduates (Croft & Grove, 2006; King, 2008). Students wishing to pursue an engineering degree course are required to be proficient in mathematics. However, many students have "no idea" what role mathematics will play in their future careers (Wood *et al.*, 2011). Most students view engineering education as further engagement in school science and mathematics (Brickhouse *et al.*, 2000). "Some see mathematics as the gateway to engineering, paving the way for sound design; others see mathematics as a gatekeeper, denying entry to otherwise talented would-be engineers" (Winkelman, 2009).

Research suggests that while professionals in numerate fields draw upon their mathematics school learning, they do so in a distinctly different manner from the way in which they experienced mathematics in school. It is reported that there is a significant difference between what a mathematician calls "doing mathematics" and what an engineer calls "doing mathematics" (Bissell & Dillon, 2000). However, in the case of engineering practice, research concerning the type of mathematics used by engineers in their work is sparse (Alpers, 2010b; Cardella, 2007; Gainsburg, 2006; Trevelyan, 2009). Gainsburg's ethnographic study contrasts school modelling problems with modelling activity of structural engineers who, due to a lack of information, experience uncertainty. In addition to mathematical analysis, practising engineers rely on mathematical thinking skills, such as skills acquired from previous experiences and situations, reasoning and justification of conclusions, to invent new designs in their work (Gainsburg, 2006).

While there are a number of studies that investigate engineers' use of mathematical thinking, most of these are conducted in academic workplaces. Difficulties associated with investigating "real" engineers' mathematics usage are that access to engineers is difficult and with many different branches and job profiles within engineering, there is no unique identity as "'the' engineer". Furthermore studies of engineers' use of mathematics have tended to take a qualitative approach that involve a small number of engineering functions and engineers and thus the findings may not represent engineers generally (Alpers, 2010a).

3 METHODOLOGY

The research methodology employed in the study described here is a sequential explanatory mixed methods design incorporating a survey followed by interview analysis (Figure 1). The sample of 365 engineers who participated in the survey is broadly representative of the professional engineering population across industry sector, engineering discipline, gender and geography and the survey sample size is satisfactory for 95% confidence that the findings represent the population of Chartered Engineers in Ireland. Chartered Engineers are professional engineers registered with Engineers Ireland[1] and they are required to have a minimum of a level 8 academic qualification (honours bachelor degree) and four years of relevant professional experience. Following the survey, interviews were conducted with 20 engineers representing low, mid and high *curriculum mathematics* users (Goold & Devitt, 2012a). *Curriculum mathematics* is the term devised in this study to represent the spectrum of school and university mathematics, ranging from secondary school to university level 8 (Goold & Devitt, 2012b).

The study investigated: engineers' *curriculum mathematics* usage; *mathematical thinking and* engagement with mathematics. The methodology used to measure engineers' *curriculum mathematics* usage is based on de Lange's pyramid of mathematics assessment (Goold & Devitt, 2012b). In de Lange's assessment model, mathematics questions are located in a pyramid according to the mathematical content; the degree of difficulty and the level of thinking. A balanced test includes questions in all content domains, of varying degrees of difficulty and at all levels of thinking. As the level of thinking required increases, it becomes harder to distinguish mathematical content domains and also the range between easy and hard questions becomes smaller, hence the model is pyramidal (De Lange, 1999; De Lange & Romberg, 2004).

---

[1]Engineers Ireland is the professional body representing engineers in Ireland.

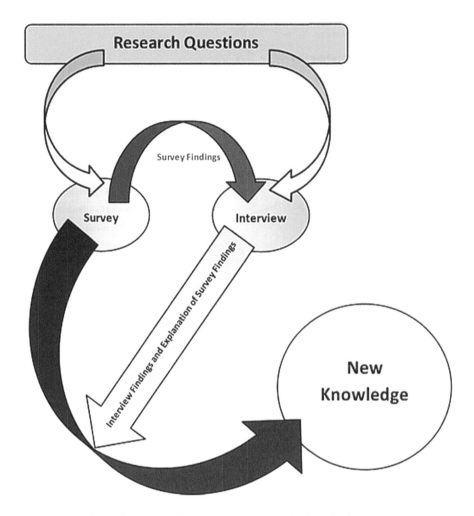

Figure 1 Sequential explanatory strategy mixed methods design.

In this study, de Lange's pyramid is adapted to measure *curriculum mathematics* in three dimensions: domain; academic level; and usage type (Figure 2).

1   Domain refers to the five mathematics content areas specified in the school mathematics syllabi in Ireland. The content areas are: statistics and probability; geometry and trigonometry; number; algebra; and functions.
2   Level refers to academic levels and ranges from Junior Certificate[2] ordinary level, which is the first formal examination taken at age 15 by students in Ireland, to

---

[2]Junior Certificate examination is the first formal examination taken at age 15 by students in Ireland. Leaving Certificate examination is taken at the end of secondary school (age 18). In both examinations subjects are offered at either ordinary level or the more advanced level.

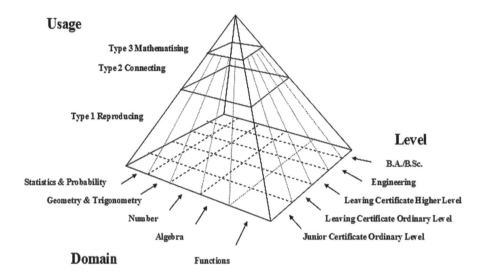

Figure 2 Curriculum mathematics assessment pyramid.

honours degree courses in both engineering and non-engineering (e.g. B.A./B.Sc.)
mathematics in level 8 academic programmes.

3   Usage type includes de Lange's three levels of thinking: type 1 (reproducing) is
usage of mathematics through knowledge of facts and concepts; type 2 (connecting) is usage of mathematics by making connections within and between different
mathematics topics and integrating information in order to solve problems where
there is a choice of strategies and mathematical tools; and type 3 (mathematising) is
usage of mathematics by extracting the mathematics embedded in a situation and
using mathematics to develop models and strategies and translating mathematical
models into real world solutions.

Two other types of mathematics usage of interest in this study are *thinking usage*
and *engaging usage*. *Thinking* usage is usage of mathematical modes of thinking
learned and practised through mathematics study and application, e.g. methods of
analysis and reasoning, logical rigour, problem solving strategies (e.g. problem decomposition and solution re-integration), recognition of patterns, use of analogy, and
a sense of what the solution to a problem might be (Schoenfeld, 1992). *Engaging*
usage relates to emotional relationships with mathematics. In the context of this study
engaging usage is defined as the motivation and persistence to take a mathematical
approach to a problem as a result of one's attitudes, beliefs, emotions, goals, sense of
value, interest, confidence, self-efficacy and sociocultural influences (Csíkszentmihályi,
1992; McLeod & Adams, 1989; Schunk *et al.*, 2010).

Following analysis of the survey data using Minitab statistical software, interviews were conducted with 20 engineers representing low, mid and high *curriculum
mathematics* users. Additionally these engineers included a diversity of engineering

disciplines; roles; sectors, organisations; urban and rural backgrounds; Leaving Certificate mathematics levels; and engineering education routes (direct entry into level 8 degree courses or progression from level 7 to level 8). 25% of the interviewees were female; 25% were less than 35 years of age. A manual data analysis process was employed (Goold & Devitt, 2012a, 2012b).

4   FINDINGS

4.1   First finding

- While almost two thirds of engineers use high level *curriculum mathematics* in engineering practice, *mathematical thinking* has a greater relevance to engineers' work compared to *curriculum mathematics*

In the survey engineers rate their mean mathematics usage for the 75 domain-level-usage combinations of *curriculum mathematics* as 2.73 Likert units[3] which is in the range "very little" to "a little". Of the five domains, *number* has the highest usage at 3.07 Likert units (Figure 3).

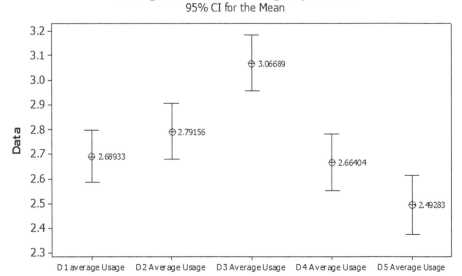

Average Mathematics Usage By Domain
95% CI for the Mean

Answer options: 1 = Not at all; 2 = Very little; 3 = A little; 4 = Quite a lot; 5 = A very great deal

Figure 3   Confidence interval: Engineers' mean curriculum mathematics usage by domain.

---

[3]Likert units: Units on 5 point Likert scale, 1 = "not at all", 2 = "very little", 3 = "a little", 4 = "quite a lot", 5 = "a very great deal".

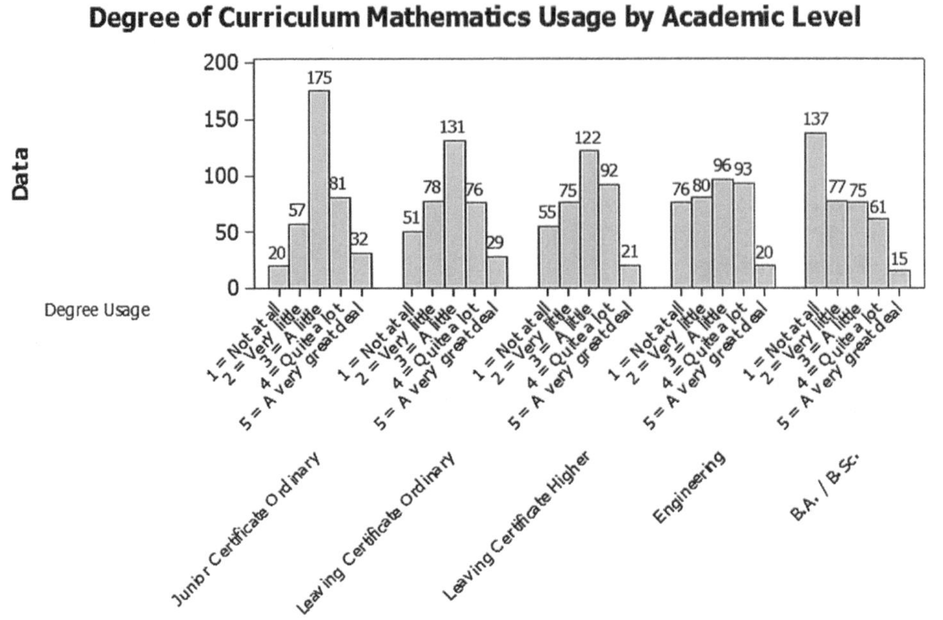

Degree of Curriculum Mathematics Usage by Academic Level

Figure 4 Histogram: Engineers' mean curriculum mathematics usage by academic level.

Survey analysis shows that almost two thirds of engineers (64.4%) use higher level Leaving Certificate mathematics in their work either "a little", "quite a lot" or "a very great deal". 57.3% of engineers use engineering mathematics and 41.4% of engineers use B.A./B.Sc. mathematics to the same degree (Figure 4).

Engineers rate their *mathematical thinking* usage as 4.02 Likert units which is considerably higher than their overall mean *curriculum mathematics* usage (2.73 Likert units). *Thinking* usage is highest (4.19 Likert units) when engineers are within 2 years of graduation and it reduces thereafter (Figure 5).

In the qualitative interviews, engineers articulate why, in their work, *thinking* usage is greater than *curriculum mathematics* usage:

– One electrical engineer, working as head of operations for a leading telecommunications company's public sector contracts, maintains that it is his experience and maturity rather than any technical qualification that qualifies him for this position. He believes that, in his current job, he could quite satisfactorily proceed without higher level mathematics. While he has responsibility for a "highly sophisticated system", this system "was designed, implemented, tested and brought into service by some technical engineers" and another team of engineers are responsible for the maintenance of the system. In terms of *curriculum mathematics*, his use of mathematics is limited to the number domain at Leaving Certificate ordinary level and particularly involves interpreting financial and statistical reports which he classifies as reproducing and connecting type usage.

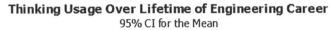

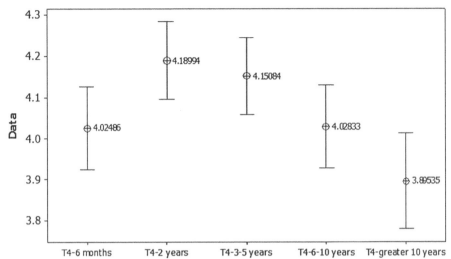

Answer options: 1 = Not at all; 2 = Very little; 3 = A little; 4 = Quite a lot; 5 = A very great deal

Figure 5 Confidence interval: Engineers' mean thinking usage.

– As a manager of a team of engineers, one mechanical engineer contrasts school mathematics where "there must be one answer" with his current job whereby if he came "to one solution ... that would be a disaster"; instead he must look at how his "design fits in" with the other disciplines. For mathematical calculations, he has "set up the computer to do all of that" and for any new calculation, he gets "the graduates to do it because they are closer to college". When hiring graduates, higher level mathematics is not necessary, instead he looks for "the guy that likes engineering" and "would fit in" with his team.

– A project manager with a multi-disciplinary engineering firm that develops solutions for complex capital projects serving multinational clients worldwide only uses mathematics that he is "confident about". He says his work is "about having the principles right and conclusions right from a good understanding of the problem with some checking by maths rather than doing a big long calculation". While "statistics and probability" at engineering level is the *curriculum mathematics* most relevant to his work, he maintains that engineering is the "bigger picture", it "is learning to understand engineering principles, experience, experimentation, the various tools you use, all of the whole lot, maths is in there but maths is only a tool ... it comes down to problem solving". He also maintains that as one progresses in engineering, less mathematics is required and his "whole way of analysing things and reasoning and organising got better as time went on; that was just experience".

– A civil engineer who is a projects manager with a state agency where her responsibilities include design, tender, implementation and construction of rail traffic infrastructure, maintains that "as an engineer you don't sit in front of the computer and do maths all day". In her job the "technical solution is the first step in about ten to get things done". She says, while "there are programs that will do a lot of the groundwork", she needs to know "how to use the programs and interpret the results". A major part of her work is "contracts" and the "commercial implications" of decisions. She states that she is "at the stage where common sense applies more than the maths". Furthermore her "grounding in maths" helps her to "look at the figures very quickly and make decisions". She says that "in order to have common sense I think you need to understand ... the effect of one piece of work on another part of our system". She rates her current *curriculum mathematics* usage in work as mainly of reproducing type and of higher level Leaving Certificate standard. She uses statistics and probability in a project about "noise monitoring" on a rail line and she uses "basic geometry and trigonometry to work out site levels". Her work involves "huge amounts of interaction with people" as she has to explain to her colleagues how she reached her conclusions. She also says that when one progresses to management "you need to know enough to know when someone is pulling your leg".

- Much of an information technology consultant's current work is "mathematical problem decomposition and restructuring and sequencing". For the 10% of the time that he needs mathematics in his job, "it is very valuable". He relishes mathematics, he sees it as a challenge and he would prefer a role that required greater mathematics usage. He uses statistics and probability, algebra and functions, at a level between higher level Leaving Certificate and engineering mathematics to determine "the most economically advantageous tender" for public sector contracts and to present civil servants with "a rock solid argument". He describes his usage type as mathematising. He believes that many people in his company have "no interest in presenting an argument or presenting something through mathematics and they would use whatever shortcuts they can or get other people to do it". He states that there is a "confidence" issue attached to using mathematics and "if you get it wrong it can look very bad".

- An electrical engineer with a multinational consulting engineering practice believes that, apart from statistics, he will never, in his work, use the level of mathematics he studied in university. However he maintains that "the sort of person who does honours maths ends up thinking and acting in a certain way" and this is what "engineers bring to their work". He notes that while "hand calculations" dominated engineering work "ten or fifteen years ago", this has changed; "now, it is all done by computer modelling". When approaching "problems that mightn't be mathematical by nature", because he is "comfortable with maths" he approaches them with a "mathematical logic".

- A manufacturing engineer who is a "programme manager" with a large U.S. multinational company maintains that while "higher level Leaving Certificate maths isn't necessary" in his work, he has brought the "discipline" he established from doing higher level Leaving Certificate mathematics into his "working life". His work involves "looking at data, making decisions and giving directions" and he uses "mathematical templates" to assist him "make the best decision for the business" and to "stand over the decisions" he makes. While he chooses the "quickest way to solve that problem", a mathematical approach is often required in complex cases; otherwise he might choose "the wrong solution". Much of the data used in high volume manufacturing is "outputs from the machines, it is pre-done for you" and has the advantage that "everybody is using the same system". Statistics and probability is a big part of the production process. He estimates that only "10% of the engineers on site would need some of the learning from higher level Leaving Certificate maths".

– A manager with a multinational software organisation is of the view that "Leaving Certificate ordinary level is enough maths for me to get by". He describes his current mathematics usage as "a lot of implicit number work ... a very basic bit of algebra ... some statistics". He states that, while ordinary level Leaving Certificate mathematics is "plenty" for his work, he does "need the thought process" developed when learning mathematics. He maintains that the "practice" of getting the "brain going in different ways" when solving mathematical problems is "good" and this gives him the ability "to reason out problems" outside of mathematics. He states that in his work "estimating" is "powerful" when "a reasonably good answer" is required "quickly".

– A recently retired electrical engineer whose career progressed from engineering work "into general management" believes "that a good grasp of maths is essential to being a good engineer" and that engineers have varying "degrees of involvement with mathematics". He is of the view that as engineers develop "the relevance and importance of mathematics" declines to the point where "experience replaces mathematics" and "for most engineering problems there is a myriad of strategies". He says he has used "algebra, functions, numbers, statistics and probability" at a minimum level of higher level Leaving Certificate in his career. Early on in his career the usage would have been type 3, mathematising. Over time his work was "primarily about decision making"; this he describes as representing "the available information" mathematically or making "the best possible estimate". He maintains "there is always an acceptable level of imperfection" and, unlike school mathematics, "there is very seldom a unique right answer in engineering challenges". He rates "problem solving against a background of incomplete information" as "one of the biggest challenges in engineering".

– For a civil engineer who is a senior area manager with a local authority, her work moved back to mathematics, equations and designs after a long period of "no maths". She claims to be the only one in her staff of fifty two people "that can actually do something from first principles". She uses functions, geometry and trigonometry when designing storm water pipes. Statistics and probability is essential in "traffic management" and numbers are required for "managing budgets". In her current work concerning "unfinished housing estates", her "main mode of *thinking*" is how to "get two hundred housing estates completed". This she describes as "horse trading" with estate developers, where "you have to give something, get something, threaten something, make a stand but whatever you do say, you have to be able to stand over it". She says her *thinking* is how to "figure out what is the optimum" she can get from the €82 million developer bonds she retains. This, she says, is "not straight black and white, it is a logical analysis". Following the collapse of "the tiger economy" and with "reduced staff" and "a lot of recent floods", there is "a completely different mode of thought" in her work. She maintains that, due to her experience, she has acquired an "automatic thinking" capability whereby she could produce "a solution immediately". She says she has "a feel for what's going" on but she "would always do a quick calculation just to make sure" her answer was correct. In managing people, she describes her role as "getting other people to think and getting other people to develop the solutions".

– An electronics engineer working as an "educator, university lecturer and researcher" maintains that "a good rigorous mathematical ability is an advantage" in engineering especially when "trying to push frontiers". However he asserts that there is a "broadness" associated with engineering roles and that many engineers work in roles "where maths isn't central to their everyday activity". He maintains that engineers need an "ability to think" and he believes that "mathematical training is good for your brain". He also argues that "while the maths is very useful for elements of problems particularly in engineering, it is not necessarily the full solution" and that some problems "mightn't need maths". He is of the view that engineers "who grew up loving maths, are probably more in danger of being the ones less focussed on what the customer really wants". Very often engineers come up with a "mad sophisticated solution that is not related to any real problem" or they might "shy away from a question which can't be formulated mathematically". He maintains that "engineering should be about trying to identify the right question". He presents that "real world" engineers "have to frame the problem correctly and maybe express it in maths, then they have to solve it and then they have to interpret the solution and communicate that to the decision maker". He asserts that without this full solution "the decision makers" might "ignore the engineer" and instead they "use their own intuition".

– A sub-station designer with an electricity generating company believes that while "a lot of engineers end up working in the social side" of engineering and do not require higher level Leaving Certificate mathematics, she "definitely" requires it in her job. She describes her work as a mixture of "design" and "project management". She says that while "she could do ninety per cent of her job without maths", engineering is that "extra ten per cent I get paid for". In her job, she uses algebra, geometry, "a lot of calculus" and "very little statistics" at engineering level in reproducing, connecting and mathematising ways. She states than in addition to "doing direct mathematics", there is also "indirect" usage which she describes as both "a logical way of thinking" and a "way of working". This indirect usage "comes from having done maths". She describes mathematics as "clean", "logical", "totally transparent" and "a good way to justify an argument" because "nine out of ten times" she is "dealing with engineers" who "understand the logical approach".

– An engineering manager with a major telecommunications company requires "an understanding of mathematics and a mathematical view" in "everything" he does at work and very often he needs to be able to do mathematics "at speed". Every day, he uses statistics, geometry, trigonometry, numbers, algebra and functions and sometimes these domains are required at engineering level and in reproducing, connecting and mathematising ways. Additionally his company is "riddled" with "black box software solutions", some of which he made himself. These software solutions standardise solutions across groups of engineers. When designing networks or solving "synchronisation" problems, he needs to "convert boxes and

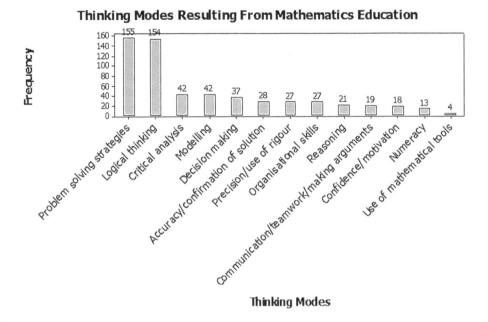

Figure 6 Engineers' modes of thinking.

widgets into euros", turn "things like man hours into megabits per second" and create documents for the "finance people to provide" the money. These documents have to be "double" checked and while mathematically based they have to be put into "a form that a non-engineer will understand". He states that because "there are so many different layers in telecommunications networks" eight of his team of ten engineers are continuously "thinking and problem solving". When managing the "capacity in the network", he has to make decisions "based on problem solving and logical thinking". He has to do a "certain amount of estimation" to predict future network usage, for this he relies on both his "experience" and knowledge of "statistics". While his usage of *curriculum mathematics* is quite high, he maintains that his *mathematical thinking* usage is higher because he has "to apply the maths" not just to engineering but to "all the aspects of his company".

The modes of *thinking* resulting from mathematics education which influence engineers' work performance are: problem solving strategies (identified by 26.4%), logical thinking (26.2%); critical analysis (7.2%); modelling (7.2%); decision making (6.3%); accuracy/confirmation of solution (4.8%); precision/use of rigour (4.6%); organisational skills (4.6%); reasoning (3.6%); communication/teamwork/making arguments (3.2%); confidence/motivation (3.1%); numeracy (2.2%); and use of mathematical tools (0.7%) (Figure 6).

Interview data confirms that both higher level Leaving Certificate mathematics and engineering level mathematics are required in many engineers' work and that much of engineers' mathematics usage is at the higher types of connecting and mathematising.

There is a common view that individual engineers in general use no more than ten per cent of the mathematics learnt in university and the difficulty for engineering education is "figuring out which ten per cent for each individual".

All engineers interviewed rate their *mathematical thinking* usage higher than their *curriculum mathematics* usage in their work. For one engineer *thinking* usage is the "value" he brings to his job and another engineer says that *thinking* usage is "where it's all at ... to me this is absolutely critical". Engineers present that their *thinking* usage comprises: problem solving; "big picture thinking"; decision making; logical thinking; estimation and confirmation of solution. Problem solving is a major part of engineers' mathematics *thinking* usage. Engineers say that engineering problems have multiple answers and that their job is to determine "what the answer means", which is "the best answer for all participants" and what "is the knock on effect" of the answer. 'Big picture thinking' is the term engineers associate with *mathematical thinking* in "real world" engineering where engineers need to "have a real tangible understanding of the effect of one piece of work on another part of the system". It is defining a problem or identifying a question that meets the overall "objective" and "the overall concept of a situation". According to the engineers, "engineering should be about trying to identify the right question, because a lot of the times, people are obsessing over the wrong question". These findings are similar to the findings in a study of new engineers where the new engineers describe their work as a "problem-solving process or way of thinking"; where they try to "organise, define, and understand a problem; gather, analyse, and interpret data; document and present the results; and project-manage the overall problem-solving process" (Korte *et al.*, 2008).

Engineers view the association between mathematics and *thinking* as "indirect" where thinking is how engineers use mathematics rather than the actual mathematics they use. They say that, when learning mathematics: doing "things in a particular order ... teaches logical thinking"; the practice of working around a problem and getting "your brain going in different ways ... transfers into other things that you do"; the emphasis on getting the right answer teaches one to "double check on everything"; and the discipline of "organising your study and the time it took to do your honours Leaving Certificate maths" is "something you bring through college and into to your working life".

Early in the engineers' careers, *curriculum mathematics* usage is higher and mathematics *thinking* usage is lower. *Thinking* usage increases for technical, commercial and management roles over the course of engineering careers. One engineer, whose *curriculum mathematics* is highest of all the engineers interviewed, says that his *thinking* usage is "probably higher than his *curriculum mathematics* usage because his role is management orientated and he has "to apply the maths not just to engineering, but also to finance, to manpower and to people".

Engineers maintain that graduate engineers with their "black and white solutions" are not ready to engineer. An ability to do engineering work comes from the "experience of working in an engineering environment" watching other engineers estimate and work out real problems and seeing how they view "the bigger picture". One engineer claims it took her four years to become an "independent thinker". This is consistent with the views in research literature where newly graduated engineers are not ready to engineer (Korte *et al.*, 2008; Trevelyan, 2011).

The interview data also shows that computer solutions are widely used in modern engineering practice. Engineers say that computational tools have many advantages in engineering practice because they bypass the need to write down the fundamental engineering equations and solve them and they offer a standard methodology for developing solutions within organisations. Most engineers say they use Excel. Engineers note that results produced by computational tools can easily be misinterpreted. One engineer presents that using computational tools is "a different type of mathematics" and he is of the view that "the engineer should understand how the program is solving the equations and what it is doing, because it is always dangerous not to".

4.2 Second finding

–   Engineers show high affective engagement with mathematics and their usage of mathematics in engineering practice is influenced by the value given to mathematics within their organisation.

Almost three quarters (74.0%) of the engineers who participated in the survey say that they enjoy using mathematics in their work either "quite a lot" or "a very great deal". Over 80% (80.6%) of the engineers surveyed feel confident dealing with mathematics in their work either "quite a lot" or "a very great deal". The survey data shows that engineers "love the challenge in solving problems mathematically", they enjoy "the satisfaction of a result", they find it easier to communicate using mathematics compared to words and they prefer "a 100% right answer rather than the ambiguity of non-mathematical solutions". For the engineers who enjoy using mathematics in work, there is a sense that mathematics is "part of who" they are. Memories of school mathematics are the main reason engineers do not enjoy using mathematics in work. For example, one engineer who has an "in built hatred of mathematics from secondary school" avoids mathematics in his work. Engineers' "grounding" in mathematics and subsequent usage are two major confidence influencers. For many engineers high mathematical self-efficacy develops in school where engineers learn to check their answers and where they are "in the habit of getting 100% in maths and maths-based exams". Engineers who have high confidence in using mathematics also show high confidence in mathematics solutions and in the "logical and objective nature of maths". These engineers note the need to "check if a solution is correct" and they are also of the view that "there is no reason for ambiguity in maths; there is only a right or wrong answer". Low confidence mathematics engineers avoid mathematics in their work while high confidence mathematics engineers readily "revise and brush up" on the required mathematics.

Interview analysis also shows that engineers' confidence in their mathematical ability grew from recognition of success in school mathematics such as their latest test grades, getting top marks or being the best in the class. For one engineer the "sense" of getting "the answer right" and knowing that he had "the right answer" was "very direct gratification". Another engineer asserts that the key to mathematics learning is "finding that you are able to do it" and this "unique skill doesn't come up much in any of the other subjects. A further engineer says "I got confidence in the fact that I was getting good results in mathematics and then I realised this is something that I could be good at". Due to "the very poor grounding" one engineer "got in maths" he says he "was afraid of some of" the mathematics he encountered in engineering practice and he has "a nagging fear that" he has "got something wrong" in his work. When he encounters a mathematics problem, he "refers" to his work colleagues.

Engineers' confidence in mathematical solutions in work is very evident in the interview data. Engineers like getting an "exact solution" and they tend to "double check" the mathematics before presenting a solution to co-workers. For one engineer mathematics is "a safety valve" in his work. Another engineer always chooses the "maths way" of doing things because mathematics is "very easy to reference and verify". Another engineer says that mathematics "is clean ... it is completely logical, ... it is totally transparent and basically once you are happy with it yourself, no one else can really question the validity of it".

Almost two thirds (64.6%) of the engineers who participated in the survey are of the view that a specifically mathematical approach is necessary in engineering practice either "quite a lot" or "a very great deal". Similarly, almost two thirds (63.3%) of engineers say that they actively seek a mathematical approach either "quite a lot" or "a very great deal". The value of engineers' engagement with mathematics in their work includes the usefulness of mathematics "for explaining results to others" and engineers' confidence in mathematics solutions. Costs of their mathematics engagement are the availability of sufficient ready-made solutions and "taking a mathematical approach may be risky and slow".

Only 3.9% of the engineers who participated in the survey say that they had a negative experience when using mathematics either "quite a lot" or "a very great deal". The majority of engineers, due to confidence in their mathematical ability and mathematical solutions, say that they did not have any negative experience using mathematics in the previous six months. However the "quirkiness of computational tools" and their "lack of understanding" and "over reliance on computer analysis" sometimes generate errors. For some engineers, mathematics consumes too much time, for example one engineer says "occasionally I have spent a long time trying to shoehorn something into mathematical language and failed, which was frustrating". The commonest reason attributed by the engineers surveyed to negative experiences using mathematics relates to communicating mathematics and the negative feelings resulting from their colleagues' lack of understanding and consequently engineers' difficulty in influencing business decisions. It is interpreted that when graduate engineers make the transition from an education environment where mathematics has high importance to engineering practice where many of their work colleagues do not understand mathematics and where there is less time to engage in mathematics that graduate engineers experience a reduction in motivational influences to use mathematics.

Interview analysis also shows that confidence in mathematical ability and in mathematical solutions are the main motivators for engineers to use mathematics in their work. However engineers say that engineering is much more than mathematics. They say that there are tiers of mathematics requirements in engineering practice that range from a majority of engineers who "need to understand" mathematics to a minority of engineers who "require a very high standard of maths". Given the diversity of their work, engineers estimate that mathematics is "valuable" in only ten per cent of their work. While a majority of the engineers interviewed are of the view that a specifically mathematical approach is not necessary in their work, at the same time a majority of these engineers say they use aspects of either higher level Leaving Certificate mathematics or engineering level mathematics in their work and they also use *curriculum mathematics* in either connecting or mathematising ways. From the interview analysis it is interpreted that *curriculum mathematics* is a small proportion but necessary part of engineers' work and engineers view mathematics as *curriculum mathematics* usage and not mathematics *thinking* usage which is significantly greater than *curriculum mathematics* usage for all engineers interviewed.

One explanation for the gaps between engineers' confidence dealing with mathematics in their work and the degree they actively seek a mathematical approach in their work and their overall *curriculum mathematics* usage is given in the interview analysis. There, engineers suggest that the necessity of a specifically mathematical approach in engineers' work is related to the value given to *curriculum mathematics* in engineering practice. For example, in one engineer's company, it is "more cost effective" not to use mathematics and in another company engineers don't have "time" to "actually use mathematics". A further engineer claims that he "wouldn't be thanked" for using mathematics. Colleagues' respect for mathematics is a factor in the value of mathematics in engineering practice. For example, one engineer says that in his company there is a respect for "maths only to the extent that it is useful". Another engineer is of the view that the "respect for mathematics" in his company "seems to change as the management changes . . . the emphasis is on sales and marketing and away from the maths right now". Difficulty communicating mathematics reduces the value of mathematics in engineering practice. Engineers say there is "skill in communicating maths". It is the "craft" of putting the mathematics "into a form that a non-engineer will understand". Consequences of poor mathematics communication skills are that calculations are "meaningless" and the message can be "biased" or "abused". Compared to other professions, engineers say they are not good communicators and a consequence of poor mathematics communications is that engineers are left in the "background" One engineer asserts that "if engineers are to survive then they need to somehow harness communication skills" Another engineer asserts that if one does not "bring the problem and the solution to people in their language", mathematics becomes "elitist". Ernest has a similar view, he states that the perception of mathematics "in which an elite cadre of mathematicians determine the unique and indubitably correct answers to mathematical problems and questions using arcane technical methods known only to them" puts "mathematics and mathematicians out of reach of common-sense and reason, and into a domain of experts and subject to their authority. Thus mathematics becomes an elitist subject of asserted authority, beyond the challenge of the common citizen" (Ernest, 2009).

5   CONCLUSIONS

In both the survey and interview data it is clear that engineers perceive mathematics as a highly "affective subject" where motivational beliefs such as affective memories (previous emotional experiences with mathematics), task value (why should I do mathematics?) and expectancy (am I able to do mathematics?) influence their engagement with mathematics. Throughout the engineers' education, the task value of mathematics is mainly derived from engineers' feeling of success when they get the correct answer. Costs of learning mathematics include: the wrong answer; time requirements; lack of relevance/usefulness; lack of respect for mathematics shown by peers and society and poor mathematics communication skills. Mathematics education that neglects the affective domain has consequences for both mathematics learning and for engineering career choice. Furthermore, when engineering graduates move from education to work environments they encounter difficulties communicating mathematics to non-mathematically competent people and mathematical solutions are consequently bypassed in decision making. While engineers say that the ability to communicate mathematics is an important skill for engineers themselves, they also maintain that it is the predominant characteristic of good mathematics teachers. Engineers hold mathematics teachers accountable for the lack of relevance in mathematics teaching to everyday life.

From both sets of data in this study it is apparent that *curriculum mathematics* is different to mathematics used in engineering practice. Solving real world engineering problems is about how engineers use mathematics rather than the actual mathematics they use. According to one engineer, an engineers' role is "to frame the problem correctly and maybe express it in maths, then they have to solve it and then they have to interpret the solution and communicate that to the decision maker". Engineers have a view that an ability to do engineering work comes from the "experience of working in an engineering environment", watching other engineers estimate and work out real problems and seeing how they view "the bigger picture". Graduate engineers lack this tacit knowledge. This view is reinforced in the research literature (Korte, *et al.*, 2008; Trevelyan, 2011).

This study informs mathematics teachers, engineering educators, practising engineers, students, parents and society. For each of these groups this study gives an insight into engineering practice and how mathematics is used in the workplace. This study also illustrates that feelings about mathematics are an important factor in mathematics learning and usage. One implication for mathematics curricula development and assessment is that mathematics learning generally focuses on objective analysis while *thinking* usage, subjective analysis and communicating mathematics are also required in engineering practice and possibly in other numerate professions such as economics.

Another implication for educators, parents and society, arising from this study, is that mathematics is a highly affective subject where student feelings about mathematics are a major influence on their engagement with the subject.

The findings from this study have particular implications for teaching mathematics and for engineering education. Teachers' own attitudes about mathematics are a big influence on students' relationship with mathematics. The study shows that students develop mathematical self-efficacy in school when they discover that they are able to do mathematics and they bring this confidence with them to university, work and into society. Engineers maintain that teachers should "emphasise more the applications of maths ... say that this is why we are doing it, the place of maths in the world and make that part of the taught and examined subject". There is strong evidence that mathematics learning requires a social environment whereby students benefit from group discussion and peer learning. The ability to communicate mathematics and its relevance is the predominant characteristic of good mathematics teachers. Teachers need to help students acquire a task value of mathematics and they need to engage with students in mathematics discussions and subjective analysis. According to Vygotsky's social constructivist mathematics learning theory, a teacher's role is to provide scaffolding on which students construct their learning. Scaffolding is a means whereby a more skilled person imparts knowledge to a less skilled person and discussion between teacher and students and amongst students themselves enhances students' mathematical thinking and communication (Vygotsky, 1978 pp. 79–91). A social mathematics learning environment enables students to enhance their tacit knowledge and this type of knowledge is required in workplace situations (Ernest, 2011).

The key message for engineering education, arising from the study, is that building a mathematics curriculum that more closely represents the way mathematics is used in engineering practice will strengthen it. This study provides evidence that while a majority of engineers in Ireland use both higher level Leaving Certificate mathematics and engineering level mathematics in their work, *curriculum mathematics* is different to much of the mathematics used in engineering practice. In engineering practice, mathematics is used primarily as a tool to estimate and confirm multiple solutions to real problems while, in engineering education, mathematics is about deriving a unique and exact solution to theoretical problems from first principles. Data analysis, which is often neglected in engineering education, is required in all engineering areas to inform engineering decisions.

A significant difference between engineering practice and engineering education is practising engineers' reliance on tacit knowledge while engineering education is based on explicit knowledge. Workplace problems often lack data and are more ambiguous compared to problems encountered in engineering education and an engineer's job is to determine "what the answer means", "which is the best answer for all participants" and "what is the knock on effect" of the answer. Engineers have particular difficulty interpreting computer solutions which have become a significant part of modern engineering practice. They say that the "quirkiness of computational tools" and their "lack of understanding" and "over reliance on computer analysis" sometimes generate errors.

Another difference between engineering education and practice is the social aspect of work compared to education. Tackling workplace problems is usually a team effort while in engineering education problem solving is mostly an individual effort. Graduate engineers' difficulty communicating mathematics is a significant weakness of engineering education and consequently, when engineers move from engineering education into engineering practice where mathematics is given a lower value than in education environments, they do not realise their mathematical ability.

To better prepare engineering students for engineering practice, they need to engage in "real world" practicality where "speed of response" and cost, subjective analysis and group work are important factors. Many engineers have an opinion that an ability to do engineering work comes from the "experience of working in an engineering environment", watching other engineers estimate, work out real problems and how they view "the bigger picture". Big picture thinking is taking the "the real world" into consideration where engineers need to "have a real tangible understanding of the effect of one piece of work on another part of the system" and "engineering should be about trying to identify the right question, because a lot of the times, people are obsessing over the wrong question".

The findings in this study suggest that engaging in active or social learning environments that emulate engineering practice would benefit engineering education. This type of learning environment would provide a greater focus on: engineering practice; real world applications of mathematics; working with tacit knowledge; teamwork; communicating mathematics; data analysis and decision making; and interpreting computer solutions. Students would be required to present and defend their mathematical solutions to both their peers and their lecturers. Based on the findings in this study, it is anticipated that this type of learning environment would develop students' mathematics communications skills and would also enhance their mathematics *thinking* and confidence.

It is concluded that the focus on "objective" solutions in mathematics education at the expense of subjective analysis, tacit knowledge and positive affects contributes to engineers' poor communication skills and reduces the value of mathematics in engineering practice. Solely focusing on "objective" solutions creates an affective hurdle for graduate engineers to overcome when they begin working as engineers. It could be argued that engineers' confidence in mathematical solutions restricts their vision of engineering solutions. For example, one engineer offers the insight that she enjoys using mathematics in work because "it is clean … it is completely logical … it is totally transparent and basically once you are happy with it yourself, no one else can really question the validity of it". However engineers maintain that "real life" engineering problems are "bigger" than mathematics, they have multiple answers and an engineer's job is to determine "what the answer means", which is "the best answer for all participants" and what "is the knock on effect" of the answer. This is supported in the research literature where it is maintained that "the unique charm of mathematics in engineering lies in the many levels and forms in which it is evoked, revoked, used, abused, developed, implemented, interpreted and ultimately put back in the box of tools, before the final engineering decision, made within the allotted resources of time, space and money, is given to the end user" (Chatterjee, 2005). In both the survey and interview data analysis, a diversity of practising engineers highlight the importance of *mathematical thinking* usage in their work compared to *curriculum mathematics*. *Mathematical thinking* knowledge is a type of tacit knowledge, this is "unwritten know-how carried in the minds of engineers developed through practice and experience" (Trevelyan, 2010a) and it differs from school mathematics (Ernest, 2011; Schoenfeld, 1992; Trevelyan, 2010a, 2010b).

Engineers' task value of mathematics developed in school where the feelings of success associated with getting "the correct answer" made "quantitative" subjects more enjoyable than "qualitative" subjects. Engineers bring their confidence in mathematical solutions with them into the world of engineering practice where many engineers are also motivated to get the "exact solution" at the expense of engaging in *mathematical thinking* and effective mathematics communications. However, according to practising engineers, in engineering practice mathematics is required to estimate and confirm multiple solutions to real problems, unlike engineering education where mathematics is about deriving unique and exact solutions to theoretical problems from first principles. Engineers demonstrate an over-attachment to "objective" solutions at the expense of "real world" solutions. "Objective" solutions have limited value in engineering practice particularly when engineers have difficulty communicating mathematics. However while there is "seldom a unique right answer in engineering", engineers prefer "a one hundred percent right answer rather than the ambiguity of non-mathematical solutions". This suggests an important finding that the focus on "objective" solutions at the expense of tacit knowledge in mathematics education reduces the value of mathematics in engineering practice. This finding has consequences, for both mathematics education in secondary schools and engineering education, where tacit knowledge is neglected at the expense of "objective" knowledge at both levels. There is evidence in the research literature that learning mathematics in a social context enables students to enhance the tacit knowledge required in workplace situations (Ernest, 2011). It is concluded that the mathematics taught pre- and during engineering education could be better matched to the mathematics required in engineering practice.

ACKNOWLEDGEMENTS

We wish to thank all the study participants for their time and contributions to this study. We are also grateful for the assistance given by Damien Owens, Registrar, Engineers Ireland and James Reilly, Statician, Institute of Technology Tallaght, Dublin.

REFERENCES

Alpers, B. (2010a) Methodological Reflections on Capturing the Mathematical Expertise of Engineers. *Proceedings of the Educational Interfaces Between Mathematics and Industry, 19–23 April 2010, Lisbon, Portugal.*

Alpers, B. (2010b) Studies on the Mathematical Expertise of Mechanical Engineers. Journal of Mathematical Modelling and Application, 1(3), 2–17.

Bissell, C. & Dillon, C. (2000) Telling Tales: Models, Stories and Meanings. For the Learning of Mathematics, 20(3), 3–11.

Brickhouse, N. W., Lowery, P. & Schultz, K. (2000) What Kind of a Girl Does Science? The Construction of School Science Identities. Journal of Research in Science Teaching, 37(5), 441–458.

Cardella, M. (2007) What Your Engineering Students Might Be Learning From Their Mathematics Pre-Reqs (Beyond Integrals and Derivatives). *Proceedings of the 37th American Society for Engineering Education (ASEE) / Institute of Electrical and Electronics Engineers (IEEE) Frontiers in Education Conference Milwaukee, WI.* Available from: http://www.scopus.com/inward/record.url?eid=2-s2.0-50049097209&partnerID=40&md5=ab67eb605d70f6af714deadc95277ba4 [acessed 20th February 2013].

Chatterjee, A. (2005) Mathematics in Engineering. Current Science, 88(3), 405–414.

Croft, T. & Grove, M. (2006) Mathematics Support: Support for the Specialist Mathematician and the More Able Student. MSOR Connections, 6(2), 1–5.

Csíkszentmihályi, M. (1992) The Flow Experience and its Significance for Human Psychology. In M. Csíkszentmihályi & I. S. Csikszentmihalyi (eds.) Optimal Experience: Psychological Studies of Flow in Consciousness. pp. 15–35. New York: Cambridge University Press

De Lange, J. (1999) Framework for Classroom Assessment in Mathematics. Utrecht, The Netherlands, Freudenthal Institute.

De Lange, J. & Romberg, T. A. (2004). Monitoring Student Progress. In T. A. Romberg (Ed.), Standards-Based Mathematics Assessment in Middle School: Rethinking Classroom Practice pp. 5–24. New York, Teachers College Press.

Elliott, L. (2009) UK 'must find 600,000 new engineers in seven years', The Observer. Available from http://www.guardian.co.uk/technology/2009/nov/29/manufacturing-engineering-recession-recovery [acessed 20th February 2013].

Ernest, P. (2009) Values and the Social Responsibility of Mathematics. In: P. Ernest, B. Greer & B. Sriraman (eds.) Critical Issues in Mathematics Education. pp. 207–216. Charlotte, NC, Information Age Publishing.

Ernest, P. (2011) The Psychology of Learning Mathematics: The Cognitive, Affective and Contextual Domains of Mathematics Education. Saarbrücken, Germany, Lambert Academic Publishing.

Forfás. (2008) The Expert Group on Future Skills Needs Statement of Activity 2007. Dublin, Department of Enterprise, Trade and Employment.

Gainsburg, J. (2006). The Mathematical Modeling of Structural Engineers. Mathematical Thinking and Learning 8(1), 3–36.

Goold, E. & Devitt, F. (2012a) Engineers and Mathematics: Engineers' Stories on Career Choice and Professional Practice. Saarbrücken, Germany, Lambert Academic Publishing.

Goold, E. & Devitt, F. (2012b) Engineers and Mathematics: The Role of Mathematics in Engineering Practice and in the Formation of Engineers. Saarbrücken, Germany, Lambert Academic Publishing.

King, R. (2008) Addressing the Supply and Quality of Engineering Graduates for the New Century. University of Sydney.

Korte, R., Sheppard, S. & Jordan, W. (2008) A Qualitative Study of the Early Work Experiences of Recent Graduates in Engineering. *Proceedings of the American Society for Engineering Education (ASEE) Annual Conference & Expositio. 22–25 June 2008, Pittsburgh, PA.*

McKinsey. (2011) Growth and Renewal in the United States: Retooling America's Economic Engine. The McKinsey Global Institute.

McLeod, D. B. & Adams, V. M. (1989) Affect and Mathematical Problem Solving. New York, Springer-Verlag.

Organisation for Economic Co-Operation and Development. (2010) The High Cost of Low Educational Performance. The Long-Run Economic Impact of Improving PISA Outcomes. Paris, OECD.

Schoenfeld, A. H. (1992) Learning to Think Mathematically: Problem Solving, Metacognition, and Sense-Making in Mathematics. In: D. A. Grouws (ed.) Handbook for Research on Mathematics Teaching and Learning. New York, Macmillan.

Schunk, D. H., Pintrich, P. R. & Meece, J. L. (2010) Motivation in Education: Theory, Research, and Applications. Upper Saddle River, NJ, Pearson Educational International.

Trevelyan, J. (2009) Steps Toward a Better Model of Engineering Practice. *Proceedings of the Research in Engineering Education Symposium Palm Cove, REES 2009, 20–23 July 2009, Queensland, Australia.*

Trevelyan, J. (2010a) Mind the Gaps: Engineering Education and Practice. *Proceedings of the Australasian Association for Engineering Education (AAEE) Conference, 2010 AaeE Conference, 5–8 December 2010, Sydney, Australia.*

Trevelyan, J. (2010b) Reconstructing Engineering from Practice. Engineering Studies, 2(3), 175–195.

Trevelyan, J. (2011) Are We Accidently Misleading Students about Engineering Practice? *Proceedings of the Research in Engineering Education Symposium, 4–7 October 2011, Madrid, Spain.*

Vygotsky, L. S. (1978) Mind in Society: The Development of Higher Psychological Processes. In M. Cole, V. John-Steiner, S. Scribner & E. Souberman (eds.) Cambridge, MA Harvard University Press.

Winkelman, P. (2009) Perceptions of Mathematics in Engineering. European Journal of Engineering Education, 34(4), 305–316.

Wood, L. N., Mather, G., Petocz, P., Reid, A., Engelbrecht, J., Harding, A., Houston, K., Smith, G. & Perrett, G. (2011) University Students' Views of the Role of Mathematics in Their Future. International Journal of Science and Mathematics Education 10(1), 99–119.

Chapter 12

Engineers' professional learning: Through the lens of practice

Donna Rooney[1], Keith Willey[2], Anne Gardner[2], David Boud[1],
Ann Reich[1] & Terry Fitzgerald[1]
[1]*Faculty of Arts and Social Sciences*
[2]*Faculty of Engineering and Information Technology, University of Technology,*
Sydney, NSW, Australia

1 INTRODUCTION

The focus of this theoretically motivated chapter is on the learning of experienced engineers throughout their professional work. It draws heavily on literature from workplace learning research. In particular, it takes up practice theorisations (Hager *et al.*, 2012a and b) to present an alternative to narrow, but widely held, contemporary disciplinary views of how engineers learn. Firstly, the chapter begins by drawing attention to the complexities of professional learning. Secondly, it outlines some developments in research in the area of workplace learning and presents an argument for the utility of a socio-material 'practice' lens to investigate professional learning. Thirdly, it provides a brief introduction to a small empirical study. Fourthly, the utility of a practice-based approach is illustrated by way of a brief discussion of the civil engineering practice of 'site walks' taken from the empirical study. Finally, the chapter concludes with some comments regarding the potential usefulness of a practice lens and its possible implications for engineers, organisations and professional bodies and for university providers of engineering education.

2 BACKGROUND

The demand for professionals, including engineers, to learn continuously in their work is well recognised. There is an expectation by organisations that the professionals they employ engage in ongoing learning in order to meet the demands of continuing change. Many large engineering firms provide comprehensive graduate programs specifically to support new engineers' entry into the profession. These programs recognise the importance of providing new graduates with a broad range of learning experiences including structured activities and exposure to various facets of engineering work. The programs are relatively well documented, scrutinised and evaluated by employers and researchers, and encouraged by professional associations. However, for experienced engineers the situation is typically less comprehensive or explicit, even when the demand for continuing professional learning remains. Experienced engineers, while still required to meet demands of ongoing change, may not attract the same organisational learning emphasis as their less experienced counterparts.

Further, this organisational emphasis on professional learning is often focused on attendance at a host of formal courses, seminars, conferences or other identified educational activities (e.g., Engineers Australia, 2012). These might be provided in-house, or by professional bodies, or by other interested parties. However, while structured educational activities like these serve an important purpose, they do not account for the full gamut of professional learning. Researchers in workplace learning have drawn attention to the significant amount of learning that happens as part of work itself (for example, Billett, 2004; Boud & Middleton, 2003; Boud et al., 2009; Eraut, 2004; Scheeres et al., 2010). This research suggests that accounts of professional learning that focus solely on participation in structured learning activities miss potential aspects of learning and serve to limit the potential for making the most of professional learning opportunities.

While engineering employers, associations, and engineers themselves, have tended to place less explicit emphasis on learning through the challenges of everyday work, we suggest there are problems with the privileging of structured educational activities. One problem is that structured learning activities are normalised and have become embedded in professional learning systems. These systems have consequences for professionals' employment as well as, in some cases, their continued registration and their right to practice their profession. Responsibility for engineers' professional development has been gradually subject to more rigorous surveillance, and indeed self-surveillance. Furthermore, because it is easier to measure attendance than almost anything else, professional learning is too often synonymous with participation in courses or seminars; equating attendance with the assumption that the participants have acquired a skill or learnt a thing and can use it. This is counterproductive because it shifts focus from the outcomes of learning (development itself) to the input (the activity that is supposed to lead to development) (Boud & Hager, 2012).

Large organisations, whose competitive edge is linked to the continual learning of its staff, can also conflate professional learning with workshops, seminars and courses. This is seen in performance management systems that endorse employees' attendance as evidence of professional learning. Similarly, even when professionals themselves acknowledge that they learn on the job, the privileging of systematised educational activities can mean that they are more likely to seek these out when a learning need arises rather than actively seek out other kinds of learning opportunities in their work. A comparable privileging can be noted in professional associations, like Engineers Australia, who also accept that professional learning occurs as part of work, but impose significant limits on the amount of such learning permitted as part of their continuing professional development requirements.

Overall, this situation sustains the endurance of training, courses and workshops as the default response to professional learning despite potential alternatives. Not only is this an issue for organisations who might benefit from knowing more about how professionals learn, but it also sustains a myopic view of continuing professional learning that helps to maintain and justify an entrenched system where training interventions might be offered by some professional associations without also recognising their own interests for doing so. Further still, professionals themselves may not recognise, and hence seek out, learning-rich activities in their work when the dominant understanding is that legitimate learning only takes place in training settings. Finally, those involved in entry-level professional university programs may be doing a disservice if their programs play down the importance of other activities where students can satisfy future (and ongoing) professional learning requirements.

Writers on workplace learning (e.g., Hager & Hodkinson, 2009; Fenwick, 2009) also argue that many traditional assumptions about professional learning are problematic. One difficulty has been that within some understandings, learning is seen as something that *individuals* do. For instance, it is the individual engineer's attendance in a course that is acknowledged and rewarded. This obscures professional learning resulting from teams of engineers working together, or from projects where multidisciplinary stakeholders converge to create or undertake something together. Much workplace learning research, and indeed in other chapters in this volume, demonstrates how the social dimensions of work provide a rich context for professional learning. Even more particularly, some of these studies show that the work is not only a context, or backdrop, but is fundamentally implicated in learning (Billett 2001, 2004; Skule & Reichborn, 2002). Yet despite some acceptance of the social and situated nature of work, professional learning is still typically accounted for, and rewarded, in individualistic terms.

A further difficulty arises from the commonplace metaphors evoked within contemporary understandings of professional or workplace learning. Two of these metaphors are 'acquisition' and 'transference'. It is believed that when professionals learn they gain (i.e., acquire) knowledge as a 'thing' that they thereby possess to apply as needed at some later date (i.e., transfer). The terms 'acquiring' and 'transferring' learning are used without recognition that they are metaphors and not descriptions of concrete reality (Boud & Hager, 2012). These metaphors assume a limited appreciation of knowledge exchange, and in doing so avoid complex understandings where new knowledges may be co-produced as a result of collective endeavours with heterogeneous others (Hager *et al.*, 2012).

We suggest that these traditional metaphors need to be replaced with more generative ones that represent the social and situated nature of professional learning that can lead to fresh understandings. Of importance here are different metaphors invoked by newer theories of learning that draw on ideas of *participation, construction* and *becoming* (Hager & Hodkinson, 2009). Here 'learning as participation' focuses on learning as a daily social practice, intertwined with work practices, and not one that is limited to transference of content acquired from 'experts' in training events. It emphasises the social nature of learning at work rather than assumptions about purely individual endeavour: a view shared by most chapters in this volume. Construction draws attention to the messiness of professional practice and the heterogeneity of those involved in it. It refers to complex knowledges being constructed in networks constituted by a variety of actors; a more social take on Itabashi-Campbell and Gluesing's notion of cognitive convergence (discussed in this book). Rather than unproblematic skills and knowledges being 'transferred' from one site to another, the notion of construction appreciates that the utilisation of knowledge and skills constitutes a new instance that is dependent on the context in which it is realised. Thus knowledges are continuously constructed (and reconstructed) in daily practice. Finally, the idea of becoming recognises the enduring and continuing character of learning and professional work. In doing so, it accepts the demand for continual learning in its acknowledgement that even the most experienced professional learns as they engage in their practice. Together, these new understandings of learning challenge the ways in which professional learning is understood and evaluated. They provide us with alternatives for theorising and assessing professional learning by shifting the focus from the individual learner to the practice itself. A new approach to professional learning of experienced engineers is therefore needed. As we suggest in the following section, this approach should reflect the social, material and situated nature of their learning.

3 TURNS TOWARDS PRACTICE THEORISATIONS

The work of researchers interested in the general field of workplace learning has moved well beyond behavioural and/or cognitive understandings of learning (see Hager, 2011). Without giving a history, it is sufficient to say that it has been influenced by several arguable 'turns' in social theory over the last two decades. A few in particular warrant a brief explanation here – although we refrain from claiming chronological or foundational accuracy given the certainty of contestation should we do so. Rather, our point in discussing these turns is to highlight some of the major influences of workplace learning research, and in doing so provide explanation for how we are led to the practice lens adopted later. One final note before proceeding is that these turns are not mutually exclusive: they share features with other turns.

A recent influence in workplace learning research comes from a 'spatial' turn, originating largely within cultural geographies (for example Cresswell, 2004; Soja, 1996, 2001). This invited consideration of the significance of physical, historical and socio-political spaces of learning (Billett, 2001; Edwards & Usher, 2008; Scheeres *et al.*, 2010; Solomon *et al.*, 2008; Usher, 1996). No longer could workplaces or the spaces within them be treated as mere backdrops or stages where learning was occurring. Several studies thus demonstrated how work itself is not a context or backdrop but is fundamentally implicated in learning (Billett, 2001, 2004; Skule & Reichborn, 2002). Indeed, even the term workplace learning emphasises the spatial dimensions of learning. Work configures opportunities for learning. As Hager and others point out, learning for work purposes has escaped the confines of the classroom. Indeed, much contemporary workplace learning literature has focused on informal learning (Eraut, 2004; Gallacher *et al.*, 2006; Hager, 1998; Hager & Halliday, 2006; Marsick & Watkins, 1990) – which itself draws research focus to less systematised and concrete spaces.

What might arguably be a 'socio-cultural' turn influenced researchers to look towards particular groups and relationships between people where learning at and for work were occurring, rather than take up individualistic notions from formal education. Perhaps one of the most influential ideas in this regard has been Lave and Wenger's 'communities of practice' (Lave & Wenger, 1991; Wenger, 1996; Wenger & Snyder, 2000). While this has since been critiqued on a variety of matters, it retains a strong appeal in contemporary organisations as it gives a readily identifiable account of working relations. These sociocultural notions of workplace learning remain well represented in contemporary literature.

Yet another 'turn' has been the linguistic turn. This literature prompted closer attention to the role of discourse and/or language in investigations of workplace learning (Edwards *et al.*, 2004). Talk between various people at or about work constitutes moments when workplace knowledges are coproduced, articulated and circulated.

A related 'epistemological' turn incited questions of knowledges themselves (Usher, 1996). Rather than knowledge simply being transferred or applied, some contemporary research illustrates how knowledge is coproduced, created and/or even performed in the practices of communities (Gherardi 2009a). Another significant turn to be presented here is a 'socio-material' turn and here actor network theory (Latour, 1999; Law, 1986, 1992; Law & Hassard, 1999) has been a powerful influence in drawing attention to the mediating influences of non-human actors or things in regard to learning (Fenwick & Edwards, 2012).

While our own work has responded to each of these influences in various ways, our current research efforts are part of what Schatzki and others term the 'practice turn' in contemporary theory (Schatzki *et al.*, 2001). Many ideas from the turns outlined above are encompassed in practice theory or, more correctly, practice theories. There are a variety of approaches on which to draw, to enable a focus on practice to be conceptualised. From ideas commonly found in the organisational literature, Gherardi (2006, 2008) identified several streams of research in what she terms 'practice-based studies': cultural and aesthetic approaches, situated learning theory, activity theory, actor network theory, and workplace studies. These streams of research have in common a focus on practice as situated, mediated and relational. Many of these theories move to a quite different kind of configuration from the conventional educator's focus on the competences of individuals and the knowledge and skills they can acquire. Such theories allow for, or indeed insist on, a variety of relational features to be considered together. Practice is an integrating idea that links thinking with doing and people with contexts.

Practice theories complement our understandings that learning in professional practice involves, *inter alia*, interactions with a variety of other people and things in a wide range of contexts. Tangible issues commonly drive such interactions. Much of the learning arises from the exigencies of work and the challenges encountered there. What holds together all the sites, purposes and relationships is that they cohere around the notion of practice. It is what professionals practise that we take as the cornerstone of work itself and the learning challenges that are part of it: they undertake practice, they extend their practice, and they take up new practices. We also believe that they learn through their practice – and if this is the case, then the participation metaphor referred to above appears to be more aligned to professional learning than acquisition.

We are finding that Schatzki's work is useful, first because of the idea that "bodies and activities are 'constituted' within practices" (Schatzki, 2001, p. 11). Practices provide units of analysis that bring together the practitioner, the material objects with which they work, their relations with others and the context in which they operate. A practice view seeks to avoid separating these as variables that might be separately manipulated apart from the whole practice. This is not to say that particular features of practices might not be foregrounded, but that they are never treated independently.

A second way Schatzki's work is helpful is that there is an emphasis on "know-how, skills, tacit understanding, and dispositions" (Schatzki, 2001, p. 16). The focus is on what is directly connected to the 'doings' of a practice and the 'sayings' that are part of the practice. Here practice is inherently dialogical and activity-focused. The organisation of practices is social, being expressed in the connections of doings and sayings that compose them, as opposed to an individual's doings and sayings. Doings are what people do with others and with things, and these interact with what people say. Both doings and sayings are done with people's bodies. However, while sayings do consist of speech, they are more than this. Sayings are differentiated from doings of practice in that a saying conveys meaning and has a semantic function whereas doings may not (Schatzki, 2002). In this sense sayings are anything that is communicative including, for example, drawings.

Schatzki also emphasises action and structure. He defines practices as "structured spatio-temporal manifolds of action ... that have two basic components: action and structure" (Schatzki, 2006, pp. 1863–1864). Structural elements include know-how concerning the actions or the 'how to' of practice; rules that specify guidance or instructions; teleo-affective structuring that explains the purposes or emotions that cause people to act towards possible ends and goals; and general understandings that may be relevant, for example, about the nature of a particular kind of work. Practice actions are performances of people that are embedded within these structural elements. Practices thus entwine people, technologies, spaces, time and artefacts. Through these embedded structures and material arrangements, practices frame future action possibilities for both individuals and the organisation.

3.1   Features of practice

There are theorists interested in a practice approach other than Schatzki. While they may disagree on some features of practice, five partly overlapping elements of practice are typically seen across various bodies of practice-related studies (Hager et al., 2012b):

1   Practices are *embodied*. Kemmis (2009, p. 23) argues that practice is "always embodied (and situated)" encapsulating "what particular people do, in a particular place and time, and it contributes to the formulation of their identities as people of a particular kind, and their agency and sense of agency";

2   Practices are *materially mediated*. When practice is undertaken, it occurs in conjunction with material arrangements in the physical world. These may include objects such as raw materials, resources, artefacts and tools, physical connections, communication tools, and material circumstances (Kemmis, 2009; Schatzki, 2005);

3   Practices are *relational*. People, artefacts, social groups and networks develop characteristics in relation to other subjects, social groups or networks such that they are formed and structured socially, (Kemmis 2009; Østerlund & Carlile, 2005);

4   Practices are *situated*. They are situated in particular settings, "in time, in language and in the dynamics of interactions" (Gherardi, 2008, p. 521);

5   Practices are always *emergent*. That is, they evolve over time and over contexts; they change in the light of circumstances.

Practices are therefore more than simple activities or actions undertaken by workers. Rather, practices are enacted; workers engage in "doings and sayings" (Schatzki, 2002, p. 81) that bring together combinations of know-how, rules, purposes, personal investments and general understandings relevant to their job (Price et al., 2012). In researching practices, an emphasis on this nexus of doings, sayings and relatings focuses the researcher on ways of uncovering the practices of a professional group as a precursor to identifying the learning opportunities inherent in these practices.

4 FINDING GENERATIVE ENGINEERING PRACTICES

The exploration of professional learning reported here originates from a small empirical study undertaken by a team of researchers from the University of Technology, Sydney (UTS). The team consisted of researchers from two Faculties: Engineering and Information Technology, and Arts and Social Sciences. This interdisciplinary team worked in partnership with one of Australia's largest engineering organisations. The organisation's remit includes work in the areas of physical infrastructure and building, civil engineering, water and environment, rail, aviation, tunnelling, mining, communication and energy. While this organisation invests considerable time and money in comprehensive graduate programs for new engineers, it was its concern with the professional learning of its more experienced engineers that prompted its involvement in this project.

The study was prompted by a need to develop new understandings of experienced engineers' professional learning and to provide an empirical base for these new understandings. It sought to locate where fruitful learning was occurring in the normal work of professional engineers. To this end the project was conceived in stages ensuring our methodology was both iterative and generative. The first was the identification phase where we analysed a range of documents from our partner organisation, held focus groups with selected teams of their experienced engineers, and followed up with semi-structured interviews with focus group participants. In both the interviews and the focus groups, participants were asked to talk about, and provide detailed examples of, their day-to-day work. These methods enabled us to elicit 'rich descriptions' (Denzin & Lincoln, 2000) of professional engineers' practices. The transcripts of the focus groups and interviews were subjected to thematic analysis, with the following themes identified: communication – talking to others/gathering information; planning; dealing with challenges; risk management/avoidance; and record keeping – diaries, meeting notes etc. The strength of the multidisciplinary team was evident in this phase with the engineers being sensitive to some of the more latent themes in the transcripts and the social scientists sensitive to the kinds of language used. While this process allowed us to get to know our data, a simple listing of themes was insufficient for identifying practices. We then examined the contexts in which these professional engineers performed their communicating, planning, dealing with challenges etc., with the intention of creating an initial list of practices that constitute the work of the experienced engineers who work for our partner organisation.

In the second stage we sought elaboration of the recognised practices, as well as identifying any further practices. To do this we carried out unstructured observations (walking around with the engineers and asking them what they do), as well as a second round of focus groups and semi-structured interviews of experienced engineers who were specifically asked for descriptions of the previously identified practices. These methods yielded further qualitative data that was examined through a practice theory lens, with the results highlighting key aspects of practices as embodied; materially mediated; relational, situated and emergent. It is these aspects of practice that inform the remainder of this chapter.

5   ENGINEERING – A BUNDLE OF PRACTICES

With the conceptualisations mentioned in section 3 in mind, we suggest that professional engineers' work when viewed through a practice lens can elucidate new understandings of professional learning. In short, we understand that a profession (engineering in this case) consists of bundles of interrelated practices and material arrangements. These include the rules of the organisation and professions, the shared practical understandings (how to carry out the basic doings and sayings) and the general understandings that are shared among those who carry out that profession (Schatzki, 2012). While some practices may be dispersed (in that they can be seen in other professions) there are many that are unique to or characteristic of a given profession. Each practice consists of activities, material 'things' and bodies and is purpose-focused. Within a given part of the profession of engineering there is a shared understanding of the practice. These practices are prefigured in that they have a history in the profession, yet each instantiation keeps alive the possibility for reconfigurement (i.e., the possibility of change).

For example the bundles of practices that make up an engineer's work might include managing projects, developing budget plans and acceptance testing. Most engineers would recognise these practices, and even though there will be differences in how they are carried out, including aspects idiosyncratic to a particular organisational culture or to an individual engineer's, or a sub-group's, preferred way of practising them, there remains enough similarity to make them recognisably germane to the profession. Along with these specifically 'engineering' practices, engineers' work would also include other practices like participating in meetings and undertaking performance reviews. While these latter practices are common in many professionals' work, they would take on specific forms that make them meaningful to the field of engineering practice. In the following section we take a common engineering practice, a site walk, and demonstrate the explanatory utility of the practice lens.

5.1   Site walk – an illustration of an engineering practice

A common practice in a number of branches of engineering is the site walk, though different names are sometimes given to it. It typically involves an inspection of a particular terrain, plant, equipment, and so on, by a number of engineers (and others). As a practice, a site walk is constituted by the activities and events in particular sites and times, and by the doings, sayings and relatings of those who undertake them. A first 'doing' is movement, as the name 'site walk' suggests, walking is one way movement occurs. However, sometimes, based on the size of the project, this practice might be even better described as a site drive or site flight because the movement involves travel over larger distances. Site walks, by nature, are not physically static. A further associated 'doing' is observing, seeing or looking. Engineers observe physical features of the site in order to locate features not represented in drawings. They observe how work is being done, possibly making suggestions about how to better follow processes and/or standards, and they observe whether all prescribed intermediate steps are undertaken or completed according to the design drawings and specifications. This type of 'doing' is mediated by a disciplinary understanding of standards, engineering principles, clients' input, and so on. These understandings are likely articulated in the 'sayings' of site-walks among the 'relatings' of the people involved.

The site walk exemplifies the key elements of practices. It is an embodied practice, by this we mean that these specific engineering practices encapsulate what professional experienced engineers routinely do. For instance, the site walk is a routinised organisational practice that occurs in specific times and locations. Those involved bring knowledges, dispositions, working histories and their bodies. It is the whole person who engages in practice, not just intellect and skills. Each instantiation of the site walk is shaped by the collective embodiment of its participants. The professional identities of participants are confirmed and reaffirmed within the practice, as are participants' sense of agency (Kemmis, 2009, p. 23).

A site walk is a practice with a purpose; or more correctly: with multiple purposes. As one engineer told us, the site walk is *"... you go out there for a purpose [...] you look at a certain thing, discuss a certain thing, resolve an issue"*. Other engineers we spoke with noted how through the site walk they would find out what was going on, check the progress of various facets of the job, and ensure the job is compliant to a myriad of standards. Another important purpose for the site walk is to identify potential problems or hazards and/or identify problems that have already occurred, and then to find solutions to or mitigate further problems.

The site walk practice is materially mediated and brings together human and non-human actors (Gherardi, 2009b). For instance: the engineers we spoke with suggested that compilations of the following human actors might undertake a site walk: the foreman, service managers, the 'guy with the shovel', digger or leading hand, a blast crew, project manager, production superintendent, senior supervisor, various consultants, an environmental team or scientists, construction manager, client etc. The coming together of these various people and the disciplines they bring not only contributes to the unfolding of events but also provides opportunities for professional learning. Such opportunities are exemplified by the comment below:

> So we know all the technical procedures and all that ... but actually doing the work? I mean, I didn't know anything about boiler making, welding or any of that until I actually interacted with some of these workers. The boilermakers and the welders [on the site walk], they show you, this is what's done, this is how you do it and this is what happens. So I think, even though as engineers, we know the technical side of it, they show us the practical side of doing the work.

Another engineer suggested that site walk provided the participants opportunities to 'bounce ideas off' each other. It is reasonable to presume that he, and others, developed new understandings as a result of this.

Furthermore, along with human actors the non-human things also shape a site walk. First there are tangible things that accompany engineers on the site walk such as motor vehicles, cameras, mobile phones, Blackberries, diaries, and pens. These non-human things enable associated activities (driving, taking pictures, looking up something on the internet, writing etc.). Then there are also the less palpable aspects like standards, engineering histories, ideologies, cultures, wider socio-political arrangements, organisational policies and procedures. All of these, while less obvious than the accompanying material things, work to constitute the practice in very particular ways. For instance, codes of practice and mandatory standards influence the gaze of site walk participants and therefore the decisions they make and the future actions they undertake. All of these influence the practice in particular ways. While engineers seek to shape materials, materials hold potential to shape practice. For instance, site walk participants react to a sloping soil surface in ways laypeople would not. For the layperson it might appear as a simple pile of dirt, for the professional it can mean risk and/or the need for further action.

Such professional practices are also understood as situated practices, and there are many ways in which a practice is situated. It is situated in particular settings, in time, in language and in the dynamics of interactions (Gherardi, 2008). For Kemmis (2009), practice "has aspects that are 'extra-individual' in the sense that the actions and interactions that make up the practice are always shaped by mediating conditions that structure how it unfolds". These may include cultures, discourses, social and political structures, and the material conditions in which a practice is situated. Participation in a practice requires ongoing development of a shared understanding of language and standards; for example, agreed levels of quality. This understanding is developed through practice and the inherent feedback involved in undertaking a practice with others.

While the practice we call site walks appears as a discrete action, the site walk itself is one of many activities that coalesce to constitute the work of professional engineers. For instance, engineers told us how something noticed/said/discovered on a site walk might then be taken to a design review, or constitute the topic of discussion at a toolbox talk. The site walk, therefore, is not independent of the other practices that constitute engineering work. Rather, it supports and informs these other practices. In other words, a site walk is a practice that is both situated and relational.

For instance, the site walk is related to networking and the building of reciprocal relationships that are essential to engineering work. Despite engineers not often physically constructing anything themselves, the quality of the finished project will be influenced by the strength of their relationships with the people who do the actual work.

Just as each instantiation of the site walk is shaped by previous site walks, so too is each engineer's knowledge. Take, for example, the engineer above who learned about boiler-making. A second example can be noted when considering designing. Designing is an ongoing activity for most large engineering jobs, yet a final design is rarely available from the onset. The site walk provides opportunities for construction and design engineers to develop the design along with the clients and, in some cases, the users or their proxies. Site walkers co-produce understandings of the work with each observation.

These practices are also relational. They occur in relation with others who practise, and in relation to the unique features a particular practitioner brings to a situation. Practice is thus embedded in sets of dynamic social interactions, connections, arrangements and relationships. Schatzki suggests that, "[o]ver any period of time, human practices link and form gigantic nets, just as arrangements are connected into immense material structure and practices and arrangements relate in myriad ways" (Schatzki, 2010, p. 209).

Practices such as the site walk are thus not static but emergent: that is, they have a history within a profession but they change over time, with contexts, and in the light of circumstances. New challenges require new ways of practising. Practices emerge in unanticipated and sometimes unpredictable ways, and in this the practitioners' knowledge and skills are open to continual construction (and reconstruction) (Johnsson & Boud, 2010). A useful example of site walks as an emerging practice is through considering how 'safety' changed site walks. Two decades ago, an engineer on a site walk may not have paid as much attention to safety and risk management. The older engineers we spoke with talked about a time when engineers actually factored in a number of deaths over the duration of a job. Today public opinion and rigorous occupational health and safety regulations would not allow such views. Therefore 'safety' is now commonplace, if not central, to engineers' work and, as such, has reshaped how site walks are undertaken.

Furthermore, it is here that we can link practices to learning. Safety has been 'learned' by professionals and become integrated into their day-to-day practices (site walks included). Moreover, each time a site walk is undertaken an engineer's knowing about safety is newly situated and mediated by the material arrangements specific to the site. Thus, rather than thinking of the site walk as an event where knowing is 'applied', it is more the case that this knowing is enacted in a uniquely situated manner (Gherardi, 2009b).

Further still, in addition to safety becoming mainstream for most engineers, there is an emergent focus on how the environment is reshaping engineering work. The growing contemporary awareness of environmental concerns has seen the advent of environmental scientists collaborating on projects along with engineers and others. When probed about an on-site explanation of earthwork activities, one participant told us how he has "... started to think a little bit more like [an environmental scientist]". This increased awareness can be understood as learning.

5.1.1   Learning and practice

The practice of the site walk described, above, along with others not described is among the 'bundle of practices and material arrangements' (Schatzki, 2012) that collectively constitute the work of professional engineers. As exemplified by our data and theorisations, we contend that learning occurs as an integral part of the practice of site walks. Engineers do not learn from site walks as they would learn from a structured course, but through engagement in a practice that is constituted as both work and learning. To discern a difference requires comparison to previous experience—not simply seeing it, being told about it, but actually by participating in (practising) the activity. Similarly, learning is both enhanced and limited by the variety of challenges that participants are confronted with during different site walks. Indeed, this preliminary account of practices suggests that practice is not only the site of learning but learning itself. Thus, we can see how engineers might extend the scope of their learning through participating in site walks.

Learning is integral to the practices described above even if learning is not necessarily how practitioners conceptualise what they do when they practise them (Boud & Solomon, 2003). Rather than understanding practice as simply 'just doing the job' we conclude by suggesting that these practice-based insights can help all engineering stakeholders think about professional learning differently. So how might stakeholders use these insights?

- A practice lens could potentially invite organisational stakeholders to reconsider traditional mechanisms that capture, acknowledge and reward professional learning based on static notions of learning where skills and knowledge are *acquired, transferred* and/or *operationalized* towards the demonstration of learning and development through a variety of instances of participation in engineering practice. This could involve a focus on practice opportunities, ranges of practices and the roles played within each, and deliberate involvement in practices beyond those of the immediate position occupied, etc.
- Professional engineers might be encouraged to use a practice theory lens through which to assess and plan their on-going professional development. Practice theory provides an alternative way for them to identify and view the learning opportunities available within their everyday work, to assess their own strengths and weaknesses in relation to these practices, and subsequently address their weaknesses through their planned participation in and interaction (including coproduction, dialogue and observation) with others participating in these same practices.
- Engineers might want to plan their careers and their moves within an organisation in the light of the particular practices and the unique challenges offered by different kinds of work. The idea of 'rotation' is common in the training of medical practitioners; medical consultants often choose work opportunities in the light of the challenges they will provide to extend their repertoire of capabilities. For example, newly employed engineers might elect to view practices such as site walks as learning opportunities, electing to undertake them with as many different stakeholders as feasible and engage in dialogue in order to learn the language, be challenged by diverse opinions and gain an appreciation of: What does the design engineer look for? What does the estimator look for? What is the environmental scientist looking for? What is the client checking? Accessing the perspectives of the range of people that may participate in a site walk can then influence how the site engineer conducts their own site walk in the future: What do they look for now? What do they note down? What do they communicate and to whom at the conclusion of the site walk that they may not have before? What do they see now that they are looking with other peoples' eyes in addition to their own? A design office equivalent might be the work-in-progress meeting that is ostensibly about managing the projects involved, but provides opportunity for each engineer to learn about the problems faced and how their colleagues solved them, or to contribute to their solution.
- A more nuanced understanding of learning might aid engineering educators' exploration of how students' learning can further benefit from authentic practice (Boud, 2012). For instance, engineering educators might explicitly identify learning opportunities that engage students in practices required of professional engineers and audit involvement in these as part of work placements. A further example might be to focus on developing teamwork activities in order to promote learning as participation and reflect the social nature of workplace learning.
- An understanding of learning through practice could enable engineering academics to develop scaffolding opportunities that assist students to both understand and

make the most of the practice-oriented learning opportunities they design, project-based learning or more general collaborative activities (Willey & Gardner, 2012). These learning 'activities' can be framed to foreground the practices that students will participate in, and how these practices relate to those they will engage in at work. Describing these practices can provide the professional context for developing those graduate attributes, often generic, that engineering students (and some engineering academics) see as irrelevant to their course. A typical example would be communication skills. Engineering students are often reported as being resistant to learning activities designed to develop their communication skills, yet effective communication (listening, questioning, explaining, sketching) is required to participate in a site walk or a design review meeting.

– Practice theory might be used to explain to students the benefits of any work placement/internship experiences they undertake as part of their studies and help them plan, evaluate and gain the most from these experiences. Students could be asked to identify the practices of the workplaces in which they are employed, and how they might learn through participating in them.

– Finally, practice theory might prove useful in helping students to conceptualise, understand and plan their own learning, particularly within collaborative activities, by enabling them to see the value of participating in the process of collaboration with a diverse group of peers as opposed to focusing on producing an artefact to achieve a desired grade. This process may assist them both to transition from a 'grading' to a 'learning' focus (Willey & Gardner, 2012) and to begin to identify as a learner and an engineer.

6 CLOSING REMARKS

We close this chapter by proposing that significant continuing professional learning results from engaging in the collective practices that constitute professional work. We suggest that learning through practice is essential to the development of engineers and, through extension, to the competitiveness and success of engineering firms.

This chapter presented a practice approach to investigating professional learning of experienced engineers. It framed an investigation of experienced engineers' work within notions of practice. Through first presenting developments in workplace learning literature, and then practice theorisations in particular, followed by a brief example of the site walk, the chapter points to the challenges and opportunities that this approach could lead to in regard to alternative understandings of learning for engineering companies and professional associations; for professional engineers themselves in relation to assessing and planning their own development; and for engineering educators and their students in relation to effectively preparing students for professional engineering practice. Further research is needed to identify, characterise and describe other practices, and the inherent learning opportunities, that professional engineers engage in as part of their everyday work. Viewing learning through a practice lens opens up the possibility of shifting educational activities away from a focus on deconstructed knowledge and skills towards a conception of them more aligned with the normal complexity and generativity of engineering practice.

REFERENCES

Billett, S. (2001) Learning through work: Workplace affordances and individual engagement. *Journal of Workplace Learning*, 13 (5), 209–214.

Billett, S. (2004) Workplace participatory practices: Conceptualising workplaces as learning environments. *The Journal of Workplace Learning*, 16 (6), 312–324.

Boud, D. (2012) Problematising practice-based education. In: Higgs, J., Barnett, R., Billett, S., Hutchings, M. & Trede, F. (eds.) *Practice-Based Education: Perspectives and Strategies.* Rotterdam, Sense Publishers.

Boud, D. & Hager, P. (2012) Re-conceptualising continuing professional development through changing metaphors and locations in professional practices. *Studies in Continuing Education*, 34, 17–30.

Boud, D. & Middleton, H. (2003) Learning from others at work: Communities of practice and informal learning. *Journal of Workplace Learning*, 15 (5), 194–202.

Boud, D. & Solomon, N. (2003) "I don't think I am a learner": Acts of naming learners at work. *Journal of Workplace Learning*, 15, 326–331.

Boud, D., Rooney, D. & Solomon, N. (2009) Talking up learning at work: Cautionary tales in co-opting learning. *International Journal of Lifelong Education*, 28 (3), 323–334.

Cresswell, T. (2004) *Place: A short introduction.* 2nd edition. Maldon, MA, Blackwell Publishing.

Denzin, N. K. & Lincoln, Y. S. (2000) Introduction: The discipline and practice of qualitative research. In: Denzin, N. K. & Lincoln, Y. S. (eds.), *Handbook of qualitative research.* Thousand Oaks, CA, Sage. pp. 1–28.

Edwards, R. & Usher, R. (2008) *Globalisation and pedagogy: Space, place and identity.* 2nd edition. London, Routledge.

Edwards, R., Nicoll, K., Solomon, N. & Usher, R. (2004) *Rhetoric and educational discourse: persuasive texts?* London, RoutledgeFalmer.

Engineers Australia (2012) Professional development courses. [Online] Available from: http://www.eeaust.com.au/Professional-Development-Courses-Category/ [Accessed 2nd August 2012].

Eraut, M. (2004) Informal learning in the workplace. *Studies in Continuing Education*, 26 (2), 247–272.

Fenwick, T. (2009) Making to measure? Reconsidering assessment in professional continuing education. *Studies in Continuing Education*, 31, 229–244.

Fenwick, T. & Edwards, R. (eds.) (2012) *Researching education through actor-network theory.* London, Wiley.

Gallacher, J., Edwards, R. & Whittaker, S. (2006) *Learning outside the academy: International research perspectives on lifelong learning.* London, Routledge.

Gherardi, S. (2006) *Organizational Knowledge: The texture of workplace learning.* Oxford, Blackwell.

Gherardi, S. (2008) Situated knowledge and situated action: What do practice-based studies promise? In: Barry, D. & Hansen, H. (eds.) *The Sage handbook of new approaches in management and organization.* London, Sage Publications. pp. 516–525.

Gherardi, S. (2009a), Communities of practice or practices of communities. In: C. Fukami (ed.) *The SAGE handbook of manaement learning, education and development*, SAGE, Los Angeles, pp. 414–530.

Gherardi, S. (2009b) Introduction: The critical power of the 'practice lens'. *Management Learning*, 40, 115–128.

Hager, P. (1998) 'Recognition of informal learning: challenges and issues', *Journal of Vocational Education and Training*, 50 (4), 521–535.

Hager, P. (2011) Historical tracing of the development of theory in the field of workplace learning. In: M. Malloch, L. Cairns, K. Evans & B. O'Connor (eds.) *The SAGE handbook of workplace learning*. [Online] Available from: http://knowledge.sagepub.com.ezproxy.lib.uts. edu.au/view/hdbk_workplacelearning/n2.xml [Accessed 20th December 2012].

Hager, P. & Halliday, J. (2006) *Recovering informal learning: Wisdom, judgement and community*. Dordrecht, The Netherlands Springer.

Hager, P. & Hodkinson, P. (2009) Moving beyond the metaphor of transfer of learning. *British Educational Research Journal*, 35 (4), 619–638.

Hager, P., Lee, A. & Reich, A. (2012) Problematising practice, reconceptualising learning and imagining change. In: Hager, P., Lee, A. & Reich, A. (eds.) *Practice, learning and change: Practice-theory perspectives on professional learning*. Dordrecht, Springer.

Johnsson, M. & Boud, D. (2010) Toward an emergent view of learning work. *International Journal of Lifelong Learning*, 29 (3), 355–368.

Kemmis, S. (2009) Understanding professional practice: A synoptic framework. In: Green B. (ed.) *Understanding and researching professional practice*. Rotterdam, Sense Publishers. pp. 19–38.

Latour, B. (1999) On recalling ANT. In: J. Law & J. Hassard (eds.) *Actor network theory and after*. Oxford, Blackwell. pp. 15–25.

Lave, J. & Wenger, E. (1991) *Situated learning: Legitimate peripheral participation*. Cambridge: Cambridge University Press.

Law, J. (ed.) (1986) *Power, action and belief. A new sociology of knowledge?* London, Routledge & Kegan Paul.

Law, J. (1992) *Notes on the theory of actor network: Ordering, strategy and heterogeneity*. Science Studies Centre, Lancaster University. [Online] Available from: http://link.springer. com/article/10.1007%2FBF01059830?LI=true#page-1 [Accessed 17th December 2012].

Law, J. & Hassard, J. (eds.) (1999) *Actor network theory and after*. Oxford, Blackwell.

Marsick, V.J. & Watkins, K.E. (1990) *Informal and incidental learning in the workplace*. London, Routledge.

Østerlund, C. & Carlile, P. (2005) Relations in practice: Sorting through practice theories on knowledge sharing in complex organizations. *Information Society*, 4, 91–107.

Price, O., Johnsson, M., Scheeres, H., Boud, D. & Solomon, N. (2012) Learning organisational practice that persist, perpetuate and change: A Schatzkian view. In: Hager, P., Lee, A. & Reich, A. (eds.) *Practice, learning and change: Practice-theory perspectives on professional learning*. Dordrecht, Springer.

Schatzki, T. (2001) Introduction: Practice theory. In: Schatzki, T., Knorr Cetina, C. & von Savigny, E. (eds.) *The practice turn in contemporary theory*. London, Routledge. pp. 10–23.

Schatzki, T. (2002) *The site of the social: A philosophical account of the constitution of social life and change*. University Park, Pennsylvania State University Press.

Schatzki, T. (2005) Peripheral vision: The sites of organizations. *Organization Studies*, 26, 465–484.

Schatzki, T. (2006) On organizations as they happen. *Organization Studies*, 27, 1–14.

Schatzki, T. (2010) *The timespace of human activity: On performance, society and history*. Lanham, MD, Lexington.

Schatzki, T. R. (2012) A primer on practices: Theory and research. In: Higgs, J., Barnett, R., Billett, S., Hutchings, M. & Trede, F. (eds.) *Practice-Based Education: Perspectives and Strategies*, Rotterdam, Sense Publishers.

Schatzki, T., Knorr Cetina, C. & von Savigny, E. (eds.) (2001) *The practice turn in contemporary theory*. London, Routledge.

Scheeres, H., Solomon, N., Boud, D. & Rooney, D. (2010) When is it OK to learn at work? The learning work of organisations. *Journal of Workplace Learning*, 22 (1 & 2), 13–26.

Skule, S. & Reichborn, A. (2002) *Learning-conducive work: A survey of learning conditions in Norwegian workplaces*. Luxembourg, CEDEFOP.

Soja, E. (1996) *Thirdspace: Journeys to Los Angeles and other real-and-imagined places*. Malden, MA, Blackwell Publishers Ltd.

Soja, E. (2001) *Postmodern geographies: The reassertion of space in social critical theory*. London, Verso.

Solomon, N., Boud, D. & Rooney, D. (2008) The in-between: Exposing everyday learning. In: Hall, K., Murphy, P. & Soler, J. (eds.) *Pedagogy and practice: Culture and identities*. London, Sage Publications. pp. 75–84.

Usher, R. (1996) A critique of the neglected epistemological assumptions of educational research. In: D. Scott & R. Usher (eds.) *Understanding educational research*. London, Routledge. pp. 9–32.

Wenger, E. (1996) Communities of practice: The social fabric of a learning organization. *Healthcare Forum Journal*, 39 (4), 20–26.

Wenger, E. & Snyder, W. (2000) Communities of practice: The organizational frontier. *Harvard Business Review*, 78 (1), 139–145.

Willey, K. & Gardner, A. (2012) Collaborative learning frameworks to promote a positive learning culture. *Proceedings of the 2012 Frontiers in Education Conference: Soaring to New Heights in Engineering Education*, 3–6 October 2012, Seattle, Washington, USA.

Glossary

Some chapters address specific terminology mainly emanating from Science, technology and society studies (STS) and in particular Actor-network Theory (ANT). We provide this glossary to help clarify the use of these terms, especially when their sense is different from the common use of the terms.

Actor-network theory (ANT) is an approach that avoids essentialist explanations of science (those which explain the success of a theory by its epistemic superiority) or innovation (those explaining successful innovations by their technical or economical superiority). It suggests following the actors and the associations they construct, without any a priori assumptions as to their nature. They can be material, social, or semiotic. Generally actors are hybrids and can comprehend humans, rules, laws, organisations and software. ANT is considered a completely distinctive constructivist material-semiotic approach, describing things in action (i.e. theories and technologies) as sociotechnical networks. It is an actor-based approach that takes into account the agency of humans and nonhumans. ANT grounds its situational stand on semiotics and ethnomethodology (Latour, 1987). However, in a recent book, Modes of Existence, Latour proposed a new approach to ANT (renamed simply Networking) where he describes it as a way of travelling through heterogeneous domains (Latour, 2012).

Equipping is a process through which something (e.g. a number, a technical device) is added to an entity (an object, a person). For instance, equipping of intermediary objects (e.g. industrial drawing) is claimed to be a central concern of engineers and technicians in their design process. It is a process of collective work whereby new properties (e.g. to be a validated design) are conferred to intermediary objects shaping the design space (Vinck, 2011).

Holonic diagrams originate from the word 'holon', from the Greek ὅλον meaning 'whole', which was coined by Arthur Koestler in his book The Ghost in the Machine (1967) to express something that harmoniously contains the whole and the parts. Traditional Cartesian theories believe that a whole is no more than the sum of the parts. However, more recent systems theories show that the dynamic relationships between the parts lead to the emergence of properties that do not exist in the parts and only appear in the whole: the whole is greater than the sum of the parts. The word holon is used today in biological and social systems theories and in philosophy. It is also used in architecture, where one speaks of holonic structures and of the philosophy of the holon. The term holonic diagram is used in the chapter *On the historical nature of engineering practice* to express the organic, parts-whole, philosophy behind the diagram, which attempts to represent both the parts and the whole.

Inscription is a way a making things stable, or durable. To inscribe is to influence in a specific way, like 'inoculating' a sense of action in an actor, or for example inscribing scripts in technological artefacts. Inscribing induces behaviour into things, and can be seen as a way of making society more durable. But actors are not passive things, they act and react, so to inscribe is always an attempt and is not of itself a result. "An inscription is the result of the translation of one's interest into material form" (Callon 1991, 143). Societal inscription is inspired by Foucault and refers to actions that integrate something into society in a way that would appear as natural.

Intermediary objects are material entities (i.e. an industrial design, a sample, or a prototype) circulating between actors or through and around which actors interact. Following intermediary objects helps to describe scientific cooperation networks or design processes (Vinck and Jeantet, 1995; Vinck, 2012). They can be a boundary objects in the sense used by Susan Leigh Star (1991). Intermediary objects emerge and evolve through interactions, always situated in multiple settings. Objects in ANT are always quasi-objects (Serres, 1982, 1995); only at the end of a chain (infinite, in principle) of translations do we have 'an object'. So, all objects will be 'intermediary objects'. We could say that 'intermediary objects' are quasi-objects in a particular dense or thick (in terms of meaning) chain.

Mediation is an active translation. ANT can represent action in programs of action where two operations are used: 'AND' and 'OR' (the semiotic turn of ANT). AND means coalition, grabbing support from others, with no new propositions, OR means alternatives of action away from the typical patterns. Mediators are the actors that in a translation create these alternatives or creative disruptions. Moreover a mediator is not a permanent label and an actor can act as mediator in some translations and not in others.

Obligatory passage point is a central concept in ANT. Actor-networks are spaces of translation and transformation, where actors negotiate meanings and alignments. In this process some actors try to establish themselves as obligatory passage points in order to frame ideas or goals in particular ways, speaking on behalf of the others. These passage points frame or situate the ongoing process of translation. Obligatory passage points are able to mobilize effective immutable mobiles.

Problem setting is a concept coined by Donald Schön in terms of his 'knowing-in-action' approach. Schön formulated his view on design in terms of 'reflective activity' and related notions, especially 'reflective practice,' 'reflection-in-action' and 'knowing-in-action'. Schön reasoned that problem solving is generally considered as handling problems as 'given', whereas the process of 'problem setting' is neglected. With problem setting we formulate the problem, framing what needs to be considered and forgetting the rest. "Problem setting is a process in which, interactively, we name the things to which we will attend and frame the context in which we will attend to them" (Schön, 1983, pp. 39–40).

Translation is the interaction in a common language between two or more actors changing one another and creating links. In ANT terms, translation is seen as the process by which a relation is created between two entities (which can be human or nonhuman). Translation is a complex interaction that can support different moments (problematisation, interessement, enrolment, mobilization) exploring different alignments and strategies.

Shaping refers to the social construction of technology (Bijker & Law, 1992), and to how the design and implementation of technology is patterned by a range of 'social' and 'economic' factors as well as narrow 'technical' considerations (MacKenzie & Wajcman, 1985; Williams & Edge, 1996).

REFERENCES

Actor Network Resource http://www.lancs.ac.uk/fass/centres/css/ant/antres.htm

Bijker, Wiebe. & Law, John, 1992, Shaping technology/building society: studies in sociotechnical change. Cambridge, MA, London, MIT Press.

Callon, Michel 1991, Techno-economic networks and irreversibility, pp. 132–165 in A Sociology of Monsters: Essays on Power.

Garfinkel, Harold, 1967, Common sense knowledge of social structures: The documentary method of interpretation in lay and professional fact finding, in H. Garfinkel (Ed.), Studies in ethnomethodology. Englewood Cliffs, NJ: Prentice-Hall.

Koestler, Arthur, 1967, The Ghost in the Machine (Arkana), reprinted in 1990, Penguin Books.

Latour, Bruno, 1987, Science in Action: How to Follow Scientists and Engineers through Society. Cambridge, MA: Harvard University Press.

Latour, Bruno, 2012, An Inquiry Into Modes of Existence, Harvard University Press, Cambridge, Mass. Translation: Cathy Porter.

Mackenzie, Donald, & Wajcman, Judy (eds), 1985, The social shaping of technology. Milton Keynes, Open University Press.

Schön, Donald, 1983, The reflective practitioner?: how professionals think in action?, Basic Books.

Star, Susan Leigh, 1991, 'Power, Technologies and the Phenomenology of Conventions: on being Allergic to Onions', pp. 26–56 in John Law (ed.), A Sociology of Monsters? Essays on Power, Technology and Domination, London, Routledge.

Serres, M., 1982, Gene'se, Paris, Bernard Grasset.

Serres, M. (with Bruno Latour), 1995, Conversations on Science, Culture, and Time, Ann Arbor, University of Michigan Press.

Vinck, D. & Jeantet, A. 1995, Mediating and Commissioning Objects in the Sociotechnical Process of Product Design: a conceptual approach, In: Designs, Networks and Strategies, Maclean, D.; Saviotti, P. & Vinck, D. (Eds), pp. 111–129, vol. 2, COST A3, Social Sciences, EC Directorate General Science R&D, Bruxelles.

Vinck, Dominique, 2011, Taking intermediary objects and equipping work into account when studying engineering practices, Engineering Studies, 3 (1), pp. 25–44.

Vinck, Dominique, 2012, Accessing Material Culture by Following Intermediary Objects, An Ethnography of Global Landscapes and Corridors, book edited by Loshini Naidoo InTech. Rijeka (Croatia). Available from: http://www.intechopen.com/books/an-ethnography-of-global-landscapes-and-corridors/natural-interactions-in-artificial-situations-focus-groups-as-an-active-social-experiment-

Williams, Robin, & Edge, David, 1996, "The Social Shaping of Technology", Research Policy, 25, pp. 856–899.

Index

academic studies 34, 42, 52–3
Actor-Network Theory (ANT) 162–3,
  179–80, 185–6, 206–7, 281
aesthetics 13–14, 15
analogy in design 71–2, 75–6
architect-engineer divide 208–9
architekton 10–11, 12
Aristotle, knowledge types 129, 142–3
Arup case study 208–20
autonomy 135, 150, 228–9; *see also*
  constraints

ba 147–52
boundaries: problem solving 135, 145;
  professional 205, 209
Bucciarelli's object worlds 207, 209
business and management 12, 15, 25–6;
  see also management processes

Callon's actor-network theory 162–3,
  179–80, 185–6, 206–7
civil engineering 17, 18, 251–2, 253; *see also*
  structural design
client relationship 173–4
closure in problem solving 138
Code of Hammurabi 9
cognitive convergence 140–6
collaborative working 63–4, 79–80, 109–12,
  124–5; *see also* interactive processes;
  team design processes
collective learning *see* organisational learning
commercial contexts 42–3
common frameworks 74
communication processes 63–4, 165–7
communication skills 37, 116
communities of practice 151
company reputation 174
competencies 38–9, 96, 103, 207

constraints: problem solving 135, 145;
  professional boundaries 205, 209
contextual competence 82, 96
continuing professional learning *see*
  professional learning
coordination 61–2, 147, 164, 169–70,
  171–3, 177
core competencies 39
craftsmanship 9–10, 12, 14, 20–1
creativity 26–7
cross-disciplinary practice 101–27;
  challenging and transforming practice
  119–24; education for 80; intentional
  learning 112–15; multidisciplinary,
  interdisciplinary and transdisciplinary
  103, **104**; preparing professionals for
  125–6; problem solving 150; strategic
  leadership 116–19; working together
  109–12, 124–5
cultural differences *see* organisational
  culture

definitions 25, 35
democratic discourse 7–8
design 12, 15, 16, 22–4, 28
design constraints 209
design processes 79–99; contextual issues
  84–5, 85–8; studies 53, 82–5; team
  design research 85–97; timelines
  82–4
design team meetings 85–97
developing countries *see* low income
  countries (LICs)
digital humanities (DH) 61–78; case studies
  65–75; innovation processes 64
disintegrated engineering profession 7, 27, 29
document management 233–5
drawings 9, 15, 16